Lecture Notes in Computer Science   2373

Edited by G. Goos, J. Hartmanis, and J. van Leeuwen

Springer
Berlin
Heidelberg
New York
Barcelona
Hong Kong
London
Milan
Paris
Tokyo

Alberto Apostolico   Masayuki Takeda (Eds.)

Combinatorial
Pattern Matching

13th Annual Symposium, CPM 2002
Fukuoka, Japan, July 3-5, 2002
Proceedings

 Springer

Series Editors

Gerhard Goos, Karlsruhe University, Germany
Juris Hartmanis, Cornell University, NY, USA
Jan van Leeuwen, Utrecht University, The Netherlands

Volume Editors

Alberto Apostolico
University of Padova
Department of Electrial Engineering and Computer Science
Via Gradenigo 6/A, 35131 Padova, Italy
E-mail: axa@artemide.dei.unipd.it
Masayuki Takeda
Kyushu University
Department of Informatics
Fukuoka 812-8581, Japan
E-mail: takeda@i.kyushu-u.ac.jp

Cataloging-in-Publication Data applied for

CR Subject Classification (1998): F.2.2, I.5.4, I.5.0, I.7.3, H.3.3, E.4, G.2.1

Cataloging-in-Publication Data applied for

Die Deutsche Bibliothek - CIP-Einheitsaufnahme

Combinatorial pattern matching : 13th annual symposium ; proceedings / CPM
2002, Fukuoka, Japan, July 3 - 5, 2002. Alberto Apostolico ; Masayuki Takeda
(ed.). - Berlin ; Heidelberg ; New York ; Barcelona ; Hong Kong ; London ;
Milan ; Paris ; Tokyo : Springer, 2002
    (Lecture notes in computer science ; Vol. 2373)
    ISBN 3-540-43862-9

ISSN 0302-9743
ISBN 3-540-43862-9 Springer-Verlag Berlin Heidelberg New York

Springer-Verlag Berlin Heidelberg New York
a member of BertelsmannSpringer Science+Business Media GmbH

http://www.springer.de

© Springer-Verlag Berlin Heidelberg 2002
Printed in Germany

Typesetting: Camera-ready by author, data conversion by DA-TeX Gerd Blumenstein
Printed on acid-free paper   SPIN 10870392   06/3142   5 4 3 2 1 0

Foreword

The papers contained in this volume were presented at the 13th Annual Symposium on Combinatorial Pattern Matching, held July 3–5, 2002 at the *Hotel Uminonakamichi*, in Fukuoka, Japan. They were selected from 37 abstracts submitted in response to the call for papers. In addition, there were invited lectures by Shinichi Morishita (*University of Tokyo*) and Hiroki Arimura (*Kyushu University*).

Combinatorial Pattern Matching (CPM) addresses issues of searching and matching strings and more complicated patterns such as trees, regular expressions, graphs, point sets, and arrays, in various formats. The goal is to derive nontrivial combinatorial properties of such structures and to exploit these properties in order to achieve superior performance for the corresponding computational problems. On the other hand, an important goal is to analyze and pinpoint the properties and conditions under which searches cannot be performed efficiently.

Over the past decade a steady flow of high-quality research on this subject has changed a sparse set of isolated results into a full-fledged area of algorithmics. This area is continuing to grow even further due to the increasing demand for speed and efficiency that stems from important applications such as the World Wide Web, computational biology, computer vision, and multimedia systems. These involve requirements for information retrieval in heterogeneous databases, data compression, and pattern recognition. The objective of the annual CPM gathering is to provide an international forum for research in combinatorial pattern matching and related applications.

The first twelve meetings were held in Paris, London, Tucson, Padova, Asilomar, Helsinki, Laguna Beach, Aarhus, Piscataway, Warwick, Montreal, and Jerusalem, over the years 1990-2001. After the first meeting, a selection of papers appeared as a special issue of *Theoretical Computer Science* in volume 92. Selected papers of the 12th meeting will appear in a special issue of *Discrete Applied Mathematics*. The proceedings of the 3rd to 12th meetings appeared as volumes 644, 684, 807, 937, 1075, 1264, 1448, 1645, 1848, and 2089 of the Springer LNCS series.

The general organization and orientation of the CPM conferences is coordinated by a steering committee composed of Alberto Apostolico (*Padova and Purdue*), Maxime Crochemore (*Marne-la-Vallée*), Zvi Galil (*Columbia*), and Udi Manber (*Yahoo!*).

March 2002   Alberto Apostolico and Masayuki Takeda

Program Committee

Tatsuya Akutsu, *Kyoto Univ.*
Amihood Amir,
    Bar Ilan & Georgia Tech
Alberto Apostolico, co-chair
    Univ. of Padova & Purdue Univ.
Paolo Ferragina, *Univ. of Pisa*
Hiroshi Imai, *Univ. of Tokyo*
Tao Jiang, *Univ. of California*
Rao Kosaraju, *Johns Hopkins Univ.*
Gonzalo Navarro, *Univ. of Chile*
Kunsoo Park, *Seoul National Univ.*

Pavel Pevzner, *Univ. of California*
Marie-France Sagot,
    Inria Rhône-Alpes
Jim Storer, *Brandeis Univ.*
Masayuki Takeda, co-chair,
    Kyushu Univ.
Uzi Vishkin, *Univ. of Maryland*
Ian Witten, *Univ. of Waikato*
Hidetoshi Yokoo, *Gunma Univ.*
Kaizhong Zhang,
    Univ. of Western Ontario

Local Organization

Local arrangements were coordinated by Ayumi Shinohara, Kyushu Univ. The conference Web site was created and maintained by Shunsuke Inenaga. Organizational help was provided by Kensuke Baba and Takuya Kida.

List of Additional Reviewers

Julien Allali
Gerth Brodal
Jeremy Buhler
Alberto Caprara
Tieling Chen
Pierluigi Crescenzi
Lancia Giuseppe
H. Kaplan

Uri Keich
Moshe Lewenstein
GuoHui Lin
Bin Ma
Michael Mitzenmacher
Alistair Moffat
Nadia Pisanti
Gianluca Rossi

Kunihiko Sadakane
Thomas Schiex
Dana Shapira
Sing-Hoi Sze
E. Verbin
Zhuozhi Wang
Shaojie Zhang

Table of Contents

Practical Software for Aligning ESTs to Human Genome

Jun Ogasawara and Shinichi Morishita

[1] Department of Computer Science, University of Tokyo
[2] Department of Complexity Science and Engineering, University of Tokyo
{jun,moris}@gi.k.u-tokyo.ac.jp
http://grl.gi.k.u-tokyo.ac.jp

Abstract. There is a pressing need to align growing set of expressed sequence tags (ESTs) to newly sequenced human genome that is still frequently revised, for providing biologists and medical scientists with fresh information. The problem is, however, complicated by the exon/intron structure of eucaryotic genes, misread nucleotides in ESTs, and millions of repeptive sequences in genomic sequences. Indeed, to solve this, algorithms that use dynamic programming have been proposed, in which space complexity is $O(N)$ and time complexity is $O(MN)$ for a genomic sequence of length M and an EST of length N, but in reality, these algorithms require an enormous amount of processing time. In an effort to improve the computational efficiency of these classical DP algorithms, we develop software that fully utilizes the lookup-table that stores the position at which each short subsequence occurs in the genomic sequence for allowing the efficient detection of the start- and endpoints of an EST within a given DNA sequence, and subsequently, the prompt identification of exons and introns. In addition, high sensitivity and accuracy must be achieved by calculating locations of all spliced sites correctly for more ESTs while retaining high computational efficiency. This goal is hard to accomplish in practice, owing to misread nucleotides in ESTs and repeptive sequences in the genome, but we present a couple of heuristics effective in settling this issue. Experimental results have confirmed that our technique improves the overall computation time by orders of magnitude compared with common tools such as sim4 and BLAT, and attains high sensitivity and accuracy against datasets of clean and documented genes at the same time. Consequently, our software is able to align about three millions of ESTs to a draft genome in less than one day, and all the information is available through the WWW at http://grl.gi.k.u-tokyo.ac.jp/.

A. Apostolico and M. Takeda (Eds.): CPM 2002, LNCS 2373, pp. 1–16, 2002.
© Springer-Verlag Berlin Heidelberg 2002

1   Introduction

The Human Genome Project is an international collaboration, designed to investigate the genetic complexity of humans. Initially, the roughly three billion nucleotides of the human genome were elucidated (Celera Genomics [1],ü@ International Human Genome Sequencing Consortium [2]). The second step involves the interpretation of the encoded sequences. For the purpose of identifying the coding regions, i.e., regions containing exons and introns, of any given DNA sequence, the alignment of many ESTs to genomic DNA is helpful to reveal these complex structures while verifying the alternative spliced transcripts. The alignment of full-length cDNAs gives a clue to some regulatory elements in its upstream regions, and further the annotations of the upstream regions using Transfac data line up candidates of cis-elements. The alignment both of sequences associated with expression pattern and of sequences from dbSNP with identifying locations of SNPs around the gene make it possible to step forward more in function analysis.

Indeed, a variety of sequence alignment algorithms have been proposed. One technique for computing the similarity between two sequences is to assign penalties, designated by letters, to insertions, deletions and substitutions present in one sequence, but not in the other (Needleman and Wunsch [3], Smith and Waterman [4]). Recently, heuristic algorithms such as FASTA [5] and BLAST [6] have been used because of their higher speed compared to dynamic programming methods.

These algorithms however require a very large amount of time for processing or they fail to align ESTs to genome that are known to be encoded, because they are designed only to solve the similarity between two sequences, but not to decide eucaryotic gene structure (exon/intron structure) through the identification of spliced sites between exons and introns. Figure 1 illustrates our problem that we need to decompose an EST into exons and then to align each exon onto the DNA sequence while preserving the order of exons. The difficulty with the problem is that there are potentially a huge number of ways to decompose the EST. To settle this problem, it is reasonable to define scores of matching and penalty for introducing introns and then to select the optimal decomposition as that with the best score.

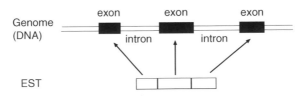

Fig. 1. Decomposing and mapping an EST

For this optimization problem, many dynamic programming algorithms that consider exon/intron structure have been developed. Gotoh's algorithm [7] defines the affine gap penalty for introns to identify very long introns, since introns correspond to long insertions. Although Gotoh's algorithm runs in $O(MN)$-time complexity and requires $O(MN)$ space for a genomic sequence of length M and for an EST of length N, the space complexity can be reduced to $O(\min(M, N))$ by Hirschberg's technique [8]. There have been developed several software tools that utilize the idea of this dynamic programming technique [9,10,11], but in practice, these tools are computationally infeasible to apply to long genomic sequences of length greater than one million. One the other hand, Sim4 [12] and BLAT [13] are also improved methods based on BLAST, which extend not only single exon but also multiple exons. Although they are able to decompose a given sequence into its exons, they are not designed to compute the optimal alignment.

Therefore, since high performance software is needed to solve this problem, we invented software that shortens the calculation time, while retaining sensitivity and accuracy. This software is able to align more than 20 ESTs per second on average to a human draft genome by using a single processor, but in practice it can process more than 100 ESTs per second on several processors, and hence about three millions of ESTs in less than half a day. With regarding sensitivity and accuracy, special care has to be taken to identify spliced sites in the final step of the software. Clean datasets of spliced sites have been collected from various species to derive statistically confirmed rules for improving gene finding algorithms [14,15,16,17,18]. These clean datasets are also valuable to evaluate sensitivity and accuracy of alignment software, and we use the HMR195 dataset [18], a collection of Mammalian sequences. We will demonstrate high sensitivity and accuracy of our method against human genes in HMR195.

2   Algorithm

In this section, we present our algorithm, and we particularly keep in mind that biologists or medical scientists who might not be familiar with computer science are also able to understand the behavior of the algorithm. We start with the introduction of basic data-structures and simpler algorithms, and move to elaborated one by improving algorithms step by step.

Mapping of the millions of ESTs in the GenBank and EST databases (dbEST) to the human genomic sequence is an arduous task. ESTs are a maximum of tens of thousands of bases long, while genomic sequences are about 3 billion bases long. In order to shorten the overall calculation time, we pre-process the genomic sequences. A DNA sequence of length L is defined as a *primary key*. Our idea is to create an auxiliary look-up table called *MapTable* that stores the position at which each primary key occurs in the genomic sequence. Figure 2 shows how to generate a MapTable when the primary key length is 2, for simplicity.

Referring to the MapTable, it is obvious that "TA" exists at the 9th position and that "GC" occurs at the 3rd and 12th positions in the genome sequence.

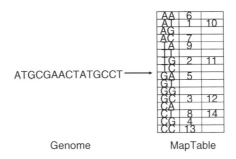

Genome   MapTable

Fig. 2. How to generate a lookup table (MapTable)

When considering a particular EST sequence, the position of the L length prefix and suffix of the gene is inferred by referring to the MapTable in the main memory. Assuming that the four characters (nucleotides) appear at random in the genome sequence, we can deduce the position from the MapTable by accessing the main memory about $M/4^L$ times (M is the length of the genome sequence). Care had to be taken in selecting appropriate values for L. Smaller L values, say 5, aligned an enormous number of positions for each 5-mer, while a larger L value increased the number of L-mers, yielding a huge index. We will mention this issue of selecting an appropriate value for L.

2.1   Simple Algorithm Assuming No Mismatches or Gaps

First we present a slow but simple algorithm which assumes that an EST sequence exactly maps to the genomic sequence with 100% matching ratio. We then improve the algorithm in a stepwise manner to accelerate performance while allowing mismatches and gaps.

In the following discussion, let us denote the genome sequence G by the sequence $g_1, g_2, ..., g_M$ ($g_i \in \{A, T, G, C, N\}$), and the EST sequence E by the sequence $e_1, e_2, ..., e_N$ ($e_i \in \{A, T, G, C, N\}$). $G_{[i,j]}$ represents a substring: g_i, g_{i+1}, $..., g_j$, so does $E_{[i,j]}$. We express the alignment of the i-th nucleotide in the EST with genome G as position $f(i)$ in genome G. We here assume that each nucleotide e_i maps to a location in G, but in general e_i might be skipped, requiring its position $f(i)$ undefined. This case will be considered later in this section. The simplest version of the algorithm (Figure 3) can be written as:

Step 1. (Detection of start- and endpoints)

Let L be the length of a primary key in the MapTable and consider a prefix of length L ($E_{[1,L]}$) and a suffix of length L ($E_{[N-L+1,N]}$) in EST E. Align the prefix and suffix with the genome by accessing the MapTable, i.e., associate $f(i)$ for each $i = 1, ..., L, N - L + 1, ..., N$.

Step 2. (Alignment by dynamic programming)

Align unassigned interval of the EST $E_{[L+1,N-L]}$ with the genome interval $G_{[f(L)+1,f(N-L+1)-1]}$ using Gotoh's dynamic programming, which yields $f(i)$ for each $i = L + 1, ..., N - L$.

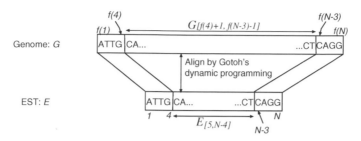

Fig. 3. Simplest algorithm with a MapTable($L = 4$)

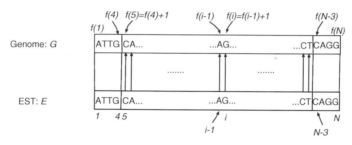

Fig. 4. Fast algorithm with a MapTable (Case of a single exon, $L = 4$)

In the case of human genome, we assign 14 to L, because most first exons and last exons are of length more than 14, though it is known that some internal exons are of length less than 14. In Step 1, there could be multiple locations for $E_{[1,L]}$ and $E_{[N-L+1,N]}$, which will be discussed later in this section. In Step 2, Gotoh's dynamic programming [7] is the algorithm that finds the optimal alignment of EST that has long introns with genome. Smith-Waterman and Needleman-Wunsch do not work well for this problem, because they pose high penalty on long introns and miss alignments with long introns. To overcome this issue, Gotoh's method allows the assignment of a small constant or an affine gap penalty to introns that could be very long, and it can output the optimal solution.

Although the detection of the start- and endpoints of a given EST in the genome sequence in Step 1 works very efficiently because it can be achieved by accessing the main memory, Step 2 could sometimes require a large amount of execution time when the length of the genome interval $G_{[f(L)+1,f(N-L+1)-1]}$ is still long for Gotoh's dynamic programming.

2.2   Fast Algorithm Assuming No Mismatches or Gaps

To accelerate the dynamic programming in Step 2, we determine $f(i)$ by elongating the exon and skipping long intron using MapTable. This greatly increases

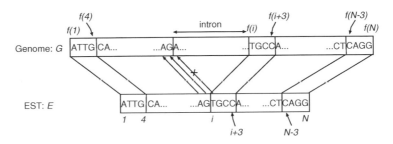

Fig. 5. Fast algorithm with MapTable (Case of multiple exons, $L = 4$)

performance because of its practicability in considering exon/intron structure. The algorithm for aligning an EST with a single exon is shown.

Step 2 (Identification of single exon)
    For each i from $L + 1$ to $N - L$, set $f(i - 1) + 1$ to $f(i)$ (see Figure 4).

We call Step 2 the *Elongation Step*, because this step adds one nucleotide to the end of the exon per one iteration.
    To handle ESTs with multiple exons, we incorporate the process that skips an intron when extension of the exon fails (Figure 5).

Step 2.1 (Identification of one exon)
    While the i-th nucleotide in E coincides with the $f(i)$-th nucleotide in G, set $f(i) = f(i - 1) + 1$ and increment i.
Step 2.2 (Search for the next exon)   Since Step 2.1 confirmed that the exon terminates at the i-th nucleotide in the EST, detect the position of the next exon by referring to the MapTable with $E_{[i,i+L-1]}$ as a primary key. After determining the locus of the next exon, to which $f(j)$ $(j = i, ..., i+L-1)$ is set, increment i by L and return to Step 2.1.

2.3   Allowing Mismatches and Gaps in Alignment

In practice, any EST cannot be fully aligned with 100 % identity, resulting in mismatches or gaps in the alignment. To allow for these mismatches and gaps, we here revise the algorithm in Section 2.3. We allow that e_i is mismatched with a different nucleotide at the $f(i)$-th position on the genome. Or, $f(i)$ is undefined when e_i is skipped and is associated with a gap. In this general setting, the start- and endpoints of the EST in the genome sequence cannot be detected if the prefix $E_{[1,L]}$ (or the suffix $E_{[N-L+1,N]}$) might contain mismatches or gaps. To resolve this problem, we scan the EST sequence E from the start until the position of a subsequence of length L is found in the MapTable, and scan E from the end in the same way (Figure 6). This method is described below.

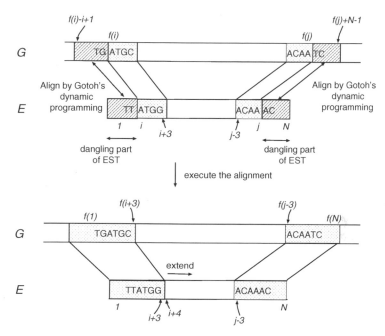

Fig. 6. Approximation of start- and endpoints ($L = 4$)

Step 1.1 (Approximation of the startpoint of an EST)
Initialize $i = 1$, and increment i until the position of $E_{[i,i+L-1]}$ is found
in the MapTable. After this, $E_{[1,i-1]}$ is called the *dangling part* of the EST
that still remains to be aligned. Then align this dangling part of EST $E_{[1,i-1]}$
with genome $G_{[f(i)-i+1,f(i)-1]}$ to decide $f(h)$ for each $h = 1, ..., i-1$ using
Gotoh's dynamic programming.

Step 1.2 (Approximation of the endpoint of an EST)
In a similar way, initialize $j = N$, and decrement j until the position
of $E_{[j-L+1,j]}$ is found in the MapTable. After this, $E_{[j+1,N]}$ is called the
dangling part of the EST that still remains to be aligned. Then align this
dangling part of EST $E_{[j+1,N]}$ with genome $G_{[f(j)+1,f(j)+N-j]}$ to decide $f(h)$
for each $h = j + 1, ..., N$ using Gotoh's dynamic programming.

Mismatches or gaps in an EST also make it more complicated to extend
an exon and skip an intron. While the elongation stops only when the exon
terminates in the case of no mismatches, elongation of the exon fails when it
reaches the end of the exon or encounters a mismatch or a gap in the alignment.

Step 2.1 (Identification of one exon)
Initialize i is the smallest position in the EST that is not aligned (for instance, (i+4)-th position of E in Figure 6) and while the i-th nucleotide in E
coincides with the $(f(i-1)+1)$-th nucleotide in G, set $f(i) = f(i-1) + 1$

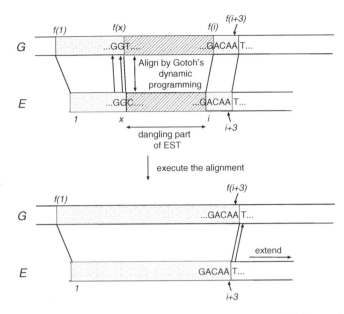

Fig. 7. Alignment of the dangling part of an EST($L = 4$)

and increment i. Then, set $x = i - 1$ to memorize the position $i - 1$ in the EST where the elongation ends.

Step 2.2 (Search for the next exon)

Increment i until the position of $E_{[i,i+L-1]}$ is found in the MapTable. After this, $E_{[x+1,i-1]}$ is called the dangling part of EST that remains to be aligned.

Step 2.3 (Alignment of the dangling part of an EST)

Align the dangling part of EST $E_{[x+1,i-1]}$ with the part of genome $G_{[f(x)+1,f(i)-1]}$ using Gotoh's dynamic programming. Figure 7 illustrates the case when no intron is detected after the dynamic programming, and Figure 8 shows when an intron is observed.

2.4   Further Acceleration by Preprocessing Long Introns

In practice, an intron could be thousands of base pairs long, while the dangling part is at most hundreds of base pairs long. Naive application of dynamic programming to the alignment of the dangling part of EST $E_{[x+1,i-1]}$ with $G_{[f(x)+1,f(i)-1]}$ is computation intense when $G_{[f(x)+1,f(i)-1]}$ contains an intron in Step 2.3. To accelerate this step, we examine whether an intron is included in the part of genome $G_{[f(x)+1,f(i)-1]}$ by comparing the length of $E_{[x+1,i-1]}$ with the length of $G_{[f(x)+1,f(i)-1]}$ before applying Gotoh's dynamic programming. If $G_{[f(x)+1,f(i)-1]}$ is much longer than $E_{[x+1,i-1]}$, we assume that $G_{[f(x)+1,f(i)-1]}$ contains an intron in it. In this case, since $E_{[x+1,i-1]}$ should be aligned to $G_{[f(x)+1,f(x)+(i-x)]}$ and $G_{[f(i)-(i-x),f(i)-1]}$, we align the dangling part of EST

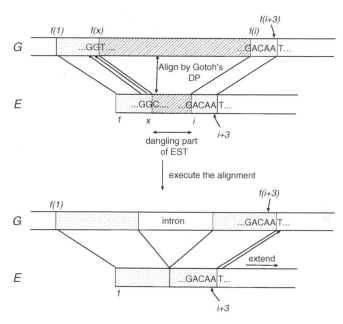

Fig. 8. Alignment of the dangling part of an EST ($L = 4$) when part of the genome contains an intron

$E_{[x+1,i-1]}$ with the concatenation of two sequences of length $i - x$, which is $G_{[f(x)+1,f(x)+(i-x)]} + G_{[f(i)-(i-x),f(i)-1]}$ (See Figure 9).

2.5 Detecting Spliced Sites

We have presented how to determine the approximate exon and intron regions. However, decisions regarding exon/intron boundaries need rigorous investigation, because most boundaries follow the GT-AG rule, but some other patterns such as GC-AG and AT-AC are also observed. In the literature, considerable efforts have been made to comprehend variants of the GT-AG rule statistically [14,15,16,17] and to make a comparative analysis of gene-finding programs [18]. For instance, Thanaraj [15] derives decision trees for inferring human exon-intron junctions from a number of EST-confirmed splice sites. Burset et al.[17] also present statistical rules for mammalian spliced sites, which also confirms that most sites obey GT-AG rule or its variants such as GC-AG and AT-AC.

Figure 10 illustrates some alternative solutions in deciding spliced sites, and the lower alignment ought to be selected according to the GT-AG rule. To this end, as shown in Figure 11, we shift the intron frame locally along the genome if no more mismatches are introduced in the alignment, for the purpose of detecting the GT-AG or its slight variants. In more precise, the intron frame is moved

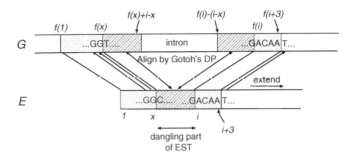

Fig. 9. Acceleration of dynamic programming

locally so that the number of matches between GT-AG and the two plus two letters at the boundary of the intron is maximized. We will demonstrate the high sensitivity and accuracy of this method later.

2.6   Matching Ratio

Having determined the intron regions, the matching ratio between the genome and the mapped EST sequence is examined. Since the matching ratio between the genome and a coded EST is 99.9% if the EST sequence is read precisely (the residual 0.1% is the difference between each human genome, i.e., SNP), we assume that an EST with a low matching ratio is not encoded. In effect, an EST whose matching ratio is low contains many misread nucleotides, or is not encoded in the genome sequence. Depending on our implementation result, the lower boundary for the matching ratio was set at 90%.

2.7   Solution for Two or More Alignments

In this section, we consider cases when there are two or more distinct alignments of an EST in a genome sequence, because an EST is often aligned to many different regions in a chromosome because of retro-transposition or gene-duplication (Figure 12). Furthermore, millions of repeptive sequences in human genome disturb correct identification of start-points and end-points.

Although the start- and endpoints of an EST are inferred in Step 1.1 and Step 1.2, the start- and endpoints is not determined uniquely if the MapTable has many candidates for the primary key. Let *StartSet* denote the set of candidate startpoints calculated in Step 1.1, and *EndSet* the set of candidate endpoints in Step 1.2. We then need to compute alignments for all the pairs in *StartSet* × *EndSet*. In practice, however, the size of *StartSet* or *EndSet* could be often more than one thousand owing to repeptive sequences. To avoid such difficult cases, we focus on the fact that the number of 14-mers appearing no more than ten times in the human genome is about 130 millions, which is about 4.3% of all the 14-mers in the human genome. In addition, sub-sequences in exons are

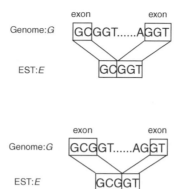

Fig. 10. Alternative solutions for spliced sites

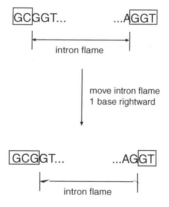

Fig. 11. Shifting the intron frame to detect GT-AG or its slight variants

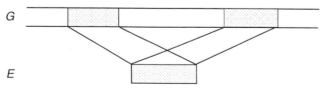

Fig. 12. EST Alignment to plural regions in genome

typically less frequent in the human genome than other sub-sequences in non-coding regions. These facts imply that scanning one hundred bases in a human EST could find, with a high probability, such a 14-mer of frequency no more than ten. We therefore revise our algorithm to search a subsequence of length L until the size of $StartSet$ (or $EndSet$) becomes no more than ten. Consequently our algorithm is described as:

Step 1.3 (Solution for many start- and endpoints)
   Assume that Steps 1.2 and 1.3 output such $StartSet$ and $EndSet$ that are of size no more than ten. Solve the alignment of $E_{[i,j-L]}$ with $G_{[start,end]}$ ($start \in StartSet$, $end \in EndSet$) by executing Step 2.1-2.3, if the distance between $start$ and end is smaller than 3,000,000 bp.

This technique seems to involve brute force, but it solves the all alignments of a given EST reliably.

3   Execution Time, Sensitivity, and Accuracy

The algorithm described here has been implemented in C++. In what follows, we call our software *Squall*. We installed and evaluated performance of our software Squall, sim4, and BLAT on a single processor of PrimePower 1000 with a clock rate of 450 MHz, 32 Gbytes of main memory, and running Solaris 8.

3.1   Execution Time Comparison with sim4 and BLAT

The performance of Squall was compared with sim4 and BLAT using the Chr.22 of the NCBI draft human genome (Build 28) . We randomly selected one thousand ESTs in the Unigene database and aligned these ESTs. Table 1 presents average time in seconds to align an EST, which makes it clear that our method has improved the computation time by orders of magnitude. Both BLAT and sim4 initially feed genomic sequences, and then perform alignments of ESTs. In Table 1, we excluded execution times of the former feeding step, so that this comparison was fair to both sim4 and BLAT, because sim4 and BLAT required much more execution time to feed the genome than our software Squall did.

3.2   Execution Time to Align Millions of ESTs with Genome

Using Squall, we evaluated average time of checking whether or not each EST in the UniGene database was aligned with the NCBI draft human genome (Build 28) and calculating the alignment when the EST mapped to the genome. Table 2 illustrates average time in seconds, and observe that average time to process one chromosome is almost proportional to the size of the chromosome. The total of average time is 0.0504 seconds. Thus, for instance, aligning three millions ESTs can be done in about 150,000 seconds on a single processor, but in practice, we can typically complete this task in less than half a day with using several processors.

Table 1. Performance comparison between Squall, BLAT, and sim4. One thousand ESTs randomly selected from UniGene are aligned with human Chr. 22 by using these three software

	Average time to align an EST (sec)
Squall (our aoftware)	0.001
BLAT	1.0
sim4	12

3.3   Sensitivity and Accuracy

To validate sensitivity and accuracy of Squall, we considered to use clean datasets of spliced sites [14,15,16,17,18]. These databases have been collected from various species to derive statistically confirmed rules for improving gene finding algorithms. Among these, we used HMR195 [18], a collection of 195 Mammalian genomic sequences that are annotated with exact locations of spliced sites, which are available at http://www.cs.ubc.ca/~rogic/evaluation/. HMR195 is useful for our experiment, because it includes 103 human genomic sequences, from which we extracted 103 mRNA sequences by consulting the locations of spliced sites. We then attempted to align these 103 sequences to the NCBI draft human genome (Build 28) by using Squall and BLAT [13]. We abandoned the same task by executing sim4 and Spidey because the task was very time-consuming and impossible to complete.

Table 3 presents the quality of each alignment by Squall and BLAT. 'o' indicates the exactly correct alignment is computed. '·' means that positions of one or two exons are incorrect, while '×' indicates that more than two exons are located incorrectly . '-' implies that no alignments are calculated for the mRNA. Table 4 summarizes the numbers of alignments in each quality category. Observe that Squall is much superior to BLAT both in sensitivity and in accuracy.

4   Graphical Viewer

We here present a graphical viewer for browsing the intron and exon structures resolved in our implementation. The viewer is available at

http://grl.gi.k.u-tokyo.ac.jp/.

This browser is called *Gene Resource Locator viewer* (*GRL viewer*, for short) [19] and it shows the alignment of each EST in the genomic sequence with a user-friendly interface. Figure 13, for instance, presents a group of ESTs mapped to the same locus. Each thick line represents the alignment of one EST (an EST alignment) in which the narrow yellow boxes are exons and the blue boxes are introns. Note that alignments share some common exons. Some alternatively spliced transcripts are also observed. The details of biological results discovered by our software can be found in [19].

Table 2. Size of chromosome and average execution time

Chr. Number	Number of bases in chromosome	Average time to align an EST (sec)
1	256,422,226	0.0044
2	241,269,072	0.0043
3	204,853,362	0.0042
4	191,388,243	0.0025
5	184,841,686	0.0031
6	178,317,572	0.0034
7	163,810,217	0.0035
8	145,730,583	0.0017
9	133,027,393	0.0017
10	142,094,824	0.0023
11	141,434,656	0.0021
12	139,632,178	0.0026
13	115,090,884	0.0010
14	106,469,138	0.0013
15	99,058,357	0.0011
16	93,800,608	0.0015
17	83,863,704	0.0019
18	81,800,002	0.0008
19	76,926,326	0.0020
20	62,963,852	0.0009
21	44,620,380	0.0004
22	47,848,586	0.0006
X	151,672,894	0.0029
Y	58,368,226	0.0002

5   Conclusion and Future Work

Our software Squall shows excellent speed and high precision, as described in Section 3. Gene structures can be resolved faster and more precisely using this algorithm than with other methods. In the field of medicine, there are many instances in which a disease-related gene is known. The precise localization of such genes will be possible in future by referring to the complete genetic map. Moreover, the generation of precise genetic maps for many creatures is a prerequisite for inter- and intra-species genome comparisons. Our algorithm can be used to generate genetic maps that combine the genomes and genes of various species.

Acknowledgements

We are grateful to Toshihiko Honkura and Tomoyuki Yamada for stimulating discussions. This research is supported by grant 12208003, a Grant-in-Aid for Scientific Research on Priority Areas from the Ministry of Education, Science and Culture, Japan.

Table 3. Quality of aligning 103 human mRNAs in HMR195 to the NCBI human draft genome (Build 28) by using Squall and BLAT. 'o' indicates the exacly correct alignment is computed for the mRNA. '·' means that positions of one or two exons are incorrect, while '×' indicates that more than two exons are located incorrectly . '-' implies that no alignments are calculated for the mRNA. The last column shows the chromosome number in which each mRNA is located

Acc numer	Squall	BLAT	Chr.	Acc numer	Squall	BLAT	Chr.	Acc numer	Squall	BLAT	Chr.
AB016625	o	×	5	U55058	o	·	1	AF043105	o	×	1
AF008216	o	o	4	AF068624	o	×	X	AF065988	o	o	12
AF092047	o	o	2	AF053069	o	·	11	AF026564	·	·	Y
AF096303	·	·	11	AF032437	o	·	12	AF037438	o	o	X
AF019563	o	·	19	AF007876	o	×	17	AF028233	o	·	17
AB012922	o	×	11	AF051160	o	·	6	AB007546	o	·	5
U25134	×	×	16	AF022382	o	×	1	AF058293	o	·	22
U17081	o	·	1	AF045999	o	×	2	AF055080	o	o	15
AB021866	-	-	?	U53447	·	o	4	AF037062	o	·	12
AF039704	×	×	11	AF009356	o	×	1	AB009589	o	·	9
AF082802	o	×	19	AF019409	o	×	11	AF047383	×	×	1
AF039954	o	·	17	AF015224	o	·	11	U31468	o	o	2
AB018249	o	·	17	AF042782	o	o	17	AF036329	o	·	20
AF099731	o	o	1	AF037207	-	-	?	AF042084	o	×	10
AF099730	o	o	1	AB016243	o	·	16	AF040714	o	·	7
AF039401	·	×	1	AF049259	o	×	17	AF039307	o	·	7
AF084941	o	o	6	AB012668	-	-	9	AF031237	o	o	3
AF059675	o	×	6	AF052572	·	·	2	AF005058	×	×	2
AF007189	o	o	7	AF042001	o	·	8	AF037372	o	o	1
AF016898	o	o	14	AF055475	o	·	X	D89060	o	×	1
AF076214	-	-	?	AF055903	o	×	11	AF032455	o	×	7
AB012113	o	·	17	AF053630	o	×	6	AB003730	o	o	12
AB019534	o	×	9	AF027148	o	·	11	AB006987	o	×	12
AF080237	o	×	16	AB007828	o	o	15	AF015812	o	×	17
AF071596	o	o	6	AF044311	-	-	?	AF015954	·	×	11
U43842	o	·	14	AF061327	o	o	19	D67013	·	×	3
AF053455	-	-	?	AF059650	o	×	5	D38752	×	o	10
AF071216	o	o	8	AB010874	·		12	D83956	o		6
AF058761	o	o	19	AF071552	o	o	8	AF016052	o	·	18
Y16791	·	×	17	AF059734	o	·	3	AB002059	o	·	22
AB016492	o	×	1	AF009962	o	o	3	AJ223321	o	o	1
AF001689	o	o	17	AB013139	o	×	8	U76254	o	o	4
AF029081	o	o	1	AF013711	o	·	11	AF017115	o	×	16
U96846	o	·	12	AF065396	o	×	6				
AF027152	o	×	12	AF058762	o	o	17				

Table 4. Statistics of alignment quality in Table 3. For instance, the number of exactly correct answers by Squall is 83 among 103 genes in HMR195

Quality	Squall	BLAT
o	83	30
·	8	32
×	5	35
—	7	6

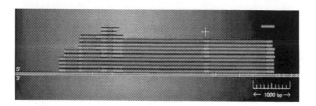

Fig. 13. GRL viewer

References

1. J. Craig Venter *et al.* The sequence of the Human Genome *Science*, 291:1304-1351 (2001) 2

2. International Human Genome Sequencing Consortium. Initial sequencing and analysis of the human genome *Nature*, 409:860-921 (2001) 2

3. S. B. Needleman and C. D. Wunsch. A general method applicable to the search for similarities in the amino acid sequence of two proteins. *Journal of Molecular Biology*, 48:443-453 (1970) 2

4. T. F. Smith and M. S. Waterman. Identification of common molecular subsequences. *Journal of Molecular Biology*, 147:195-197 (1981) 2

5. W. R. Pearson and D. J. Lipman. Improved tools for biological sequence comparison. *Proceeding of the National Academy of Sciences*, 85:2444-2448 (1988) 2

6. S. F. Altschul, W. Gis, E. W. Myers, and D. J. Lipman. Basic local alignment search tool. *Journal of Molecular Biology*, 215:403-410 (1990) 2

7. O. Gotoh. An improved algorithm for matching biological sequences. *Journal of Molecular Biology*, 162:705-708 (1982) 3, 5

8. Daniel S. Hirschberg. A Linear Space Algorithm for Computing Maximal Common Subsequences. *CACM* 18(6): 341-343 (1975) 3

9. M. S. Gelfand, A. A. Mironov, and P. A. Pevzner. Spliced alignment: A new approach to gene recognition. *Proc. Natl. Acad. Sci.* 93:9061-9066 (1996) 3

10. E. Birney and R. Durbin. Dynamite: a flexible code generating language for dynamic programming methods used in sequence comparison. *Proc. Fifth Int. Conf. Intelligent Systems Mol. Biol.* 5:55-64 (1997) 3

11. R. Mott. EST_GENOME: A program to align spliced DNA sequences to unspliced genomic DNA. *Comput. Appl. Biosci.* 13:477-478 (1997) 3

12. L. Florea, G. Hartzell, Z. Zhang, G. M. Rubin, and W. Miller. A computer program for aligning a cDNA sequence with a genomic sequence. *Genome Research*, 8(9):967-974 (1998) 3

13. W. James Kent. UCSC Human Genome Browser, http://genome.ucsc.edu/ 3, 13

14. M. Burset and R. Guigo. Evaluation of gene structure prediction programs. *Genomics*, 34:353-357 (1996). 3, 9, 13

15. T. A. Thanaraj. A clean data set of EST-confirmed splice sites from Homo sapiens and standards for clean-up procedures. *Nucl. Acids. Res.* 27: 2627-2637 (1999). 3, 9, 13

16. F. Clark and T. A. Thanaraj. Categorization and characterization of transcript-confirmed constitutively and alternatively spliced introns and exons from human. *Hum. Mol. Genet.* 11: 451-464 (2002) 3, 9, 13

17. M. Burset, I. A. Seledtsov, and V. V. Solovyev. SpliceDB: database of canonical and non-canonical mammalian splice sites. *Nucl. Acids. Res.* 29: 255-259 (2001). 3, 9, 13

18. S. Rogic, A. Mackworth and F. Ouellette. Evaluation of gene finding programs. *Genome Research*, 11: 817-832 (2001). 3, 9, 13

19. T. Honkura, J. Ogasawara, T. Yamada, and S. Morishita. The Gene Resource Locator: gene locus maps for transcriptome analysis. *Nucl. Acids. Res.*, 30(1):221-225 (2002) 13

Efficient Text Mining
with Optimized Pattern Discovery*

Hiroki Arimura[1,2]

[1] Department of Informatics, Kyushu University
Fukuoka 812-8581, Japan
[2] PRESTO, Japan Science and Technology Co., Japan
arim@i.kyushu-u.ac.jp
http://www.i.kyushu-u.ac.jp/~arim/

Abstract. The rapid progress of computer and network technologies makes it easy to collect and store a large amount of unstructured or semi-structured texts such as Web pages, HTML/XML archives, E-mails, and text files. These text data can be thought of large scale text databases, and thus it becomes important to develop an efficient tools to discover interesting knowledge from such text databases.

There are a large body of data mining researches to discover interesting rules or patterns from well-structured data such as transaction databases with boolean or numeric attributes [, ,]. However, it is difficult to directly apply the traditional data mining technologies to text or semi-structured data mentioned above since these text databases consist of (i) heterogeneous and (ii) huge collections of (iii) un-structured or semi-structured data. Therefore, there still have been a small number of studies on text mining, e.g., [, , ,].

Our research goal is to devise an efficient semi-automatic tool that supports human discovery from large text databases. Therefore, we require a fast pattern discovery algorithm that can work in time, e.g., $O(n)$ to $O(n \log n)$, to respond in real time on an unstructured data set of total size n. Furthermore, such an algorithm has to be robust in the sense that it can work on a large amount of noisy and incomplete data without the assumption of an unknown hypothesis class.

To achieve this goal, we adopt the framework of *optimized pattern discovery* [ ], also known as *Agnostic PAC learning* [ ] in computational learning theory. In optimized pattern discovery, an algorithm tries to find a pattern from a hypothesis space that optimizes a given statistical measure, such as *classification error* [ ], *information entropy* [ ], and *Gini index* [ ], to discriminate a set of *interesting* documents from a set of *uninteresting* ones. In the recent developments in computational learning theory, it is shown that such an algorithm can approximate arbitrary distributions on data within a given class of hypotheses very well in the sense of classification accuracy [,].

* This work is partially supported by the Ministry of Education, Science, Sports, and Culture, Grant-in-Aid for Scientific Research on Priority Areas Informatics (No. 14019070) 2002.

In this lecture, we present efficient and robust pattern discovery algorithms for large unstructured and semi-structured data combining the techniques from combinatorial pattern matching, computational geometry, and computational learning theory [, , ,]. We then describe applications of our pattern discovery algorithms to interactive document browsing, keyword discovery from Web and structure discovery from XML archives. We also discuss applications of optimized pattern discovery to information extraction from Web [,].

References

1. R. Agrawal, R. Srikant, Fast algorithms for mining association rules, In *Proc. VLDB'94*, 487–499, 1994.
2. H. Arimura, A. Wataki, R. Fujino, S. Arikawa, A fast algorithm for discovering optimal string patterns in large text databases, In *Proc. 9th Int. Workshop on Algorithmic Learning Theory*, LNAI 1501, 247–261, 1998.
3. T. Asai, K. Abe, S. Kawasoe, H. Arimura, H. Sakamoto, S. Arikawa, Efficient substructure discovery from large semi-structured data, In *Proc. 2nd SIAM Int'l. Conf. on Data Mining*, 158–174, 2002.
4. W. W. Cohen, Y. Singer, Context-sensitive learning methods for text categorization, J. ACM, 17(2), 141–173, 1999.
5. M. Craven, D. DiPasquo, D. Freitag, A. McCallum, T. Mitchell, K. Nigam, and S. Slattery, Learning to construct knowledge bases from the World Wide Web, Artificial Intelligence, 118, 69–114, 2000.
6. L. Devroye, L. Gyorfi, G. Lugosi, A probablistic theory of pattern recognition, Springer-Verlag, 1996.
7. U. M. Fayyad, G. Piatetsky-Shapiro, P. Smyth, and R. Uthurusamy, Advances in Knowledge Discovery and Data Mining, The MIT Press, Cambridge, 1996.
8. D. Hand, H. mannila, and P. Smyth, Principles of Data The MIT Press, Cambridge, 1996.
9. T. Kasai, G. Lee, H. Arimura, S. Arikawa, K. Park, Linear-time longest-common-prefix computation in suffix arrays and its applications, In *Proc. 12th Combinatorial Pattern Matching*, LNCS 2089, 181-192, Springer-Verlag, 2001.
10. M. J. Kearns, R. E. Shapire, L. M. Sellie, Toward efficient agnostic learning. *Machine Learning*, 17(2–3), 115–141, 1994.
11. S. Morishita, On classification and regression, In *Proc. 1st Int'l. Conf. on Discovery Conference*, LNAI 1532, 49–59, 1998.
12. L. Parida, I. Rigoutsos, A. Floratos, D. Platt, Y. Gao, Pattern discovery on character sets and real-valued data: linear bound on irredundant motifs and an efficient polynomial time algorithm, In *Proc. 11th ACM-SIAM Symposium on Discrete Algorithms*, 297–308, 2000.
13. F. Provost, M. Schkolnick, R. Srikant (eds.), Proc. 7th ACM SIGKDD international conference on Knowledge discovery and data mining ACM Press, 2001.
14. H. Sakamoto, H. Arimura, and S. Arikawa, Extracting partial structures from html documents, In *Proc. 14th Florida Artificial Intelligence Research Symposium (FLAIRS'2001)*, Florida, AAAI, 264-268, May, 2001.
15. K. Taniguchi, H. Sakamoto, H. Arimura, S. Shimozono and S. Arikawa, Mining semi-structured data by path expressions, In *Proc. 4th Int'l. Conf. on Discovery Science*, LNAI 2226, 378-388, Springer-Verlag, 2001.

16. S. Shimozono, H. Arimura, S. Arikawa, Efficient discovery of optimal word-association patterns in large text databases, *New Gener. Comp.*, 18, 49–60, 2000.

17. J. T. L. Wang, G. W. Chirn, T. G. Marr, B. Shapiro, D. Shasha and K. Zhang, Combinatorial pattern discovery for scientific data: Some preliminary results, In *Proc. SIGMOD'94*, 115–125, 1994.

Application of Lempel-Ziv Factorization
to the Approximation
of Grammar-Based Compression

Wojciech Rytter[1,2]

[1] Instytut Informatyki, Uniwersytet Warszawski, Poland
rytter@mimuw.edu.pl
[2] Department of Computer Science, Liverpool University
Chadwick Building, Peach Street, Liverpool L69 7ZF, U.K.

Abstract. We present almost linear time ($O(n \cdot \log |\Sigma|)$ time) $O(\log n)$-ratio approximation of minimal grammar-based compression of a given string of length n over an alphabet Σ and $O(k \cdot \log n)$ time transformation of $LZ77$ encoding of size k into a grammar-based encoding of size $O(k \cdot \log n)$. Computing exact size of the minimal grammar-based compression is known to be NP-complete. The basic novel tool is the AVL-grammar.

Keywords: LZ-compression, minimal grammar, AVL-tree, AVL-grammar

1 Introduction

Text compression based on context free grammars, or equivalently, on straight-line programs, has recently attracted much attention, see [, ,] and [, , ,]. The grammars give a more structured type of compression. In a grammar-based compression a single text w of length n is generated by a context-free grammar G. Assume we deal with grammars generating single words. In the paper, using ideas similar to *unwinding* from [ ] and *balanced grammars* from [ ], we show a logarithmic relation between LZ-factorizations and minimal grammars. Recently, approximation ratios of several grammar-based compression have been investigated by Lehman and Shelat in [ ]. In this paper we propose a new grammar-based compression algorithm based on Lempel-Ziv factorization (denoted here by LZ), which is a version of $LZ77$-encoding [ ]. For a string w of length n denote by $LZ(w)$ the Lempel-Ziv factorization of w. We show:

1. For each string w and its grammar-based compression G $|LZ(w)| = O(|G|)$;
2. Given $LZ(w)$, a grammar-based compression G' for w can be efficiently constaruted with $|G'| = O(\log |w| \cdot |LZ(w)|)$.

This gives $\log n$-ratio approximation of minimal grammar-based compression, since LZ-factorization can be computed effciently [ ]. The grammar-based type

A. Apostolico and M. Takeda (Eds.): CPM 2002, LNCS 2373, pp. 20– , 2002.

of compression is more convenient than LZ-compression, especially in compressed and fully compressed pattern-matching. For simplicity assume that the grammars are in Chomsky normal form. The size of the grammar G, denoted by $|G|$, is the number of productions (rules), or equivalently the number of nonterminals of a grammar G in Chomsky normal form. Grammar compression is essentially equivalent to straight-line programs. A *grammar* (*straight-line program*) is a sequence of assignment statements:
$$X_1 = expr_1; \; X_2 = expr_2; \ldots; \; X_m = expr_m,$$
where X_i are nonterminals and $expr_i$ is a single (terminal) symbol, or $expr_i = X_j \cdot X_k$, for some $j, k < i$, where $\cdot$ denotes the concatenation of X_j and X_k. For each nonterminal X_i, denote by $val(X_i)$ the value of X_i, it is the string described by X_i. The string described by the whole straight-line program is $val(X_m)$. The size of the straight-line program is m.

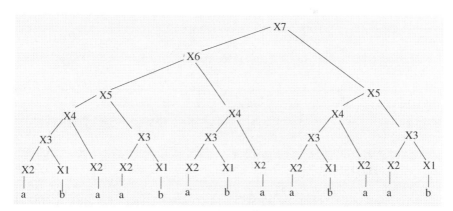

Fig. 1. $Tree(G_7)$: the parse-tree of G_7. It is a binary tree: we assume that the nonterminals generating single terminal symbols are identified with these symbols. We have: $val(G_7) = Fib_7 = abaababaabaab$

The problem of finding the smallest size grammar (or equivalently, straight line program) generating a given text is NP-complete. Hence there is an interest in polynomial-time approximation algorithms. We consider the following problem: **(approximation of grammar-based compression)**

> *Instance:*   given a text w of length n,
> *Question:*   construct in polynomial time a grammar G such that $val(G) = w$ and the *ratio* between $|G|$ and the size of the minimal grammar for w is *small*.

Example 1. Let us consider the following grammar G_7 which describes the *7th Fibonacci word* $Fib_7 = abaababaabaab$. We have $|G_7| = 7$. This is the smallest size grammar in Chomsky normal form for Fib_7. However the general test for grammar minimality is computationally hard.

$$X_7 = X_6 \cdot X_5; \ X_6 = X_5 \cdot X_4; \ X_5 = X_4 \cdot X_3;$$
$$X_4 = X_3 \cdot X_2; \ X_3 = X_2 \cdot X_1;$$
$$X_2 = a; \qquad X_1 = b;$$

If A is a nonterminal of a grammar G then we sometimes identify A with the grammar G with the starting nonterminal replaced by A, all useless unreachable nonterminals being removed. In the parse tree for a grammar with the starting nonterminal A we can also sometimes informally identify A with the root of the parse tree.

2   LZ-Factorizations and Grammar-Based Factorizations

We consider a similar version of the LZ77 compression algorithm without *self-referencing* as one used in [ ] (where it is called LZ1). Intuitively, LZ algorithm compresses the input word because it is able to discover some repeated subwords, see [ ]. The Lempel-Ziv code defines a natural factorization of the encoded word into subwords which correspond to intervals in the code. The subwords are called *factors*. Assume that Σ is an underlying alphabet and let w be a string over Σ. The LZ-factorization of w is given by a decomposition: $w = f_1 \cdot f_2 \cdots f_k$, where $f_1 = w[1]$ and for each $1 \leq i \leq k$, f_i is the longest prefix of $f_i \ldots f_k$ which occurs in $f_1 \ldots f_{i-1}$.

We can identify each f_i with an interval $[p, q]$, such that $f_i = w[p \ldots q]$ and $q \leq |f_1 \ldots f_{i-1}|$. We identify LZ-factorization with $LZ(w)$. Its size if the number of factors.

For a grammar G generating w we define the parse-tree $Tree(G)$ of w as a derivation tree of w, in this tree we identify (conceptually) terminal symbols with their parents, in this way every internal node has exactly two sons, see Figure . Define the partial parse-tree, denoted $PTree(G)$ as a maximal subtree of $Tree(G)$ such that for each internal node there is no node to the left having the same label. We define also the grammar factorization, denoted by G-factorization, of w, as a sequence of subwords generated by consecutive bottom nonterminals of $PTree(G)$, these nonterminals are enclosed by rectangles in Figure . Alternatively we can define G-factorization as follows: w is scanned from left to right, each time taking as next G-factor the longest unscanned prefix which is generated by a single nonterminal which has already occurred to the left or a single letter if there is no such nonterminal. The factors of LZ- and G-factorizations are called LZ-factors and G-factors, respectively.

Example 2. The LZ-factorization of the 7-th Fibonacci word Fib_7 is given by:

$$abaababaabaab \ = \ f_1 \ f_2 \ f_3 \ f_4 \ f_5 \ f_6 \ = \ a \ b \ a \ aba \ baaba \ ab$$

The G_7-factorization is:

$$abaababaabaab \ = \ g_1 \ g_2 \ g_3 \ g4 \ g_5 \ g_6 \ = \ a \ b \ a \ ab \ aba \ abaab$$

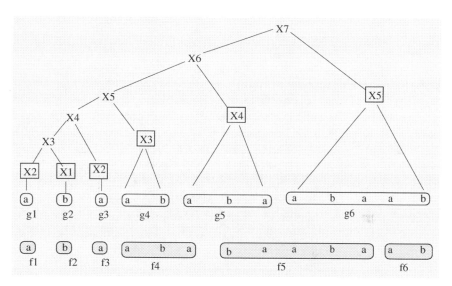

Fig. 2. $PTree(G_7)$, LZ-factorization (shaded one) is shown below G_7 factorization of Fib_7. The number of LZ-factors does not exceed the number of grammar-based factors

Lemma 1. *Let G be a context free grammar in Chomsky normal form generating a single string w. Let g be the number of G-factors of w, then $|G| \geq g$.*

Proof. The nonterminals corresponding to G-factors do not need to be distinct, however all internal nodes of the tree $PTree(G)$ have different nonterminal labels, so there are at least $g - 1$ internal nodes in this tree which correspond to nonterminals. Additionally there should be at least one nonterminal which production is of the type $A \rightarrow a$. Altogether there are at least g different nonterminals.

Theorem 1.
For each string w and its grammar-based compression G $|LZ(w)| \leq |G|$.

Proof. The number of LZ-factors is not greater than the number of G-factors. This follows from the intuitively obvious fact that L-factorization is *more greedy* than G-factorization. Let $f_1 f_2 \ldots f_k$ be LZ-factorization and $g_1 g_2 \ldots g_r$ be the G-factorization of w, then we can prove by induction on i that for each $i \leq \min(k,r)$ we have:

$$|g_1 g_2 \ldots g_i| \leq |f_1 f_2 \ldots f_i|.$$

Hence if $r \leq k$ then $|g_1 g_2 \ldots g_r| \leq |f_1 f_2 \ldots f_r|$ and $f_1 f_2 \ldots f_r = w$, since $g_1 g_2 \ldots g_r = w$. Consequently $k \leq r$. This completes the proof.

3   Efficient Concatenation of AVL-Grammars

We introduce new type of grammars: AVL-grammars. They correspond naturally to AVL-trees. The first use of of a different type balanced grammars has appeared in [ ]. AVL-trees are usually used in the context of binary search trees, here we use them in the context of storing in the leaves the consecutive symbols of the input string w. The basic operation is the concatenation of sequences of leaves of two trees. We use the standard AVL-trees, for each node v the balance of v, denoted $bal(v)$ is the difference between the height of the left and right subtrees of the subtree of T rooted at v. T is AVL-balanced iff $|bal(v)| \leq 1$ for each node v. We say that a grammar G is AVL-balanced if $Tree(G)$ is AVL-balanced. Denote by $height(G)$ the height of $Tree(G)$ and by $height(A)$ the height of the parse tree with the root labeled by a nonterminal A. The following fact is a consequence of a similar fact for AVL-trees, see [ ].

Lemma 2.
If the grammar G is AVL-balanced and $|val(G)| = n$ then $height(G) = O(\log n)$.

In case of AVL-balanced grammars in each nonterminal A additional information about the balance of A is kept: $bal(A)$ is the balance of the node corresponding to A in the tree $Tree(G)$. We do not define the balance of nodes corresponding to terminal symbols, they are identified with their fathers: nonterminals generating single symbols. Such nonterminals are leaves of $Tree(G)$, for each such nonterminal B we define $bal(B) = 0$.

Example 4. Let us consider $G = G_7$ and look at the tree in Figure . Only nonterminal nodes are considered. $bal(X_1) = bal(X_2) = bal(X_3) = 0$ and $bal(X_4) = \ldots bal(X_7) = +1$. Hence the grammar G_7 for the 7th Fibonacci word is AVL-*balanced.*

Lemma 3.   *Assume A, B are two nonterminals of AVL-balanced grammars. Then we can construct in $O(|height(A) - height(B)|)$ time a AVL-balanced grammar $G = Concat(A, B)$, where $val(G) = val(A) \cdot val(B)$, by adding only $O(|height(A) - height(B)|)$ nonterminals.*

Proof. We refer to the 3rd volume of Knuth's book, page.474, see [ ], for the description of the *concatenation algorithm* for two AVL-balanced trees $T1$, $T2$ with roots A and B. Our AVL-trees contain keys (symbols) only in leaves, so to concatenate two trees we do not need to delete the root of one of them (implying a costly restructuring), see [ ]. Assume that $height(T1) \geq height(T2)$, other case is symmetric. We follow the rightmost branch of $T1$, the heights of nodes decrease each time at most by 2. Then we stope at a node v such that $height(v) - height(T2) \in \{1, 0\}$. We create a new node v', its father is the father of v and its sons are v, $root(T2)$, see Figure .
The resulting tree can be unbalanced (by at most 2) on the rightmost branch. Suitable rotations are to be done, see Figure . The concatenating algorithm for AVL-trees can be applied to the parse-trees and automatically extended to the

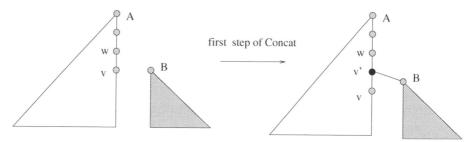

Fig. 3. The first step of $Concat(A, B)$. The edge (w, v) is split into (w, v'), (v', v), where $height(v) = height(B)$ or $height(v) = height(B) + 1$. The node v' is a newly created node. The corresponding grammar productions are added

case of AVL-grammars. The real parse-tree could be even of an exponential size, however what we need is only its rightmost or leftmost branch, which can be recovered from the grammar going top-down.

There is one more technical detail distinguishing it from the concatenation of trees. It can happen that only a constant number of rotations have been done, which is reflected by an introduction of several new productions of the grammar, see Figure . However this would imply creating copies of old terminals on the path from v' to the root, due to the change of subtrees rooted at v. However the number of affected nonterminals is only $O(|height(A) - height(B)|)$. If we change production rule for a nonterminal in $Tree(G)$ we should do it on its newly created copy, since this nonterminal can occur in other places, and we cannot affect other parts of the tree. Possibly the structure of the tree is changed in one place at the bottom of the rightmost path. However for all nodes on this path the corresponding nonterminals have to change names to new ones, since sequences of leaves in their subtrees have changed (by a single symbol). The rebalancing has to be done only on the rightmost branch bottom-up starting at v. The part of this branch is of length $O(|height(A) - height(B)|)$.

4   Construction of Small Grammar-Based Compression

Assume we have an LZ-factorization $f_1 f_2 \ldots f_k$ of w. We convert it into a grammar whose size increases by a logarithmic factor. Assume we have LZ-factorization $w = f_1 f_2 \ldots f_k$ and we have already constructed *good* (*AVL-balanced* and of size $O(i \cdot \log n)$) grammar G for the prefix $f_1 f_2 \ldots f_{i-1}$. If f_i is a terminal symbol generated by a nonterminal A then we set $G := Concat(G, A)$. Otherwise we locate the segment corresponding to f_i in the prefix $f_1 f_2 \ldots f_{i-1}$. Due to the fact that G is balanced we can find a logarithmic number of nonterminals $S_1, S_2, \ldots S_{t(i)}$ of G such that $f_i = val(S_1) \cdot val(S_2) \cdot \ldots val(S_{t(i)})$, see Figure . The sequence $S_1, S_2, \ldots S_{t(i)}$ is called the *grammar decomposition* of the factor f_i.

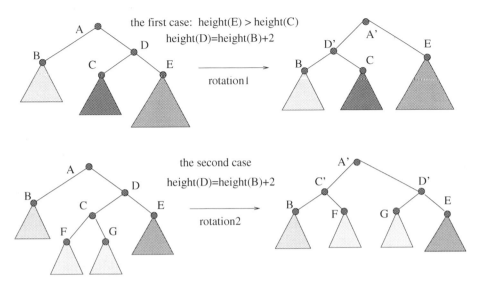

Fig. 4. All nodes are well balanced except the root A, which is overbalanced to the right. There are two cases. A single rotation in $Tree(G)$ corresponds to a local change of constant number of productions and creation of some new nonterminals. The root becoms balanced, but its father or some node upwards can be still unbalanced and the processing goes up

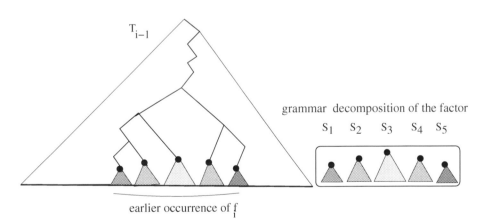

Fig. 5. The next factor f_i is split into segments corresponding to nonterminals occurring to the left. There are $O(\log n)$ segments since the height of the parse-tree is $O(\log n)$. Observe that $height(S1) \leq height(S-2) \leq height(S3) \geq height(S4) \geq height(S5)$. The sequence of heights of subtree $S_1, S_2, \ldots S_{t(i)}$ is *bitonic*

We concatenate the parts of the grammar corresponding to this nonterminals with G, using the operation *Concat* mentioned in Lemma . Assume the first $|\Sigma|$ nonterminals corresponds to letters of the alphabet, so they exist at the beginning. We initialize G to the grammar generating the first symbol of w and containing all nonterminals for terminal symbols, they don't need to be initially *connected* to the string symbol. The algorithm starts with the computation of LZ-factorization, this can be done using suffix trees in $O(n \log |\Sigma|)$ time, see [ ]. If LZ-factorization is too large (exceeds $n/\log n$) then we neglect it and write a trivial grammar of size n generating a given string. Otherwise we have only $k \leq n \cdot \log n$ factors, they are processed from left to right. We perform :

ALGORITHM *Construct-Grammar(w)*; $\{|w| = n\}$
     compute LZ factorization $f_1 f_2 f_3 \ldots f_k$
     $\{$ in $O(n \log |\Sigma|)$ time , using suffix trees$\}$
     if $k > n/\log(n)$ **then return** trivial $O(n)$ size grammar
     else
     for $i = 1$ **to** k **do**
        (1) Let $S_1, S_2, \ldots, S_{t(i)}$ be grammar decomposition of f_i;
        (2) $H := Concat(S_1, S_2, \ldots, S_{t(i)})$;
        (3) $G := Concat(G, H)$;
     return G;

Due to Lemma we have $t(i) = O(\log n)$, so the number of two-arguments concatenations needed to implement single step (2) is $O(\log n)$, each of them adding $O(\log n)$ nonterminals. Steps (1) and (3) can be done in $O(\log n)$ time, since the height of the grammar is logarithmic. Hence the algorithm gives $O(\log^2(n))$-ratio approximation.

At the cost of slightly more complicated implementation of step (2) $\log^2 n$-ratio can be improved to a $\log n$-ratio approximation. The key observation is that the sequence of heights of subtrees corresponding to segments S_i of next LZ-factor is *bitonic*, see Figure . We can split this sequence into two subsequences: height-nondecreasing sequence$i$ $R_1, R_2, \ldots R_k$, called right-sided, and height-nonincreasing sequence $L_1, L_2, \ldots L_r$, called left-sided. We can save a logarithmic factor due to the following fact.

Lemma 4. *Assume $R_1, R_2, \ldots R_k$ is a right-sided sequence, and G_i is the AVL-grammar which results by concatenating $R_1, R_2, \ldots R_i$ from left-to-right. Then*

$$|height(R_i) - height(G_{i-1})| \leq \max \{(height(R_i) - height(R_{i-1}), 1\}$$

Proof. We use the following obvious fact holding for any two nonterminals A, B. Denote $h = \max\{height(A), height(B)\}$, then we have:

$$h \leq height(Concat(A, B)) \leq h + 1 \tag{1}$$

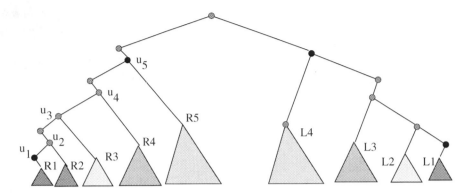

Fig. 6. An example of the grammar decomposition of the next factor f_i into a sequence of right-sided and a sequence of left-sided subtrees: $R1 \cdot R2 \cdot R3 \cdot R4 \cdot R5 \cdot L4 \cdot L3 \cdot L2 \cdot L1$ The *right-sided* sequence of subtrees is $R1, R2, \ldots R5$

Let u_i be the father of the node corresponding to R_i, see Figure . We show:

Claim. $height(G_i) \leq height(u_i)$.

The proof of the claim is by induction. For $i = 1$ we have $G_i = R_i$. In this case $height(u_1) = height(R_1) + 1 \geq height(G_1)$. Assume the claim holds for $i - 1$: $height(G_{i-1}) \leq height(u_{i-1})$. There are two possibilities.

Case 1. $height(G_{i-1}) \leq height(R_i)$.
   Then, according to Equation : $height(G_i) \leq height(G_{i-1} + 1$, and due to the inductive assumption $height(G_i) \leq height(u_{i-1}) + 1 \leq height(u_i)$.
Case 2. $height(G_{i-1}) \leq height(R_i)$.
   Then, again using Equation , $height(G_i) \leq height(R_i) + 1 \leq height(u_i)$.

This completes the proof of the claim. We go now to the main part of the proof of the lemma.

If $height(G_{i-1}) \geq height(R_i)$ then
$$|height(R_i) - height(G_{i-1}))| = height(G_{i-1}) - height(R_i)$$
$$\leq height(u_{i-1} - height(R_i) \leq 1.$$

The last inequality follows from the AVL-property.

If $height(G_{i-1}) \leq height(R_i)$ then
$$|height(R_i) - height(G_{i-1})| = height(R_i) - height(G_{i-1})$$
$$\leq height(R_i) - height(R_{i-1}),$$

since $height(G_{i-1}) \geq height(R_{i-1})$. This completes the proof.

Theorem 2. *We can construct in a $O(n \log |\Sigma|)$ time a $O(\log n)$-ratio approximation of a minimal grammar-based compression.*

Given LZ-factorization of length k we can construct a corresponding grammar of size $O(k \log n)$ in time $O(k \log n)$.

Proof. The next factor f_i is decomposed into segments $S_1, S_2, \ldots S_{t(i)}$. It is enough to show that we can create in $O(\log n)$ time an AVL-grammar for the concatenation of $S_1, S_2, \ldots, S_{t(i)}$ by adding only $O(\log n)$ nonterminals and productions to G, assuming that the grammrs for $S_1, S_2, \ldots, S_{t(i)}$ are available.

The sequence $(S_1, S_2, \ldots, S_{t(i)}$ consists of a right-sided sequence and left-sided sequence. The grammars H', H'' corresponding to these sequences are computed (by adding logarithmically many nonterminals to G), due to Lemma . Then H', H'' are concatenated. Assume $R_1, R_2, \ldots R_k$ are right-sided subtrees. Then the total work and number of extra nonterminals needed to concatenate $R_1, R_2, \ldots R_k$ can be estimated as follows:

$$\sum_{i=2}^{k} |height(R_i) - height(G_{i-1})| \leq \sum_{i=2}^{k} \max \{height(R_i) - height(R_{i-1}), 1\}$$

$$\leq \sum_{i=2}^{k} (height(R_i) - height(R_{i-1})) + \sum_{i=2}^{k} 1 \leq height(R_k) + k = O(\log n).$$

The same applies to the left-sided sequence in a symmetric way. Altogether processing each factor f_i enlarges the grammar by an $O(\log n)$ additive factor and needs $O(\log n)$ time. If our goal is $\log n$-ratio then we consider only the case when the number k of factors is $O(n/\log n)$. If LZ-factorization is computed ($O(n \log |\Sigma|)$ time using suffix trees, or $O(n)$ time for integer alphabets, see [ ])), then the total time and the size of the output grammar is $O(k \log n)$.

5 Final Remarks

The main result is $\log n$-ratio approximation of a minimal grammar-based compression. However the transformation of LZ-encodings into grammars is of the same importance (or maybe even more important). The grammars are easier to deal than LZ-encodings, particularly in compressed pattern-matching, see [ ]. Our method leads to an simpler alternative algorithm for LZ77-compressed pattern-matching. Another useful feature of our grammars is their logarithmic height. We can take any grammar G (straight-line program) generating a single text and produce the G-factorization. Then we can transform it into a balanced grammar in the same way as it is done for LZ-factorization. This gives an alterantive algorithm for balancing grammars and straight line programs, it has been originally done using the methods from parallel tree contraction.

Theorem 3. *Assume G is a grammr (straight-line program) of size k generating a single string of size n. Then we can construct in $O(k \log n)$ time an equivalent grammar of height $O(\log n)$.*

Assume we have a grammar-compressed pattern and a text, where m_1, m_2 are the sizes of their compressed versions. In [ ] an improved algorithm for fully compressed pattern-matching algorithm has been given, which works in time $O(m_1 \cdot m_2 \cdot h_1 \cdot h_2)$, where h_1, $h2$ are the heights of corresponding grammars. We can use AVL-grammar together with the algorithm from [ ] to texts which are polynomially related to their compressed versions. This gives an improvement upon the result of [ ] for LZ fully compressed matching in case when encodings are polynomially related to explicit texts (which is a typical case). Let notation $\widetilde{O}(g(k))$ stand for $O(g(k)\log^c(k))$, where c is a constant.

Theorem 4. *Given LZ-encodings of sizes m_1 and m_2 of the pattern P and a text T respectively. Assume that the original texts are polynomially related to their compressed versions. Then we can do fully compressed pattern-matching in time $\widetilde{O}(m_1 \cdot m_2)$.*

Grammar compression can be also considered for two-dimensional texts, but this case is much more complicated, see [ ].

References

1. A. Apostolico, S. Leonardi, Some theory and practice of greedy off-line textual substitution, DCC 1998, pp. 119-128
2. P. Berman, M. Karpinski, L. L. Larmore, W. Plandowski, and W. W. Rytter, On the Complexity of Pattern Matching for Highly Compressed Two-Dimensional Texts, *Proceedings of the* 8th *Annual Symposium on Combinatorial Pattern Matching* LNCS 1264, Edited by A. Apostolico and J. Hein, (1997), pp. 40–51. Full version to appear in JCSS
3. M. Crochemore and W. Rytter, *Text Algorithms,* Oxford University Press, New York (1994)   ,   ,
4. M. Farach and M. Thorup, String matching in Lempel-Ziv compressed strings, *Proceedings of the* 27th *Annual Symposium on the Theory of Computing* (1995), pp. 703–712   ,   ,
5. L. Gąsieniec, M. Karpinski, W. Plandowski and W. Rytter, Efficient Algorithms for Lempel-Ziv Encoding, *Proceedings of the* 5th *Scandinavian Workshop on Algorithm Theory.* Springer-Verlag (1996)
6. Martin Farach, "Optimal suffix tree construction with large alphabets", FOCS 1997
7. M. Hirao, A. Shinohara, M. Takeda, S. Arikawa, Faster fully compressed pattern matching algorithm for balanced straight-line programs", Proc. of 7th International Symposium on String Processing and Information Retrieval (SPIRE2000), pp. 132-138. IEEE Computer Society, September 2000   ,
8. M. Karpinski, W. Rytter and A. Shinohara, Pattern-matching for strings with short description, *Nordic Journal of Computing*, 4(2):172-186, 1997
9. J. Kieffer, E. Yang, Grammar-based codes: a new class of universal lossless source codes, IEEE Trans. on Inf. Theory 46 (2000) pp. 737-754
10. D. Knuth, *The Art of Computing, Vol. III Second edition.* Addison-Wesley (1998), page.474
11. J. K. Lanctot, Ming Li, En-hui Yang, Estimating DNA Sequence Entropy, SODA 2000

12. E. Lehman, A. Shelat, Approximation algorithms for grammar-based compression, SODA 2002
13. J. Ziv and A. Lempel, A Universal algorithm for sequential data compression, *IEEE Transactions on Information Theory* IT–23 (1977), pp. 337–343
14. M. Miyazaki, A. Shinohara, M. Takeda, An improved pattern-matching algorithm for strings in terms of straight-line programs, Journal of Discrete Algorithms, Vol.1, pp. 187-204, 2000
15. C. Nevill-Manning, Inferring sequential structure, PhD thesis, University of Waikato, 1996
16. W. Rytter, Compressed and fully compressed pattern-matching in one and two-dimensions, Proceedings of IEEE, November 2000, Volume 88, Number 11, pp. 1769-1778

Block Merging for Off-Line Compression

Raymond Wan and Alistair Moffat

Department of Computer Science and Software Engineering, The University of Melbourne
Victoria 3010, Australia
www.cs.mu.oz.au/~{rwan,alistair}

Abstract. To bound memory consumption, most compression systems provide a facility that controls the amount of data that may be processed at once. In this work we consider the RE-PAIR mechanism of Larsson and Moffat [2000], which processes large messages as disjoint blocks. We show that the blocks emitted by RE-PAIR can be post-processed to yield further savings, and describe techniques that allow files of 500 MB or more to be compressed in a holistic manner using less than that much main memory. The block merging process we describe has the additional advantage of allowing new text to be appended to the end of the compressed file.

1   Introduction

Dictionary-based compression techniques offer fast decoding rates and compression ratios close to those obtained by the best context-based schemes. Dictionary-based models can be broadly categorized as either on-line (adaptive), or off-line (semi-static).

On-line mechanisms process the input text incrementally, continually updating the dictionary during compression, without having to explicitly send it to the receiver. On the other hand, an off-line model is presented with the entire text at once and uses it to build an explicit dictionary. The dictionary and the compressed document, which is composed of indices into the dictionary, are thus two separate entities; and can be transmitted separately to the receiver. Decompression requires the decoding of the dictionary and then a symbol-by-symbol look up of each index into the dictionary.

To bound memory consumption, most compression systems provide a facility that controls the amount of data that may be processed at once. Some systems allow a command-line argument that stipulates (in megabytes) the amount of memory that may be used. Others, such as BZIP2 [Seward and Gailly, 1999], limit the amount of memory used by partitioning the document into blocks and compressing each block independently. Then, the memory space used depends on the block size.

In a block-based off-line system, each block contains everything that is required to decode that block, with no reliance on other blocks. However, as each block is segmented arbitrarily from what is often a homogeneous original text, it is clear that some overlap in information must occur, particularly in the dictionary of each compressed block. This overlap results in a loss of compression effectiveness.

In this paper we describe a mechanism to merge the compressed blocks produced by the off-line dictionary-based algorithm RE-PAIR [Larsson and Moffat, 2000]. We

A. Apostolico and M. Takeda (Eds.): CPM 2002, LNCS 2373, pp. 32–41, 2002.

seek to achieve two objectives: a reduction in the amount of memory used during compression; and the creation of a single dictionary for up to a gigabyte of text so that the underlying document can be easily browsed and searched.

RE-PAIR recursively replaces all instances of the most frequently occurring pair of adjacent symbols in the original text with a new symbol, so that each set of replacements reduces the message length by at least one. This replacement process is repeated until no pair of adjacent symbols occurs more than once in the reduced message. Two products are created during the replacement: a dictionary of phrases; and the reduced message sequence, which can be thought of as a start rule for a grammar.

The original document is assumed to be composed of *primitives*. We assume here that the primitives are the ASCII characters, but note that any set could be used. Each iteration of the recursive pairing process introduces a new *phrase* into the dictionary, or *phrase hierarchy*. Collectively, the primitives and phrases are called *symbols*. Each phrase in the phrase hierarchy is the concatenation of two symbols.

Every symbol has a *generation* associated with it. As a basis for the generations, all primitives are assigned to generation zero. Each phrase is then assigned to a generation one greater than the higher generation of its constituent symbols. Larsson and Moffat [2000] show that the phrase hierarchy can be stored compactly if it is sorted in increasing generation order, mapped to integers by a device they call the *chiastic slide*, and then encoded using *interpolative coding* [Moffat and Turpin, 2002].

The reduced message (or *sequence*) is a list of integer symbol numbers. It can be encoded using any coder, such as a minimum-redundancy coder or an arithmetic coder [Moffat and Turpin, 2002].

2   Block Merging

Larsson and Moffat [2000] describe structures that allow RE-PAIR to execute in time and space linear in the length of the original text. But the constant factor on memory space is relatively large, and when the primitives are characters, as much as $5n$ words of memory can be required to process a message of n bytes. In practical terms, this means that to process a message of (say) 500 MB, around 10 GB of random-access memory is required, a non-trivial amount even by today's standards.

One obvious work-around to this problem is to process the source text in fixed-length blocks, with the block size determined by the available memory. Each block is then compressed independently, and the compressed representations concatenated. Compared to a holistic approach, compression effectiveness is sacrificed: many of the phrases appear in the hierarchy of more than one block; and phrases that repeat at infrequent intervals may never be recognized.

The purpose of block merging is to regain some or all of that lost effectiveness. The process, dubbed RE-MERGE, combines two blocks at a time. A complete pass thus halves the number of blocks, and a compressed message of b blocks is reduced to a single block in $\lceil \log b \rceil$ passes. Each pair of blocks is merged in two separate steps. In the first step, the union of the two phrase hierarchies is formed. In the second step, the new phrase hierarchy is used to renumber the concatenated sequence, so that it refers to the common phrase hierarchy.

There are several possibilities in carrying out the second step. The simplest, which we call *Phase 1* operation, simply maps each symbol number to its replacement, so that the combined length of the merged sequence is exactly the sum of the two separate sequences. *Phase 2* operation seeks to rationalize the concatenated sequence produced by Phase 1 by searching for symbol pairs in the concatenated sequence that now exist in the combined phrase hierarchy. Two different ways of accomplishing this exist, corresponding loosely to a greedy approach versus a globally optimal approach. After Phase 2, repeated symbol pairs may exist in the merged sequence which do not appear in the phrase hierarchy. *Phase 3* addresses this problem by re-scanning the concatenated sequence to search for new candidates for replacement.

These three phases are described in this section. The number of phrases in a phrase hierarchy P is denoted as $|P|$. Likewise, the number of symbols in a sequence S is denoted as $|S|$. We use α and β to represent symbols.

2.1   Phase 1: Combining Phrase Hierarchies

P_1		P_2		P'	
$5 \rightarrow$ zi	1	$5 \rightarrow$ en	1	$5 \rightarrow$ zi	1
$6 \rightarrow$ en	1	$6 \rightarrow$ enz	2	$6 \rightarrow$ en	1
$7 \rightarrow$ enzi	2	$7 \rightarrow$ enzi	3	$7 \rightarrow$ enz	2
$8 \rightarrow$ zien	2	$8 \rightarrow$ zenzi	4	$8 \rightarrow$ enzi	2
				$9 \rightarrow$ zien	2
				$10 \rightarrow$ zenzi	3

Fig. 1. Merging phrase hierarchies P_1 and P_2 to form P'. Each phrase's generation appears to its right. In each of the three phrase hierarchies, phrases 0 to 4 map to the primitives c, e, i, n, and z, respectively

As the first step of block merging, the union of two phrase hierarchies P_1 and P_2 must be formed. By removing duplicate phrases, we expect to produce a phrase hierarchy P' which is bigger than each of P_1 and P_2, but smaller than their sum: $\max\{|P_1|, |P_2|\} \le |P'| \le |P_1| + |P_2|$.

To find duplicates, both phrase hierarchies are fully expanded to form the underlying sets of strings. Sorting the strings then brings together the pairs of duplicates. When duplicate strings are found, the one with the lower generation is retained in P', and the other is discarded. Any subsequent phrases constructed from a removed phrase must refer to the one that will be in P', producing a ripple effect that causes phrases to tend to move to lower generations.

An example of phrase hierarchy merging is shown in Figure 1. Duplicate phrases within the same generation are removed from P' (phrase 5 in P_2). The removal of phrase 7 in P_2 causes phrase 8 in P_2 to move to the third generation of the new phrase hierarchy (phrase 10 in P').

After merging the phrase hierarchies, their respective sequences (S_1 and S_2) are concatenated and renumbered into a single sequence, S'.

2.2   Phase 2: Identifying Old Phrases

During the sequence renumbering process, pairs of symbols might be encountered which correspond to a phrase in P'. For example, the adjacent symbols $\alpha\beta$ may occur only once in S_1, but multiple times in the sequence that led to S_2, in which case the combination $\alpha\beta$ appears in P' as a consequence of its inclusion in P_2. When S_1 is being renumbered, $\alpha\beta$ can be identified and replaced.

Old phrases can be identified in a number of different ways. A simple way is to pair each symbol with its predecessor in the sequence, and search for it in a table of phrases. If a match is found, then the replacement is made. The new symbol is then paired up with its preceding symbol, and a further pair of symbols searched for. This repeats until no replacement is found, at which point we move to the next symbol in the sequence. We denote this method of rationalizing a concatenated sequence as *Phase 2a*.

The replacements made by phase 2a may not be optimal, as a substitution made immediately may prohibit a replacement later that would have offered a shorter overall sequence. The alternative, *Phase 2b*, also identifies old phrases, but in an optimal manner. To achieve this, the source text is completely decompressed, and then re-compressed with reference solely to the static phrase hierarchy P', without consideration of the segmentation into phrases that was indicated by the earlier RE-PAIR process.

Phase 2b was described by Katajainen and Raita [1989] and by Klein [1997], and involves a reduction to the task of determining a shortest path in a directed acyclic graph (DAG). Each primitive in the document source text is placed in a node in the DAG, and linked to its successor by an edge. An additional edge is created from the final primitive of the text to a special sink node. That is, for a text of n primitives, the DAG contains $n + 1$ nodes, and n one-step edges. In addition, each subsequence of primitives that corresponds to any phrase in the hierarchy gives rise to a further directed edge, from the first node of the subsequence to the successor node of the last primitive in the subsequence. The least number of edges traversed from the first node of the DAG to the sink node represents the shortest parsing of S' using P'.

To allow efficient construction of the DAG, the phrase hierarchy is processed into a search structure. Each entry stores a string from the phrase hierarchy, and also notes the *longest prefix phrase* for that string, where phrase β is the longest prefix phrase for phrase α if and only if it is both a prefix of α, and there is no other phrase in the hierarchy that is a prefix of α and is longer than β. The longest prefix phrase of each of the primitives is the empty string. As a consequence of the phrase hierarchy being constructed by RE-PAIR, every non-primitive has at least one prefix that is also in the phrase hierarchy – its left component. However, there might be a longer phrase that can serve this role.

When the DAG is complete, each node contains edges representing phrases in the phrase hierarchy that start from that node. Every node will have at least one edge coming from it – the edge representing the primitive symbol of length one. Similarly, every node has at least one edge entering it. Breadth-first search is used to find the shortest path, which runs in time linear to the number of edges.

The amount of memory used can be bounded through the use of *synchronization points* (also known as *choke points*, or *cut vertices*). Figure 2 shows the DAG constructed for a string "zizenzi" using the phrase hierarchy {c, e, en, enz, enzi, i, n, z, zenzi, zi, zien}. There are no edges that bypass the third primitive (the second "z"). If we were to take a pair of scissors and cut the DAG at that point, no edges would be severed. Hence, every path from the first primitive to the sink node in this graph must pass through that second "z", including the shortest path. We can process this graph as two disjoint *segments* – first, by finding the best way to construct "zi"; and then, as an independent exercise, by finding an optimal representation for "zenzi". That is, once a synchronization point is reached, the segment which it ends can be processed to determine an optimal representation with respect to the phrase hierarchy, and that representation written to disk.

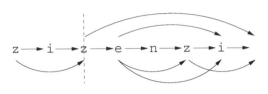

Synchronization points tend to be common, and the DAG segments relatively small. The length of each segment depends on the message and the contents of the phrase hierarchy. In our experiments on 509 MB of newspaper text, we found that the distance between synchronization points is, on average, approximately 20 characters.

Fig. 2. Possible parsings of the phrase "zizenzi"

This is rather more than the 1.46 characters reported by Katajainen and Raita [1989], but our test file is several magnitudes larger than the 43 kB they used, and our dictionary similarly bigger. Katajainen and Raita use phrases of up to 8 characters; our RE-PAIR/RE-MERGE hierarchy generated a longest phrase of 4,970 characters and an average phrase coverage of 21.5 primitives.

2.3 Phase 3: Identifying New Phrases

After merging the phrase hierarchy and the sequence, it is possible that a pair of symbols $\alpha\beta$ exists twice in S', but only once in sequence S_1 and once in sequence S_2. No phrase would have been formed to replace $\alpha\beta$, but after merging has completed, it can be replaced. We call this phase "identifying new phrases", which can be performed by applying RE-PAIR to S'. But, while S' is smaller than the original message, it is still too large to be handled as a single RE-PAIR block.

Instead, we relax the RE-PAIR requirements, and instead of searching for all possible pair replacements, and carefully processing them in decreasing frequency order, we look for *any* possible repeated pairs, and replace them in an arbitrary order. While the distinction between these two is small, it allows a considerable saving in memory space. Phase 3 proceeds as follows.

Suppose that the merged sequence S' contains n symbols. A bit-vector B of $8n$ bits (n bytes) is used to probabilistically note which of the $n-1$ possible pairs in S' reappear. Each pair $\alpha\beta$ of adjacent symbols in S' is hashed to three locations in B using three independent hashing functions. If any of the three bits is off, then $\alpha\beta$ has not appeared previously. In this case, the three indicated bits are all set to one, and processing continues with the next pair. On the other hand, if all three bits are on, then $\alpha\beta$ may have appeared previously in S'. In this case we add $\alpha\beta$ to a list L of possible replacements, and continue.

At the end of this first step, L contains every pair that repeats, and a number of *false duplicates* that do not, and were included only because of the probabilistic nature of the filter used to create L. Array L is then ordered, so that it can be searched, and a second sequential pass made through S'. In this pass, the frequency of all pairs in L is accumulated. Every pair in L has frequency at least 1, and some will have a frequency greater than 1. At the end of this second step, L is purged so that only the pairs with frequency greater than 1 are retained. These are the true duplicates.

A third pass is then made through S', checking each pair in S' for membership in the purged L. Each pair that remains in L should be replaced by a new symbol, subject to some slight complications caused by possible overlapping pairs. Because S' is processed sequentially, pairs will not be replaced in decreasing frequency order. But because S' was already created through the use of RE-PAIR, the repetitions being extracted in phase 3 are either unimportant, or very long, and it is unlikely that any high frequency repetitions occur. Hence, the difference between this final approximate process and a genuine monolithic application of RE-PAIR should be small.

Memory space is dominated by B, which requires n bytes, and L, which records symbol pairs. The size of L is determined by the number of duplicates, false and true, that appear in S'. The number of true matches is determined by the nature of S', but is expected to be small when reasonably-sized initial RE-PAIR blocks are used.

The number of false duplicates can be estimated as a function of n. The probability of any particular bit in B remaining off after $3n$ bits in B are randomly selected and turned on is $(1 - 1/8n)^{3n}$. The probability that a random selection of three bits in B are all on is thus given by

$$\left(1 - \left(1 - \frac{1}{8n}\right)^{3n}\right)^3.$$

Furthermore, this is the largest the false duplicate probability gets, as $3n$ bits can have been set only at the end of the processing of S'. Numerically, taking $n = 50 \times 10^6$ (that is, assuming an average phrase length of 20 symbols when processing a gigabyte of text) this expression gives a false match probability of a little over 3%. Multiplying by n, and allowing for the fact that each entry of L requires 8 bytes to store two symbol numbers, gives a false duplicate cost of approximately $0.25n$ bytes. It seems rather unlikely that true duplicates could add more than (say) another $0.05n$ bytes to the size of L.

Our analysis presumed that B contains $8n$ bits, and that 3 bits are checked and then set for each pair in S'. A more detailed analysis than is possible here shows that when the objective is to minimize the combined space required by B and L, these choices are a good combination. In total, approximately $1.3n$ bytes are required, or around 70 MB for the hypothetical problem involving $n = |S'| = 50 \times 10^6$. No memory space is required for S' – it is processed sequentially, and can be read the necessary three times from disk.

Note that there is also a small probability that the pair replacements performed by phase 3 might lead to further follow-on replacements being possible. If necessary, phase 3 can be iterated until a fixed point is reached, in a manner similar to the processing mechanism of the RAY system described by Cannane and Williams [2001].

3   Coding the Sequence

The relationship between RE-PAIR and the various phases of RE-MERGE is shown in Figure 3. After phase 3, an entropy coder is used to encode the sequence. The entropy coder can also be applied to the sequence produced by RE-PAIR or any phase of RE-

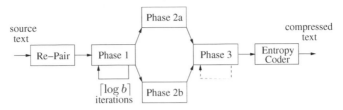

Fig. 3. An overview of how RE-MERGE's three phases interact with each other

MERGE. The entropy coder used in our investigation is SINT, a structured arithmetic coder [Moffat and Turpin, 2002] and generally yields the best compression that might be achieved from the RE-PAIR/RE-MERGE combination. It provides compression close to the self-information of the sequence, and is also sensitive to localized variations in symbol frequency, especially if there is a correlation between symbol identifier and symbol probability. Unlike RE-PAIR, SINT does have the drawback of being incremental, and not permitting random-access inspection of the compressed file.

To summarize the effect of the merging options provided by RE-MERGE, experiments were undertaken on the 509 MB collection of newspaper articles mentioned earlier. The English articles are drawn from the *Wall Street Journal* (WSJ) and embedded with SGML markup. The experimental machine was a 933 MHz Pentium III with 1 GB RAM and 256 kB on-die cache. Table 1 shows the compression performance of RE-PAIR with RE-MERGE.

Table 1. Compression ratios (bits per character, relative to the source file) of the phrase hierarchy and the sequence for WSJ after the three RE-MERGE phases

	RE-PAIR only	RE-MERGE Level 1	2a	2b	3
phrase hierarchy	0.223	0.064	0.064	0.064	0.146
sequence	1.640	1.645	1.562	1.556	1.466
total	1.863	1.709	1.626	1.620	1.611

In Table 1, the column "RE-PAIR only" shows the compression attained by RE-PAIR alone on WSJ as a sequence of 51 blocks, each 10 MB. The cost of storing the phrases is relatively high, because there are 51 separate hierarchies. Phase 1 significantly reduces this cost, a saving that is retained when phase 2 is applied. Phase 2b provides slightly better compression effectiveness than phase 2a, but at considerably increased execution cost (detailed shortly). Finally, phase 3 provides a further minute gain in compression effectiveness, but at the cost of a considerably larger phrase hierarchy – many additional phrases are identified, but each occurs only a few times. As mentioned earlier, context-based schemes tend to offer better compression than dictionary-based schemes. One of these systems, PPM [Moffat and Turpin, 2002], compressed WSJ to 1.60 bits per character using a sixth order model and 255 MB of memory.

Table 2. Time to encode and decode 509 MB of text, in CPU seconds. Two composite systems are compared, indicating the tension between execution time and compression ratio. As a reference point, while BZIP2 [Seward and Gailly, 1999] uses only a fraction of the memory used by RE-PAIR/ RE-MERGE, it does partition the input text into blocks and shows how fast a well-engineered compression system can run. BZIP2 requires 456 seconds to encode and 162 seconds to decode the same file

	Best efficiency	Best effectiveness
RE-PAIR (51 blocks)	1607	1607
phase 1 (6 iterations)	560	560
phase 2a	381	–
phase 2b	–	4160
phase 3	–	1735
entropy coding	38	260
total encoding time	2586	8322
entropy decoding	5	333
phrase substitution	52	60
total decoding time	57	393

Table 2 shows execution times, summarized in two different ways: one biased in favor of fast execution using the RE-VIEW coder [Moffat and Wan, 2001]; and one aimed at obtaining the greatest compression effectiveness using the SINT coder. The RE-VIEW coder is a two-byte-aligned coder that is used to encode the sequence with only a slight loss in compression ratio over an entropy coder such as SINT.

4   Appending Text

Section 2 discussed various phases for the merging of two phrase hierarchies (P_1 and P_2) and their sequences (S_1 and S_2). We can also generalize phase 2b, and suppose that P_1 and S_1 are known, but that P_2 does not exist and S_2 is uncompressed ASCII text.

In this case P', the final phrase hierarchy, is identical to P_1, and S_1 is used unchanged as the first component of S'. The second section of S' is formed by compressing S_2 relative to the phrases in $P' = P_1$. The only additional requirement is that RE-PAIR needs to include all of the ASCII characters as primitives.

The benefit of this approach is that text can be added to the end of a compressed file without applying RE-PAIR to the entire text. If the text used to generate P_1 and S_1 is representative of S_2, then the set of phrases available in P_1 will be good, and compression effectiveness will not suffer. More typical is for the nature of the text to slowly drift, and for compression effectiveness to slowly erode as the site of compression becomes removed from the section used to form the phrase hierarchy. When this happens, complete re-compression using RE-PAIR and RE-MERGE creates a more representative phrase hierarchy.

Figure 4 shows the cost of performing this append operation on the same WSJ source data as was used above. In each experiment an initial section was used as a seed to create a phrase hierarchy, and then the balance of the 509 MB added. Compression improves as more seed text is made available, but relatively slowly. For example, a seed text of just 30% of WSJ yields overall compression only 11% worse than when the phrase hierarchy is built from all of WSJ. Another way of looking at this is to extrapolate, and presume that the WSJ phrase hierarchy might comfortably be used to compress as much as a gigabyte of new data from the same source. The execution time also varies with the amount of seed text – with a small phrase hierarchy, all of the subsequent processing is faster, and use of a 10 MB seed text gives a total execution time approximately one third of the full RE-PAIR/RE-MERGE time.

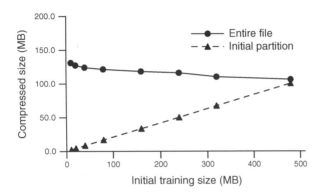

Fig. 4. Compressed size using RE-MERGE and the SINT coder, as a function of the amount of initial seed text made available to SINT when compressing the 509 MB collection WSJ

5   Related Work

There are a number of off-line compressing mechanisms that share features with RE-PAIR, and these might also be assisted by application of block merging. They include RAY [Cannane and Williams, 2001], XRAY [Cannane and Williams, 2000], and OFF-LINE [Apostolico and Lonardi, 2000]. Bentley and McIlroy [1999] describe a slightly different scheme in which long replacements are identified and used as the basis for substitutions. Early work in the area is due to Rubin [1976]. SEQUITUR [Nevill-Manning and Witten, 2000] is also a phrase-based compression scheme, but operates on-line.

In other work [Moffat and Wan, 2001], we describe RE-PHINE, a tool used to browse the phrases created from compressing a text document. Using RE-VIEW allows RE-PHINE to perform extremely fast searches in the compressed text.

An extension to SEQUITUR allows phrase browsing using a tool called PHIND [Nevill-Manning et al., 1997], and our RE-PHINE browser draws heavily on that work. Bahle et al. [2001] describe a browsing system in which successor words are stored directly in an explicit index.

Manber [1997] and de Moura et al. [2000] describe pragmatic techniques for searching compressed text. XRAY [Cannane and Williams, 2000] provides a facility similar to that of Section 4 whereby new text can be appended and compressed using a static model; a slightly different adaptive approach for appending new documents in information retrieval systems is described by Moffat et al. [1997]. Finally, as was noted in Section 2.2, Katajainen and Raita [1989] and Klein [1997] have previously examined the issue of finding good parsings once a static dictionary has been derived, and phase 2b is essentially an implementation of that approach.

6 Conclusion

Our primary motivation in this work has been to identify repeated phrases in very large texts. Improved compression effectiveness – to the level attained by a good context-based mechanism using a similar amount of main memory – has been a by-product of that quest. The result is a system that allows excellent compression effectiveness to be obtained if an arithmetic coder is used, or allows fast phrase-based browsing and searching on large documents if a simpler byte-based coder is used.

References

A. Apostolico and S. Lonardi. Off-line compression by greedy textual substitution. *Proc. IEEE*, 88(11):1733–1744, Nov. 2000.

D. Bahle, H. E. Williams, and J. Zobel. Compaction techniques for nextword indexes. In G. Navarro, editor, *Proc. 8th International Symposium on String Processing and Information Retrieval*, pages 33–45. IEEE Computer Society Press, Los Alamitos, CA, Nov. 2001.

J. Bentley and D. McIlroy. Data compression using long common strings. In J. A. Storer and M. Cohn, editors, *Proc. 1999 IEEE Data Compression Conference*, pages 287–295. IEEE Computer Society Press, Los Alamitos, California, Mar. 1999.

A. Cannane and H. E. Williams. A compression scheme for large databases. In M. E. Orlowska, editor, *Proc. 11th Australasian Database Conference*, pages 6–11, Canberra, Australia, 2000. IEEE Computer Society Press, Los Alamitos, CA.

A. Cannane and H. E. Williams. General-purpose compression for efficient retrieval. *Journal of the American Society for Information Science and Technology*, 52(5):430–437, Mar. 2001.

E. S. de Moura, G. Navarro, N. Ziviani, and R. Baeza-Yates. Fast and flexible word searching on compressed text. *ACM Transactions on Information Systems*, 18(2):113–139, 2000.

J. Katajainen and T. Raita. An approximation algorithm for space-optimal encoding of a text. *The Computer Journal*, 32(3):228–237, 1989.

S. T. Klein. Efficient optimal recompression. *The Computer Journal*, 40(2/3):117–126, 1997.

N. J. Larsson and A. Moffat. Offline dictionary-based compression. *Proc. IEEE*, 88(11):1722–1732, Nov. 2000.

U. Manber. A text compression scheme that allows fast searching directly in the compressed file. *ACM Transactions on Information Systems*, 15(2):124–136, Apr. 1997.

A. Moffat and A. Turpin. *Compression and Coding Algorithms*. Kluwer Academic Publishers, Boston, MA, 2002.

String Matching with Stopper Encoding and Code Splitting*

Jussi Rautio[1], Jani Tanninen[2], and Jorma Tarhio[1]

[1] Department of Computer Science and Engineering
Helsinki University of Technology
P.O. Box 5400, FIN-02015 HUT, Finland
{jrautio,tarhio}@cs.hut.fi
[2] Department of Computer Science, University of Joensuu
P.O. Box 111, FIN-80101 Joensuu, Finland
jtanni@cs.joensuu.fi

Abstract. We consider exact string searching in compressed texts. We utilize a semi-static compression scheme, where characters of the text are encoded as variable-length sequences of base symbols, each of which is represented by a fixed number of bits. In addition, we split the symbols into two parallel files in order to allow faster access. Our searching algorithm is a modification of the Boyer-Moore-Horspool algorithm. Our approach is practical and enables faster searching of string patterns than earlier character-based compression models and the best Boyer-Moore variants in uncompressed texts.

1 Introduction

The *string matching problem*, which is a common problem in many applications, is defined as follows: given a pattern $P = p_1 \ldots p_m$ and a text $T = t_1 \ldots t_n$ in an alphabet Σ, find all the occurrences of P in T. Various good solutions [ ] have been presented for this problem. The most efficient solutions in practice are based on the Boyer-Moore approach [ ].

Recently the compressed matching problem [ ] has gained much attention. In this problem, string matching is done in a compressed text without decompressing it. Researchers have proposed several efficient methods [, , ,] based on Huffman coding [ ] or the Ziv-Lempel family [,].

One of the most efficient approaches has been developed by Shibata et al. [ ]. They present a method called BM-BPE which finds text patterns faster in a compressed text than Agrep [ ] finds the same patterns in an uncompressed text. Their search engine is based on the Boyer-Moore algorithm and they employ a restricted version of byte pair encoding (BPE) [ ] achieving a saving of 40% in space. BPE replaces recursively the most common character pair by an unused character code. According to their experiments BM-BPE is faster than most of the earlier methods.

---

* This work has been supported by the National Technology Agency (Tekes).

A. Apostolico and M. Takeda (Eds.): CPM 2002, LNCS 2373, pp. 42– , 2002.
© Springer-Verlag Berlin Heidelberg 2002

Two other works apply the Boyer-Moore approach in compressed texts. Manber [ ] presents a non-recursive coding scheme related to BPE. No character can be both the right character of one pair and the left character of another. His method achieves a saving of 30% in space and the search speed is 30% faster than Agrep. Because of special coding, Manber's approach works poorly with short patterns. Moura et al. [ ] present a method with a better compression ratio than BPE and with a faster search than Manber, but their search works only with words.

We present a new method, which is faster than the best variation of BM-BPE with a comparable compression ratio. In our method characters are encoded as variable-length sequences of base symbols, where each base symbol is represented by a fixed number of bits. Our coding approach is a generalization of that of Moura et al. [ ], where bytes are used as base symbols for coding words. In addition, we split the base symbols into two parallel files in order to allow faster access. Our search algorithm is a variation of the Boyer-Moore-Horspool algorithm [ ]. The shift function is based on several base symbols in order to enable longer jumps than the ordinary occurrence heuristic.

We tested our approach with texts of natural language. Besides outperforming BM-BPE, our approach was clearly more efficient than the best Boyer-Moore variants of Hume and Sunday [ ] in uncompressed texts for $m > 3$. Our approach is efficient also for short patterns, which are important in practice. For example, our approach is 20% faster than BM-BPE and an efficient Boyer-Moore variant for patterns of four characters.

Our approach is not restricted to exact matching nor the Boyer-Moore algorithm, but it can be applied to string matching problems of other types as well.

2   Stopper Encoding

2.1   Stoppers and Continuers

We apply a semi-static coding scheme called *stopper encoding* for characters, where the codewords are based on frequencies of characters in the text to be compressed. The frequencies of characters are gathered in the first pass of the text before the actual coding in the second pass. Alternatively, fixed frequencies based on the language and the type of the text may be used.

A codeword is a variable-length sequence of *base symbols* which are represented as k bits, where k is a parameter of our scheme. Because the length of a codeword varies, we need a mechanism to recognize where a new one starts. A simple solution is to reserve some of the base symbols as *stoppers* which can only be used as the last base symbol of a codeword. All other base symbols are *continuers* which can be used anywhere but in the end of a codeword. If $u_1 \ldots u_j$ is a codeword, then $u_1, \ldots, u_{j-1}$ are continuers and u_j is a stopper.

Moura et al. [ ] use a fixed coding scheme related to our approach. They apply 8-bit base symbols to encode words where one bit is used to describe whether the base symbol is a stopper or a continuer. Thus they have 128 stoppers and 128 continuers.

2.2   Number of Stoppers

It is an optimization problem to choose the number of stoppers to achieve the best compression ratio (the size of the compressed file divided by that of the original file). The optimal number of stoppers depends on the number of different characters and the frequencies of the characters. Let F_i be the frequency of the ith character in the decreasing order according to frequency. When encoding with k-bit base symbols, s stoppers, and $2^k - s$ continuers, the compression ratio C for a fixed division to stoppers and continuers can be calculated with the following formulas, where $L_{s,x}$ is the number of different codewords with x or less base symbols when there are s stoppers.

$$L_{s,x} = \sum_{t=0}^{x-1} s(2^k - s)^t$$

$$Q_i = \frac{kx}{8}F_i, \quad \text{where } x \text{ is the smallest such that } i \leq L_{s,x}.$$

$$C = \frac{\sum_i Q_i}{\sum_i F_i}$$

Let us consider 3-bit base symbols as an example. Table 1 shows how many characters at most can be processed optimally with s stoppers, when the frequency distribution of the characters is uniform or follows Zipf's law.

Table 1. Optimal stopper selection

Stoppers	Uniform	Zipf
7	15	16
6	19	44
5	66	79
4	87	437
3	480	15352

As another example, let us consider the bible.txt of the Canterbury Corpus [ ]. For this text, 14 is the best number of stoppers, when base symbols of four bits are used. Then each of the 14 most common characters of the text (63.9% of all characters) is encoded with one base symbol of 4 bits and the next 28 characters with two base symbols. In this scheme, 56 of the next characters could be encoded with three base symbols, but there are only 21 of them left (0.6% of all characters.)

Moura et al. [ ] use 128 stoppers for 8-bit base symbols. This number is not optimal. More stoppers produce a better compression ratio—the gain is about 5% in the case of the words of the bible.txt. However, the difference is marginal in the case of longer texts.

Perhaps the easiest method of finding the optimal number of stoppers is to calculate first the cumulative frequencies of characters. Let $F'(i)$ be the cumulative frequency of the ith character in the decreasing order of frequency such that $F'(0) = 0$. Then $S_{s,x} = \frac{kx}{8}(F'(L_{s,x}) - F'(L_{s,x-1}))$ is the total coding space for all k-bit base symbols of width x. Then we examine which value of s minimizes the sum $\sum_{x=1}^{x_N} S_{s,x}$ where N is the number of different characters and x_N is the largest x such that $L_{s,x} \leq N$ holds.

2.3   Building the Encoding Table

After the number of stoppers (and with it, the compression ratio) has been decided, an encoding table can be created. The average search time is smaller if the distribution of base symbols is as uniform as possible. We present here a heuristic algorithm, which produces comparable results with an optimal solution in the average case.

The procedure depends on the width of base symbols. We present here the 4-bit version. Let us assume that the characters are ordered in a decreasing order of frequency. If the number of stoppers is s, we allocate the s first base symbols as one-symbol codewords for the s most common characters. The next $s(16 - s)$ characters will have two-symbol codewords, starting with a continuer (the index of the base symbol is $s + (c - s) \bmod (16 - s)$), and ending with a stopper (the index of the base symbol is $(c - s) \operatorname{div} (16 - s)$), where c is the index of the character in turn. Further symbols are encoded with more continuers and a stopper according to the same idea.

The encoding table is stored with the compressed text. We need $N + 2$ bytes to encode the table for N 8-bit characters. One byte is reserved for the number of characters and another byte for the number of stoppers. After these bytes, all the characters present in the text are given in the decreasing order of frequency. With this information, the encoding table can be reconstructed before decompression.

3   Code Splitting

We made an experiment of accessing 100 000 bytes from a long array. For each $g = 0, 1, ..., 7$ we run a test where the bytes were accessed with repetitive gaps of g, i.e. we access bytes $g \cdot i$, $i = 0, 1, ..., 99999$.

Table 2 shows the results of the experiment which was run on a 500 MHz Celeron processor under Linux. The times are relative execution times with a fixed gap width. The main task of the test program was access the array, but it did some additional computation to make the situation more realistic. This extra computation was the same for each run.

Table 2. An access experiment

Gap	0	1	2	3	4	5	6	7
Time	1.00	1.56	2.18	2.70	2.02	2.39	2.55	2.72

According to Table 2, dense accessing is clearly more efficient than sparse accessing, although the total time does not grow monotonously. This dependency is a consequence of the hierarchical organization of memory in modern computers. We tested the same program also with other processors and the results were rather similar. This phenomenon suggests that string matching of the Boyer-Moore type could be made faster by splitting the text to several parallel files.

Combining code splitting with stopper encoding. The splitting of the text could be done in many ways. We apply the following approach. Let the text be represented as k-bit base symbols. We concatenate the h high bits of the base symbols to a file and the l low bits to another file, $h + l = k$. In practice these files could be still concatenated, but here we consider two separate files for clarity. We call this method *code splitting*.

We denote stopper encoding with the division to h high bits and l low bits by $SE_{k,h}$. The version without code splitting is denoted by $SE_{k,0}$. The plain code splitting without compression is denoted by $SE_{8,h}$. Note that $SE_{8,h}$ can be seen a representative of stopper encoding: in $SE_{8,h}$ all the 256 base symbols of eight bits are stoppers.

We consider mainly three versions of stopper encoding: $SE_{4,0}$, $SE_{8,4}$, and $SE_{6,2}$. Note that $SE_{4,0}$ applies compression, $SE_{8,4}$ code splitting, and $SE_{6,2}$ both of them.

4   The Searching Algorithm

The key point of the searching algorithm is that the pattern is encoded in the same way as the text. So we actually search for occurrences of a string of base symbols, or low bits of them, if code splitting is applied. In the latter case, the search in low bits produces only potential matches which all should be checked with high bits.

After finding an occurrence of the encoded pattern, we simply check the base symbol preceding the occurrence. If the base symbol is a stopper (or the occurrence starts the text), we report a match, otherwise we ignore this alignment and move on.

4.1   Searching in an Alphabet of 16 Characters

Let us assume that we have a text with 16 or less different characters. Then all the characters of the text can be represented with four bits, and we can store two consecutive characters in one 8-bit byte.

The basis of our searching algorithm is ufast.fwd.md2, a fast Boyer-More-Horspool variant presented by Hume and Sunday [ ]. This algorithm employs an unrolled skip loop and a fixed shift in the case of the match of the last character of the pattern. The shift is based on the text character under the rightmost character of the pattern. It is straightforward to modify ufast.fwd.md2 to our setting.

Because one byte holds two characters, there are two different byte alignments of an occurrence of the pattern. Therefore there are two acceptable bytes which may start the checking phase of the algorithm, corresponding to these two alignments.

The shift is based on a character pair in the terms of the original text. This approach in ordinary string matching has been studied by Baeza-Yates [ ] and Zhu and Takaoka [ ]. Zhu and Takaoka take the shift as a minimum of shifts based on match and occurrence heuristics like in the original Boyer-Moore algorithm [ ]. However the mere occurrence heuristic is faster in practice for natural language texts.

This searching algorithm works fine with the variant $SE_{4,0}$, where 4-bit base symbols are used and thus the size of alphabet is just 16.

4.2   Searching in $SE_{8,4}$

Recall that no compression is involved with $SE_{8,4}$. All the bytes of the text are split in two parts: the four high bits to one part and the four low bits to the other. These parts are stored in separate files, where new bytes are made from two half-bytes. For example the text "Finland!", which is 46-69-6e-6c-61-6e-64-21 in hexadecimal, will have its high bits stored as 46-66-66-62 and the low bits as 69-ec-1e-41.

Now suppose we wanted to search the pattern land, which is 6c-61-6e-64, in the encoded text. We start by searching the low bits for the corresponding combination of low bits which we will call the encoded pattern, namely c1-e4. Then the pattern could start from the beginning of a byte in the low bits (c1-e4) or from the middle of a byte (*c-1e-4*), where the asterisk represents any hexadecimal digit.

Now we use some method to search the low bits for all occurrences (of both variants) of this encoded pattern. When and only when a match is found in the low bits, the corresponding high bits are checked. So if there are no matches in the low bits, the high bits can be ignored.

An advantage of this method is that only a fraction of the characters of the text are inspected. *False matches* (substrings of the text where the low bits match with the pattern and the high bits do not) are rare in most texts of natural language, so we seldom need to check the high bits at all. Another advantage is that two characters are accessed at a time while scanning the text.

4.3   Searching in $SE_{6,2}$

This 6-bit variant sacrifices space for speed. The ideal compression ratio is 75%, when there are 64 or fewer different characters in the text. Since it is difficult to store sequences of 6-bit base symbols into 8-bit bytes, code splitting is applied. We store four low bits and two high bits separately.

This allows two variations. The first variation goes through the 4-bit part and checks the 2-bit part only when a match in the 4-bit part is found. This is what we will call the 4+2 searching algorithm. The second variation, 2+4, does the same thing vice versa. The 2+4 variation is generally faster, because it only needs to take $\frac{1}{3}$ of all data into account on the first pass, while the 4+2 takes $\frac{2}{3}$ of it. However, the overhead of having to search 4 patterns simultaneously and inefficiency in the case of patterns of 7 or less characters, also make the 4+2 variation usable on the side of the 2+4 one. The best algorithm is obviously a combination. Based on our experiments, we decided to use the 2+4 variation for $m \geq 8$ and the 4+2 one for $m < 8$.

5   Experimental Results

When Boyer-Moore string searching described above is combined with stopper encoding $SE_{k,h}$, the total method is denoted by $BM\text{-}SE_{k,h}$.

We tested the performance of the algorithms $BM\text{-}SE_{4,0}$, $BM\text{-}SE_{8,4}$, and $BM\text{-}SE_{6,2}$. Recall that $SE_{4,0}$ applies compression, $SE_{8,4}$ code splitting, and $SE_{6,2}$ both of them. We compared them with four other searching algorithms. We used Tuned Boyer-Moore or ufast.fwd.md2 [ ] denoted by TBM as the searching algorithm for uncompressed texts. Three versions of the BM-BPE algorithm (a courtesy from M. Takeda) for compressed texts were tested: one with maximal compression ratio and no upper limit for the number of characters represented by a byte (max), another with optimal search speed where a byte can represent at most two characters (fast), and the third one where a byte can represent at most three characters and which was recommended by the authors (rec). All the algorithms were modified to read first the whole text to the main memory and then to perform the search. All the tests were run on on a 500 MHz Celeron processor with 64 MB main memory under Linux.

The compression ratio was measured with four texts (Table 3): the bible.txt [ ], the CIA World Factbook of 1992, Kalevala, the national epic of Finland (in Finnish), and E.coli, the genome of *Escherichia coli*, entirely composed of the four DNA symbols. As explained earlier, there is no compression involved with $SE_{8,4}$, only a different encoding. The compressed files include the encoding tables which are necessary to uncompress them. As a reference, we give also the compression ratios achieved with Gzip.

To make a fair comparison with BM-BPE, the version $BM\text{-}SE_{4,0}$ is the right choice, because its compression rate is similar to that of the fast BM-BPE.

The compression and decompression algorithms of BM-SE are very fast (17 MB/s) due to the lightweight encoding and decoding schemes.

Table 3. Compression ratio

| | bible.txt | CIA1992 | Kalevala | E.coli |
	3.86 MB	2.36 MB	0.52 MB	4.42 MB
BM-BPE max	47.8%	56.8%	51.9%	31.3%
BM-BPE fast	56.2%	63.0%	55.1%	50.0%
BM-SE$_{4,0}$	58.9%	68.2%	58.1%	50.0%
BM-SE$_{6,2}$	75.0%	75.8%	75.1%	75.0%
Gzip	29.4%	29.3%	36.3%	28.9%

We tested the search speed with two texts: bible.txt (Table 4) and E.coli (Table 5). We used command-line versions of all the algorithms. We measured the processor time in milliseconds required by the search. Although the excluding of the I/O time slightly favors poorer compression methods, we wanted to measure the efficiency of the pure algorithms without any disturbance due to buffering. The same test was repeated for 500 different strings of the same length randomly chosen in the text.

Table 4. Search times (ms), bible.txt, $3 \leq m \leq 20$

	3	4	5	6	8	10	12	16	20
TBM	53.4	47.4	42.8	40.4	37.0	35.2	35.0	33.8	31.8
BM-BPE max	68.4	66.2	63.4	61.2	57.6	55.2	53.2	39.4	38.6
BM-BPE rec	71.8	51.2	45.0	44.2	35.0	31.2	30.6	26.8	25.8
BM-BPEfast	52.4	46.4	38.2	36.2	31.0	27.8	26.4	24.2	23.6
BM-SE$_{8,4}$	59.4	38.4	32.0	26.2	22.6	20.0	18.8	17.2	17.0
BM-SE$_{4,0}$	49.0	37.2	31.6	28.0	24.2	22.2	21.0	20.0	19.4
BM-SE$_{6,2}$	66.8	38.0	32.6	26.6	18.8	15.2	13.6	12.0	10.4

In the bible.txt, the versions BM-SE$_{4,0}$ and BM-SE$_{6,2}$ of Boyer-Moore with stopper encoding are clearly faster than BM-BPE for all pattern widths shown in Table 4. However, they are also faster than TBM excluding very short patterns $m < 4$. Even the version without compression, BM-SE$_{8,4}$ is faster than TBM and BM-BPE for $m > 3$. None of the BM-SE algorithms is distinctly the fastest one. BM-SE$_{4,0}$ is the fastest for $m < 6$, BM-SE$_{8,4}$ for $m = 6$, BM-SE$_{6,2}$ for $m > 6$. The times of four algorithms are shown graphically in Figure 1.

The advantage of BM-SE is smaller in the DNA text, because the average length of shift is shorter. According to Table 5, BM-SE$_{4,0}$ is the fastest for short patterns $m \leq 12$ and BM-BPE rec for longer ones. Note that BM-BPE rec is now clearly faster than BM-BPE fast. As one may expect, BM-SE$_{6,2}$ is very poor in the DNA text and so we left it out from this comparison. Probably BM-SE$_{2,0}$ (which has not yet been implemented) will be even better than BM-SE$_{4,0}$ for DNA data.

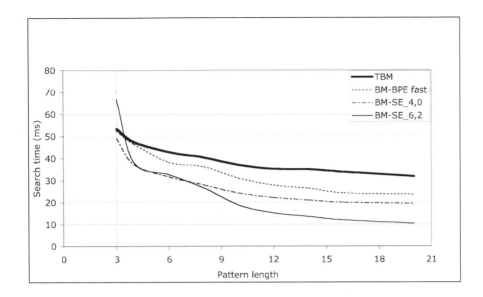

Fig. 1. Search times in bible.txt

TBM is not a good reference algorithm for DNA matching. BNDM [ ] would be more appropriate, because is the fastest known algorithm for patterns $m \leq w$, where w is the number of bits in the computer word. Other alternatives would have been ufast.rev.gd2 [ ] or algorithms based on alphabet transformations [,].

Table 5. Search times (ms), E.coli, $6 \leq m \leq 48$

	6	12	24	48
TBM	67.0	61.2	60.0	60.2
BM-BPE max	52.8	34.2	26.0	23.0
BM-BPE rec	43.2	28.0	22.0	21.0
BM-BPE fast	52.4	36.8	31.4	30.4
BM-SE$_{8,4}$	37.8	27.4	23.6	22.0
BM-SE$_{4,0}$	35.8	26.2	23.0	21.4

6   Concluding Remarks

We have presented a new practical solution for the compressed matching problem. According to our experiments the search speed of our BM-SE is clearly faster than that of BM-BPE for natural language texts. The version BM-SE$_{4,0}$

has similar compression ratio to the fast BM-BPE. In DNA texts there is no significant difference in the search speed.

Moreover our BM-SE is faster than TBM for patterns longer than three characters.

It would be interesting to compare BM-SE with Manber's method [ ], because he reports a gain of 30% in search times. It is clear that this gain is not possible for short patterns because of Manber's pairing scheme. A part of the gain is due to the save in I/O time which was excluded in our measurements.

References

1. A. Amir and G. Benson. Efficient two-dimensional compressed matching. In *Proc. DCC'92*, pages 279–288, 1992.
2. A. Amir, G. Benson, and M. Farach. Let sleeping files lie: Pattern matching in Z-compressed files. *J. of Comp. and Sys. Sciences*, 52(2):299–307, 1996.
3. R. Arnold and T. Bell. A corpus for the evaluation of lossless compression algorithms. In *Proc. DCC '97, Data Compression Conference*. IEEE, 1997.   ,
4. R. Baeza-Yates. Improved string searching. *Software – Practice and Experience*, 19(3):257–271, 1989.   ,
5. R. S. Boyer and J. S. Moore. A fast string searching algorithm. *CACM*, 20(10):762–772, 1977.   ,
6. M. Crochemore and W. Rytter. *Text Algorithms*. Oxford University Press, 1994.
7. P. Gage. A new algorithm for data compression. *C/C++ Users Journal*, 12(2), 1994.
8. R. N. Horspool. Practical fast searching in strings. *Software Practice and Experience*, 10:501–506, 1980.
9. D. Huffman. A method for the construction of minimum-redundancy codes. *Proc. of the I. R. E.*, 40(9):1090–1101, 1952.
10. A. Hume and D. Sunday. Fast string searching. *Software – Practice and Experience*, 21(11):1221–1248, 1991.   ,   ,   ,
11. U. Manber. A text compression scheme that allows fast searching directly in the compressed file. *ACM Trans. on Information Systems*, 15(2):124–136, 1997.   ,
12. E. Moura, G. Navarro, N. Ziviani, and R. Baeza-Yates. Fast and flexible word searching on compressed text. *ACM Trans. on Information Systems*, 18(2):113–139, 2000.   ,   ,
13. G. Navarro, T. Kida, M. Takeda, A. Shinohara, and S. Arikawa. Faster approximate string matching over compressed text. In *Proc. 11th IEEE Data Compression Conference (DCC'01)*, pages 459–468, 2001.
14. G. Navarro and M. Raffinot. Fast and flexible string matching by combining bit-parallelism and suffix automata. *ACM Journal of Experimental Algorithmics (JEA)*, 5, 2000.
15. G. Navarro and J. Tarhio. Boyer-Moore string matching over Ziv-Lempel compressed text. In *Proc. 11st Annual Symposium on Combinatorial Pattern Matching (CPM 2000)*, LNCS 1848, pages 166–180, 2000.
16. H. Peltola and J. Tarhio. String matching in the DNA alphabet. *Software – Practice and Experience*, 27:851–861, 1997.

52   Jussi Rautio et al.

17. Y. Shibata, T. Matsumoto, M. Takeda, A. Shiohara, and S. Arikawa. A Boyer-Moore type algorithm for compressed pattern matching. In *Proc. 11st Annual Symposium on Combinatorial Pattern Matching (CPM 2000)*, LNCS 1848, pages 181–194, 2000.
18. S. Wu and U. Manber. Agrep – a fast approximate pattern-matching tool. In *Proc. USENIX Technical Conference*, pages 153–162, Berkeley, CA, USA, 1992.
19. R. Zhu and T. Takaoka. On improving the average case of Boyer-Moore string matching algorithm. *Journal of Information Processing*, 10:173–177, 1987.
20. J. Ziv and A. Lempel. A universal algorithm for sequential data compression. *IEEE Trans. Inf. Theory*, 23:337–343, 1977.
21. J. Ziv and A. Lempel. Compression of individual sequences via variable length coding. *IEEE Trans. Inf. Theory*, 24:530–536, 1978.

Pattern Matching Problems
over 2-Interval Sets

Stéphane Vialette

Laboratoire de Génétique Moléculaire
École Normale Supérieure
46, rue d'Ulm 75230 Paris cedex 05
vialette@ens.fr

Abstract. We study the computational complexity of *pattern match-ing* problems over *2-interval sets*. These problems occur in the context of molecular biology when a structured pattern, *i.e.*, a RNA secondary structure, has to be found in a sequence. We show that the PATTERN MATCHING OVER 2-INTERVAL SET problem is NP-complete for struc-tured patterns where no pair precedes the other, but can be solved in polynomial time for several interesting special cases .

1 Introduction

The Ribonucleic acid (RNA) is a family of molecules which has several important functions in the cell. For example, the role of transfer RNA (tRNA) concerns the process of protein synthesis. The functionality of a specific RNA molecule de-pends mostly on its *secondary structure*. The RNA is a single stranded molecule viewed as a linear sequence $x_1 x_2 \ldots x_n$ where $x_i \in \{A, C, G, U\}$. RNA secondary structures refer to conformation of the single stand after it folds back on itself, by forming base pairs.

Many interesting RNAs preserve a secondary structure of base-pairing inter-actions more than they conserve their sequence. There is therefore a growing demand for general purpose search programs that take into account both se-quence and structure patterns. One way of finding such a secondary structure pattern in a RNA sequence is by *pattern matching*. The PALINGOL software [ ] provides a framework of reference for this approach. The basic idea of PALIN-GOL is a two step procedure. In the first step the sequence is scanned in order to build a set of all helices found on it. For efficiency, we may perform simple checking to avoid generating too long a set of helices. The important point is just to ensure that we get all helices that could be involved in the structure. In the second step, a pattern matching algorithm finds all occurrences of a specific structure in the set of all helices. This is usually done using a *branch and bound* like procedure [   ]. The purpose of this paper is to highlight some of the issues

---

[1] The work described in this paper was developed during the author's PhD at LIAFA laboratory - Université Denis Diderot - 2, place Jussieu F-75251 Paris Cedex 05, France.

involved in this second step taking advantage of a new representation of the problem. It should be pointed out that we are only concerned in this paper with the specific problem of searching for known *structure patterns* within a sequence and not finding the optimal folding.

Our approach in this paper is to first establish a *geometric* description of helices. The basic idea is to use a natural generalization of intervals, namely a 2-*interval*. A 2-interval is the disjoint union of two intervals on the line. The geometric properties of 2-intervals provide a possible guide for understanding the computational complexity of finding structured patterns in RNA sequences. An illustration is given in figure   .

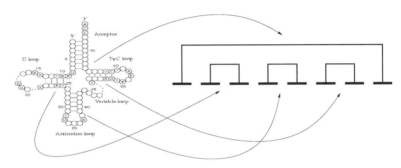

Fig. 1. A tRNA secondary structure and the associated 2-interval pattern

Using a model to represent non sequential information allows us for varying restriction on the complexity of the pattern structure. Two disjoint 2-intervals can be in precedence order, be allowed to nest and/or be allowed to cross. Furthermore, the 2-interval set and the pattern can have different restrictions. These different combinations of restrictions alter the computational complexity of the PATTERN MATCHING OVER 2-INTERVAL SET problem, and need to be examined separately. This examination produces efficient algorithms for more restrictive patterns, and hardness results for those less restrictive.

We briefly review the related terminology used in this paper. A 2-*interval* is the disjoint union of two intervals and 2-*interval graphs* are the intersection graphs of a set of 2-intervals on the real line. D.B. West and D.B. Shmoys [   ] have shown that recognizing 2-interval graphs is an NP-complete problem. Furthermore, it is shown in [   ] that the INDEPENDENT SET problem is NP-complete even when restricted to 2-interval graphs. We will usually write a 2-interval as $D = (I, I')$ with $I < I'$ where $<$ is the precedence order between intervals. Let $D_1 = (I_1, I_1')$ and $D_2 = (I_2, I_2')$ be two 2-intervals. They are called *disjoint* if they do not intersect, *i.e.*, $(I_1 \cup I_1') \cap (I_2 \cup I_2') = \emptyset$. Of particular interest is the relation between two disjoint 2-intervals. We will write $D_1 < D_2$ if $I_1 < I_1' < I_2 < I_2'$, $D_1 \sqsubset D_2$ if $I_2 < I_1 < I_1' < I_2'$ and $D_1 \between D_2$ if $I_1 < I_2 < I_1' < I_2'$. An illustration of these relations is given in figure   - (a). The *support* of a 2-interval set $\mathcal{D} = \{D_1, D_2, \ldots, D_n\}$ is defined to be set

$\{I_i, I'_i \mid D_i = (I_i, I'_i) \in \mathcal{D}\}$. In other words, the support of a set of 2-intervals $\mathcal{D}$ is the set of all (simple) intervals involved in $\mathcal{D}$. A set of 2-intervals with the property that any two distinct intervals of its support do not intersect is said to be a *linear ordered graph*. Indeed, we can associate with each 2-interval an edge of a graph for which the set of vertices is linearly ordered.

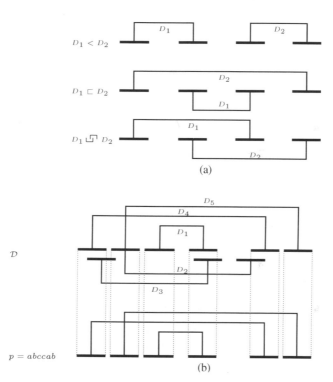

(a)

(b)

Fig. 2. (a) Relations between 2-intervals. (b) Shown here is an occurrence of the structured pattern $p = abccab$ in the set of 2-intervals $\mathcal{D} = \{D_1, D_2, D_3, D_4, D_5\}$. Observe that p is $\{\sqsubset, \between\}$-structured pattern: $aa \between bb$, $cc \sqsubset aa$ and $cc \sqsubset bb$

A *structured pattern* (or *2-interval pattern*) is a word p in which each of its letters occurs twice. Geometrically speaking, a structured pattern is merely a formal description of a set of 2-intervals such that any two of them are disjoint. Therefore, it makes sense to associate a non empty subset of $\{<, \sqsubset, \between\}$ to a given structured pattern. For example, $abaccb$ is a $\{<, \sqsubset, \between\}$-structured pattern because $aa \between bb$, $aa < cc$ and $cc \sqsubset bb$, $abbacc$ is a $\{<, \sqsubset\}$-structured pattern because $bb \sqsubset aa$, $aa < cc$ and $bb < cc$, and $abccba$ is a $\{\sqsubset\}$-structured pattern. Call a $\mathcal{R}$-structured pattern a *simple structured pattern* if $|\mathcal{R}| = 1$, *i.e.*, $\mathcal{R}$ consists of only one relation.

The pattern matching over 2-interval set problem asks to find a 2-interval subset such that any two of them are disjoint, described by a given abstract model, *i.e.*, a structured pattern. The mathematical model is best explained by referring to figure -(b). We are now in position to define the PATTERN MATCHING OVER 2-INTERVAL SET problem.

PATTERN MATCHING OVER 2-INTERVAL SET
Input : A 2-interval set $\mathcal{D}$ and a $\mathcal{R}$-structured pattern p.
Problem : Is there an occurrence of p in $\mathcal{D}$?

The PATTERN MATCHING OVER 2-INTERVAL SET problem is strongly related to the longest common subsequence problem for sequences with arc annotations [,]. Also, we show in this paper the similarity between the PATTERN MATCHING OVER 2-INTERVAL SET problem and the protein structure similarity problem (CONTACT MAP OVERLAP) [ ]. The main results of this paper are summarized in the following table.

		PATTERN MATCHING OVER 2-INTERVAL SET	
	pattern structure	support	
		no restrictions	linear ordered graph
1	$\{<,\sqsubset,\sqcup\}$	NP-complete	NP-complete
2	$\{\sqsubset,\sqcup\}$	NP-complete	NP-complete
3	$\{<,\sqsubset\}$	?	?
4	$\{<,\sqcup\}$	?	?
5	$\{<\}$	$\mathcal{O}(n \log n)$	$\mathcal{O}(n \log n)$
6	$\{\sqsubset\}$	$\mathcal{O}(n^2)$	$\mathcal{O}(n^2)$
7	$\{\sqcup\}$	$\mathcal{O}(n^2 \log n)$	$\mathcal{O}(n^2 \log n)$

Throughout the table, n denotes the cardinality of the 2-interval set. Row 1 is proved in [ ], Row 2 follows from theorem , Rows 5 and 6 follow directly from the development in section   and Row 7 is the result of a polynomial time algorithm for the CLIQUE problem restricted to *crossing circle trapezoid graphs* (proposition). The complexity of the PATTERN MATCHING OVER 2-INTERVAL SET problem is open for $\{<,\sqsubset\}$-structured patterns and $\{<,\sqcup\}$-structured patterns (Rows 3 and 4).

A *graph* G consists of a finite set $V = \{u_1, u_2, \ldots\}$ of elements called *vertices* together with a prescribed set E of undirected pair of distinct vertices of V. The number n of elements in V is called the *order* of the graph. Every unordered pair $e \in E$ of vertices u_i and u_j is called an *edge* of G, written $e = \{u_i, u_j\}$. We call u_i and u_j the *endpoints* of e and they are called *adjacent vertices*. The *open neighbor* of a vertex $u \in V$ is the set $N(u) = \{v \in V \mid \exists \{u, v\} \in E\}$. The *close neighbor* of a vertex $u \in V$ is the set $N[u] = \{u\} \cup N(u)$. The *subgraph* of G induced by a subset $V' \subseteq V$ is a graph $G(V') = (V', E')$ such that $E' = \{\{u, v\} \in E \mid u, v \in V'\}$. A *clique* is a complete graph. The CLIQUE problem is to construct for a given graph an induced complete subgraph of the maximum number of vertices.

The rest of the paper is organized as follows. In Section , we state that the PATTERN MATCHING OVER 2-INTERVAL SET problem is NP-complete even

when restricted to $\{\sqsubset,\text{⌐}\}$-structured patterns. We present in Section   polynomial time algorithms for the PATTERN MATCHING OVER 2-INTERVAL SET
problem restricted to simple structured patterns.

2   Hardness Results

Pattern matching problems over 2-interval sets are very hard decision problems. Indeed, it is first shown in [ ] that the PATTERN MATCHING OVER 2-
INTERVAL SET problem is NP-complete using a similar reduction technique as
in [ ]. We will state more in this section, namely that the PATTERN MATCH-
ING OVER 2-INTERVAL SET problem is NP-complete even when restricted to
$\{\sqsubset,\text{⌐}\}$-structured patterns.

Theorem 1. *The* PATTERN MATCHING OVER 2-INTERVAL SET *problem is* NP-
complete even when restricted to $\{\sqsubset,\text{⌐}\}$-*structured patterns.*

   The proof of theorem   consists in a polynomial time reduction from the
CLIQUE problem which is a known NP-complete problem (see for instance [ ]).
Without going into further details, it is important to point out that theorem
still holds if we assume that our 2-interval set $\mathcal{D}$ is a linear ordered graph.
   Let us now mention one important consequence of theorem   concerning
the CONTACT MAP OVERLAP problem [ ]. A *contact map* is a graph $G =
(V, E)$ such that the set of vertices $V = \{u_1, u_2, \ldots, u_n\}$ is linearly ordered,
i.e., $u_1 < u_2 < \ldots < u_n$. The CONTACT MAP OVERLAP problem is the
following optimization problem: Given two contact maps (V, E) and (V', E'),
find two subsets $S \subseteq V$ and $S' \subseteq V'$ with $|S| = |S'|$ such that the cardi-
nality $|\{\{u, v\} \in E \mid u, v \in S$ and $\{f(u), f(v)\} \in E'\}|$ is as large as possible,
where f is an order-preserving bijection between S and S'. The CONTACT MAP
OVERLAP problem is MAX SNP-complete even if both contact maps have maximum degree one or are self avoiding walks [ ]. We state a new simple spe-
cial case of this problem. Call two edges $\{u_i, u_j\}$ and $\{u_k, u_\ell\}$ of E *disjoint* if
$\max\{u_i, u_j\} < \min\{u_k, u_\ell\}$ or $\max\{u_k, u_\ell\} < \min\{u_i, u_j\}$. Define a *tangle* to be
a contact map (V, E) such that if $\{u_i, u_j\}, \{u_k, u_\ell\} \in E$ then the edges $\{u_i, u_j\}$
and $\{u_k, u_\ell\}$ are *not* disjoint.

Corollary 1. *The* CONTACT MAP OVERLAP *problem is* NP-*complete even if
both contact maps are tangles with maximum degree one.*

3   The PATTERN MATCHING OVER 2-INTERVAL SET Problem Restricted to Simple Structured Patterns

We prove in this section that the PATTERN MATCHING OVER 2-INTERVAL SET
problem restricted to simple structured patterns is solvable in polynomial time
using simple graph-based algorithms. This will be divided into two parts. For one,

we observe that the PATTERN MATCHING OVER 2-INTERVAL SET problem restricted to transitive simple structured patterns is solvable in polynomial time using standard graph theory tools, namely maximum independent set algorithm for interval graphs [ ] and maximum clique algorithm for comparability graphs [ ]. For another, we prove that the PATTERN MATCHING OVER 2-INTERVAL SET problem restricted to {⊐}-structured patterns is solvable in polynomial time using a polynomial time maximum cardinality clique algorithm for a new class of graphs ; the argument is more tricky in this case. In the sequel, n denotes the cardinality of a 2-interval set $\mathcal{D}$.

Proposition 1. *The* PATTERN MATCHING OVER 2-INTERVAL SET *problem is solvable in time* $O(n \log n)$ *when restricted to* {<}-*structured patterns.*

Proof. Define a family of intervals $\mathcal{I}$ by assigning to each 2-interval $D = (I, I')$ of $\mathcal{D}$ the least interval that covers both I and I'. Let G be the associated interval graph. Is is easily seen that there exists a bijective mapping between occurrences of {<}-structured patterns in $\mathcal{D}$ and independent sets in G. An $O(n \log n)$ algorithm for finding a maximum independent set in an interval graph given in the form of a set of intervals is presented in [ ].   □

Proposition 2. *The* PATTERN MATCHING OVER 2-INTERVAL SET *problem is solvable in time* $O(n^2)$ *when restricted to* {⊏}-*structured patterns.*

Proof. Define a graph G with vertex set $\mathcal{D}$ by choosing all those sets $\{D_i, D_j\}$ (where $i \neq j$) as edges of G for which $D_i \sqsubset D_j$ or $D_j \sqsubset D_i$ holds. Observe that G is a *comparability graph* [ ] and that the construction of G can be carried on in time $O(n^2)$. Clearly, there exists a bijective mapping between occurrences of {⊏}-structured patterns in $\mathcal{D}$ and cliques in G. It is now sufficient to use the above remark together with the fact that the CLIQUE problem is solvable in linear time when restricted to comparability graphs [ ].   □

The remainder of this section is devoted to non-transitive simple structured patterns, *i.e.*, {⊓}-structured patterns. In general terms our approach consists in using *circle trapezoids* as an alternative means of describing a 2-interval set together with a generalization of a polynomial time algorithm for the CLIQUE problem restricted to circle graphs [ ]. Let us first introduce some definitions.

A *circle trapezoid* is the region in a circle that lies between two non-crossing chords and *circle trapezoid graphs* are the intersection graphs of a family of circle trapezoids on a common circle [ ]. Two circle trapezoids are called *crossing* if they intersect in the circle **but not** on its perimeter. Call a graph $G = (V, E)$ a *crossing circle trapezoid graph* (CCT-graph) if its vertices can be put in one to one correspondence to a set of circle trapezoids such that two vertices of G are adjacent if and only if their corresponding circle trapezoids cross. An illustration of this definition is given in figure .

It is well known that *circle graphs* and *overlap graphs* are equivalent graph classes [ ]. An easy way to visualize this equivalence is by using the *projection method* suggested by F. Gavril [,] where intersecting chords of the circle correspond to overlapping intervals on the line. Following this construction, we can

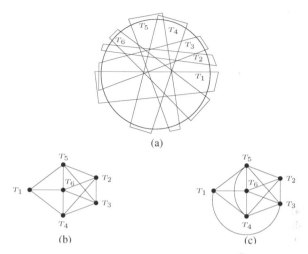

Fig. 3. (a) A circle trapezoid representation, (b) the corresponding CCT-graph and (c) the corresponding circle trapezoid graph. Observe that T_1 and T_3 intersect but do not cross. Similar remark applies to T_4 and T_5

associate a circle trapezoid representation to a given 2-interval set where intersecting circle trapezoids of the circle correspond to intersecting 2-intervals on the line. An illustration of this construction is given in figure -(a). For the convenience of the reader, the CCT-graph of a 2-interval set stands for the CCT-graph of its associated circle trapezoid representation.

Lemma 1. *Let $\mathcal{D} = \{D_1, D_2, \ldots, D_n\}$ be a 2-interval set and G be its CCT-graph. Then, there exists a subset $\mathcal{D}' \subseteq \mathcal{D}$ such that, for any two distinct D_i and D_j of $\mathcal{D}'$, either $D_i \sqsubset D_j$ or $D_j \sqsubset D_i$, if and only if there exists a clique of size $|\mathcal{D}'|$ in G.*

An illustration of the above lemma is given in figure -(b). In the following, we provide a polynomial time algorithm which solve the CLIQUE problem for CCT-graphs. The notion of *trapezoid graph* [ ] will be used. A graph is a trapezoid graph if there exists a set of trapezoids corresponding to the vertices of the graph such that two vertices are joined by an edge if and only if the corresponding trapezoids intersect. We need a new definition. Let $G = (V, E)$ be a CCT-graph and $V' \subseteq V$ be a subset of its vertices. We will denote by $\overline{G}(V')$ the graph with vertex set V' such that two vertices of $\overline{G}(V')$ are adjacent if and only if their corresponding circle trapezoid *do not cross*, i.e., $\overline{G}(V')$ is the complement of the induced subgraph $G(V')$. Let us start by characterizing the graph $\overline{G}(N[u])$ for all $u \in V$, where $N[u]$ is the close neighbor of u in the CCT-graph G.

Lemma 2. *Let $G = (V, E)$ be a CCT-graph. Then $\overline{G}(N[u])$ is a trapezoid graph for all $u \in V$.*

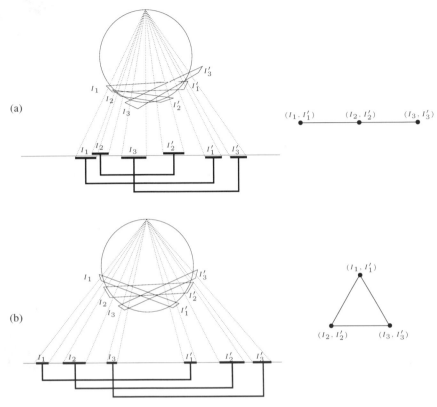

Fig. 4. Projection of 2-interval sets and the corresponding CCT-graphs

Lemma 3. *Let $G = (V, E)$ be a CCT-graph and K be a clique of G. Then K is a clique of the induced subgraph $G(N[u])$ for all $u \in K$.*

The following lemma is crucial to the proof of the theorem. Indeed, it allows us to concentrate on the INDEPENDENT SET problem restricted to trapezoid graphs.

Lemma 4. *Let $G = (V, E)$ be a CCT-graph and $u \in V$ be an arbitrary vertex. Then there exists a clique of size k in the induced subgraph $G(N[u])$ if and only if there exists an independent set of size k in the trapezoid graph $\overline{G}(N[u])$.*

An illustration of the above lemma is given in figure .

Proposition 3. *The CLIQUE problem is solvable in time $O(n^2 \log n)$ when restricted to CCT-graphs.*

Proof. The algorithm is as follows:
MAX-CLIQUE-CCT-GRAPH
Input: a CCT-graph $G = (V, E)$
Output: a maximum cardinality clique in G

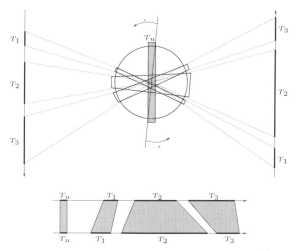

Fig. 5. A CCT-graph and the trapezoid graph $\overline{G}(N[u])$

```
1. for all  u ∈ V  do  K_u ← MAX-IND-SET-TRAPEZ-GRAPH(G̅(N[u]))
2. return the largest  K_u .
```

where MAX-IND-SET-TRAPEZ-GRAPH is an algorithm which solves the independent set problem for trapezoid graphs. Based on a geometric representation of trapezoid graphs by boxes in the plane, S. Felsner, R. Müller and L. Wernisch [ ] have designed an optimal $O(n \log n)$ algorithm for weighted independent set on such graphs. By lemmas , and , the algorithm is correct.   □

Corollary 2. *The* PATTERN MATCHING OVER 2-INTERVAL SET *problem is solvable in time $O(n^2 \log n)$ when restricted to $\{\sqcup\}$-structured patterns.*

Proposition , proposition and corollary may be summarized by the following theorem.

Theorem 2. *The* PATTERN MATCHING OVER 2-INTERVAL SET *problem is solvable in polynomial time when restricted to simple structured patterns.*

4   Conclusion and Open Problems

In the context of computational molecular biology we considered the problem of finding an occurrence of a pattern in a 2-interval set. We proved that this problem is NP-complete even when restricted to $\{\sqsubset,\sqcup\}$-structured patterns. Also, we described polynomial time algorithms which solve this problem for simple structured patterns. There are many interesting related problems arising in the above context in a natural way. Below we mention two of them.

1. What is the complexity of the PATTERN MATCHING OVER 2-INTERVAL SET problem restricted to $\{<, \sqsubset\}$-structured patterns? The rationale of this problem is that we can view a $\{<, \sqsubset\}$-structured pattern as a pseudo-knot free RNA secondary structure. Observe that this problem is solvable in polynomial time if $\mathcal{D} = \{D_1, D_2, \ldots, D_n\}$ is a 2-interval set for which D_i and D_j are $\{<, \sqsubset\}$-comparable for all $D_i, D_j \in \mathcal{D}$ using an ordered tree inclusion algorithm [ ].
2. What is the complexity of the PATTERN MATCHING OVER 2-INTERVAL SET problem restricted to $\{<, \between\}$-structured patterns?

Acknowledgments

The author is grateful to Christian Choffrut and anonymous referees for very helpful comments and suggestions. This work is partially supported by HMR-Aventis FRHMR2/9908.

References

1. B. Billoud, M. Kontic, and A. Viari. Palingol: a declarative programming language to describe nucleic acids secondary structures and to scan sequence database. *Nucl. Acids Res.*, 24:1395–403, 1996.
2. I. Dagan, M. C. Golumbic, and R. Y. Pinter. Trapezoid graphs and their coloring. *Discrete Appl. Math.*, 21:35–46, 1988.
3. P. Evans. Finding common subsequences with arcs and pseudoknots. In *Proceedings of the 10th Annual Symposium Combinatorial Pattern Matching (CPM 1999)*, volume 1645 of *Lecture Notes in Computer Science*, pages 270–280, 1999.
4. S. Felsner, R. Müller, and L. Wernisch. Trapezoid graphs and generalizations: Geometry and algorithms. *Discrete Appl. Math.*, 74:13–32, 1997.
5. M. R. Garey and D. S. Johnson. *Computers and intractability: a guide to the theory of NP-completeness*. W. H. Freeman, San Franciso, 1979.
6. F. Gavril. Algorithms for a maximum clique and a minimum independent set of a circle graph. *Networks*, 3:261–273, 1973.
7. D. Goldman, S. Istrail, and C. H. Papadimitriou. Algorithmic aspects of protein structure similarity. In *IEEE Proceedings of the 40th Annual Conference of Foundations of Computer Science (FOCS99)*, pages 512–521, 1999.
8. M. C. Golumbic. *Algorithmic Graph Theory and Perfect Graphs*. Academic Press, New York, 1980.
9. U. I. Gupta, D. T. Lee, and J.Y-T. Leung. Efficient algorithms for interval graph and circular-arc graphs. *Networks*, 12:459–467, 1982.
10. T. Jiang, G.-H. Lin, B. Ma, and K. Zhang. The longest common subsequence problem for arc-annotated sequences. In *Proceedings of the 11th Annual Symposium on Combinatorial Pattern Matching (CPM 2000)*, volume 1848 of *Lecture Notes in Computer Science*, pages 154–165, 2000.
11. P. Kilpeläinen. *Tree matching problems with applications to structured text databases*. PhD thesis, University of Helsinki, Finland, 1992.
12. E. L. Lawler and D. W. Wood. Branch and bound methods: A survey. *Operations Research*, 14:699–719, 1966.

13. S. Vialette. *Aspects algorithmiques de la prédiction des structures secondaires d'ARN*. PhD thesis, Université Denis Diderot, Paris, France, 2001. (in french).

14. D. B. West and D. B. Shmoys. Recognizing graphs with fixed interval number is NP-complete. *Discrete Appl. Math.*, 8:295–305, 1984.

The Problem
of Context Sensitive String Matching

Venkatesan T. Chakaravarthy and Rajasekar Krishnamurthy

Computer Science Department, University of Wisconsin
Madison, WI 53706, USA
{venkat,sekar}@cs.wisc.edu
http://www.cs.wisc.edu/~{venkat,sekar}

Abstract. In the context sensitive string matching problem, we are
given a pattern and a text. The pattern is a string over variables and
constants and the text is a string of constants. The goal is to find if there
is a mapping from variables to strings of constants so that on applying
this mapping to the pattern we get the given text. Languages like Perl
and Python support such a sophisticated string matching. The problem
is known to be NP-Complete. In this paper, we consider a weighted version of this problem that checks how close the pattern can be matched
with the text. We show that this variation is MAXSNP-Complete and
cannot be approximated within a factor of 3313/3312. We show that
even the restriction, where the pattern consists of variables only, is NP-
Complete and MAXSNP-Complete. When the alphabet is bounded, we
give an approximation algorithm for this restriction.

1   Introduction

String matching is a well studied problem and has applications in a wide variety
of fields. Recently, various modifications of the exact string matching problem
have been considered. Applications in computational biology and computer vision have motivated the study of approximate string matching [,], where different matching relations like swapped matching [ ], "don't cares" [ ] and overlap
matching [ ] have been proposed.

There is another interesting variation of exact string matching, which we call
context sensitive string matching. Here, the pattern is allowed to have variables
and the goal is to map variables to strings, such that, when the variables in the
pattern are replaced by the corresponding strings, we get the text. This kind of
string matching is supported by languages like Perl and Python. Moreover, some
of the complex list processing capabilities of languages like Prolog and Lisp are
captured by this problem. The problem was first considered by Angluin [ ] in
the context of finding patterns common to a set of strings. We next define the
problem formally.

Problem Definition: The input consists of a set of *constants* Σ, a set of
variables V, a *pattern* $P \in (V + \Sigma)^+$ and a *text* $T \in \Sigma^+$. An *assignment* σ
is a mapping $\sigma : V \to \Sigma^+$. $\sigma(P)$ denotes the string obtained by replacing each

A. Apostolico and M. Takeda (Eds.): CPM 2002, LNCS 2373, pp. 64– , 2002.

occurrence of a variable X in P, by $\sigma(X)$. If the strings $\sigma(P)$ and T are the same, σ is a *matching assignment*. The goal of the problem is to determine whether such a matching assignment exists. If so, then P matches T.

For example, let $\Sigma = \{a, b, c, d, e\}$, $V = \{X, Y\}$, $P = XaXY$ and $T =$ "*abaabcde*". Let σ be an assignment that maps X to "*ab*" and Y to "*cde*". Then $\sigma(P)$ is same as the text and σ is a matching assignment. On the other hand, if $P = XX$ and $T =$ "*aaa*", then it is easy to see that there cannot be any matching assignment. Note that each occurrence of a variable X in the pattern has to be replaced by the same string and no variable can be mapped to the null string.

The context sensitive string matching problem was shown to be NP-Complete, even over a binary alphabet, by Angluin [ ]. Just like how exact string matching was extended to approximate string matching, we extend context sensitive string matching to a weighted version. When there is no matching assignment between the pattern and the text, we want to find how close the pattern can be matched with the text. To explore this possibility of approximate matching, we generalize the problem to a weighted version. We show that this problem is MAXSNP-Complete and cannot be approximated within a factor of $3313/3312$. We also consider various restrictions of the problem. When the pattern consists of variables only, we show that the problem remains NP-Complete and MAXSNP-Complete. We also give a simple approximation algorithm for this restriction, when the alphabet is bounded.

We refer to the set of constants Σ as the *alphabet*. We represent the constants by small letters and special symbols, and the variables by capital letters. Throughout the paper, we refer to the context sensitive string matching problem simply as CS-Matching.

Related Work

The concept of pattern, in the above sense, has been considered in many scenarios. In [ ], the problem of finding a "good" pattern that matches a given set of strings was considered. In that context, the language $L(P)$ of a pattern P is defined to be the set of all strings that can be matched (under some assignment) to P. Given a text T and a pattern P, then the CS-Matching problem is membership testing of T in $L(P)$. This was shown to be NP-Complete in [ ]. In this paper, we consider the weighted version of the problem, where we attempt to find an assignment that matches the pattern to the text as close as possible.

In our definitions, we required that each variable is assigned to a non-null string. Such patterns are called non-erasing in literature. If we allow null-string assignments also, the patterns are called erasing. Decidability problems about the equivalence and inclusion among patterns, in both these settings, was studied in [, ,]. The generative power of pattern languages were considered in [,], from a formal language theoretic perspective. While we allow a variable to be mapped to any string (from Σ^+), they consider the power of pattern languages

---

[1] This has been referred to as *uniform substitution* in literature.

when this mapping is restricted to specific languages (like regular and context-free languages).

CS-Matching has also been considered in the context of unavoidability testing of a pattern. An infinite text T is said to avoid a pattern P, if P does not match any substring of T. A pattern is unavoidable if no infinite text avoids it. Prior work like [, ,] deals with identifying necessary and sufficient conditions for a pattern to be unavoidable. Our problem differs in that the goal here is to check how close the pattern matches a given text of finite length.

When each variable occurs only once in the pattern, the pattern can be treated as a regular expression. Thus, CS-Matching is a generalization of the well known regular expression string matching problem.

2   Weighted Context Sensitive String Matching

For a pattern P and a text T, there may not be any matching assignment. In this case, we want to see how close the pattern can be matched to the text. Towards that end, we first define a weighted version of the problem. We then show that this weighted version is MAXSNP-Complete.

There is a natural way to extend the CS-Matching problem to a weighted version. Here, apart from the usual Σ, V, P and T, the input also includes two positive numbers α, the *matching-cost*, and β, the *mismatch-cost*, with $\beta > \alpha$. We call an assignment σ to be a *feasible assignment* if $|\sigma(P)| = |T|$. The cost of a feasible assignment σ is the distance between $\sigma(P)$ and T. Let $T = a_1 a_2 \ldots a_n$ and $\sigma(P) = b_1 b_2 \ldots b_n$. The distance is given by $\Sigma d(a_i, b_i)$, where $d(a_i, b_i)$ is α, if $a_i = b_i$ and β otherwise. If σ is not a feasible assignment then its cost is ∞. We require that the input instances have at least one feasible assignment . Then the weighted CS-Matching problem is to find the optimal assignment.

As CS-Matching is known to be NP-Complete [ ], we consider approximation algorithms for the weighted version. If α and β are allowed to be arbitrary, we cannot have any constant factor approximation algorithm, unless $NP = P$. This is due to the fact that, one can reduce CS-Matching to the weighted version, by setting $\alpha = 0$ and $\beta = 1$. Now, any constant factor approximation algorithm can be used to solve the NP-Complete CS-Matching problem in polynomial time. Similarly, if we set $\alpha = 1$ and β to be a large value (like n^2), we would get the same result. So, it is interesting to consider the case when α and β are constants, with $\beta > \alpha > 0$. We discuss the case where $\alpha = 1$ and $\beta = 2$. Our results can be generalized for any constants α and β and the inapproximability bound will vary accordingly.

1,2-Matching is defined to be the weighted CS-Matching problem, where the matching cost is 1 and the mismatch cost is 2. We can define α, β-Matching similarly, for any constants, $\beta > \alpha > 0$. Let the length of the input text be $|T| = n$. Then any feasible assignment has a cost between n and $2n$. As we require that the input instance should have at least one feasible assignment, the

---

[2] In Section , we give a polynomial time algorithm to check whether a given instance of the problem has some feasible assignment.

optimal cost lies between n and $2n$. We define the difference between the cost of an assignment and n to be the *extra-cost* for that assignment.

We next prove that the 1,2-Matching problem is MAXSNP-Complete. We first prove the result when the size of the alphabet is allowed to be arbitrary. Then we adapt the proof to get the same result, even over a binary alphabet. Note that over a unary alphabet, any feasible assignment is optimal. As a corollary, we get that the α, β-Matching problem is MAXSNP-Complete.

THEOREM 1 *The 1,2-Matching problem is MAXSNP-Complete.*

Proof : In Section , we will give a 2-approximation algorithm for the 1,2-Matching problem. It is known that a problem is in MAXSNP iff it has some constant factor approximation algorithm [ ]. So the problem is in MAXSNP and we proceed to prove that it is MAXSNP-Hard.

The vertex cover problem on 3-regular graphs is known to be MAXSNP-Complete [ ]. We give an L-reduction from the vertex cover problem on 3-regular graphs to our problem. As the first step, we present an algorithm that given a 3-regular graph produces an instance of our problem. Let G be the input graph with n vertices $v_1, v_2, \ldots, v_n$ and m edges $e_1, e_2, \ldots, e_m$. Note that, as the graph is 3-regular, $m = 1.5n$. We use the alphabet $\Sigma = \{\$_1, \$_2, \ldots, \$_n, t, f\}$. For each vertex v_i of the graph, we add a variable V_i to V, the variable set. We then add $m + 1$ dummy variables $D_0, D_1, D_2, \ldots, D_m$ to V. A *dummy variable* is one that occurs exactly once in the pattern. The output pattern P has three parts, $P = P_1 P_2 P_3$. Here, $P_1 = V_1 V_2 \ldots V_n$, $P_2 = $ "$\$_1 \$_2 \ldots \$_n$" and $P_3 = D_0 s_1 D_1 s_2 D_2 \ldots D_{m-1} s_m D_m$ are called the vertex, dollar and edge segments respectively. In the edge-segment, s_i encodes the ith edge. For example, if the ith edge $e_i = (v_j, v_k)$, then $s_i = V_j V_k$. The text $T = T_1 T_2 T_3$ is also made of three segments. T_1 is the string "$fff...f$" of length n. T_2 is the same as $P_2 = $ "$\$_1 \$_2 \ldots \$_n$". T_3 is made up of m *blocks* of the string "$tfttft$". This completes the construction. Refer to Figure for an example, where the input graph is K_4, the complete graph on four nodes.

Let VC^* be an optimal vertex cover of G. We first have to show that the optimal assignment has cost, at most, $\alpha |VC^*|$, for some constant α. First note that the length of the text is $|T| = 11n$. We exhibit an assignment σ^* with cost $11n + |VC^*|$. For each vertex $v_i \in VC^*$, map the variable V_i to "t". For each vertex $v_i \notin VC^*$, map the variable V_i to "f". With this mapping the vertex segments of P and T, namely P_1 and T_1 get aligned with an extra-cost of $|VC^*|$. The dollar-segments P_2 and T_2 align with no extra-cost. We show how to map the dummy variables to strings so that there is no extra-cost in aligning the edge segments P_3 and T_3. Since the mapping is based on a vertex cover, for any edge $e_i = (v_j, v_k)$, the corresponding string $V_j V_k$ has been mapped to one of "tt", "tf" or "ft". All these are available as substrings in "$fttf$" which occurs in each block of T_3. So we can make $V_j V_k$ align with the text without any extra-cost, by making the dummies "eat" the "left-over" text in each block. Since the variables for each edge get mapped within "$fttf$" of a block "$tfttft$", each dummy variable will get mapped to a string of positive length. We exhibit this idea in Figure .

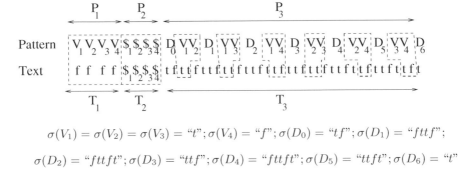

$$\sigma(V_1) = \sigma(V_2) = \sigma(V_3) = \text{``}t\text{''}; \sigma(V_4) = \text{``}f\text{''}; \sigma(D_0) = \text{``}tf\text{''}; \sigma(D_1) = \text{``}fttf\text{''};$$

$$\sigma(D_2) = \text{``}fttft\text{''}; \sigma(D_3) = \text{``}ttf\text{''}; \sigma(D_4) = \text{``}fttft\text{''}; \sigma(D_5) = \text{``}ttft\text{''}; \sigma(D_6) = \text{``}t\text{''}$$

Fig. 1. Constructing an assignment for the vertex cover $\{V_1, V_2, V_3\}$ of the graph K_4

The extra-cost for this assignment σ^* is $|VC^*|$ (the cost incurred in matching P_1 with T_1). We want to find a constant α such that, $cost(\sigma^*) \leq \alpha |VC^*|$. As G is a 3-regular graph, each vertex can cover at most 3 edges. Thus, $|VC^*| \geq n/2$. We already showed that $cost(\sigma^*) = 11n + |VC^*|$. Using these facts, with some simple arithmetic, we get that $cost(\sigma^*) \leq 23|VC^*|$.

We next give an algorithm that takes as input an assignment σ and outputs a vertex cover VC, with the property that, $|VC| - |VC^*| \leq cost(\sigma) - cost(\sigma^*)$, where VC^* is an optimal vertex cover and σ^* is an optimal assignment. To find VC, we first make certain transformations to σ to attain a feasible assignment with the following two properties:-

Property 1 : For each V_i, $\sigma(V_i) = \text{``}t\text{''}$ or $\sigma(V_i) = \text{``}f\text{''}$.

Property 2: For each edge $e_i = (v_j, v_k)$, $\sigma(V_j)$ or $\sigma(V_k)$ is "t".

We also ensure that we do not increase the cost of σ, in doing this transformation. Then, we include all the vertices that got mapped to "t" to VC.

If $|\sigma(V_i)| > 1$, P_2 will not align with T_2, incurring a minimum extra-cost of n. We can get an equally good assignment, by mapping each variable V_i to "t". The mappings for the dummy variables can be changed appropriately so that P_3 exactly matches T_3. The extra cost for this assignment is n and we can take this to be σ.

We can now assume that, for all V_i, $|\sigma(V_i)| = 1$. If for some V_i, $\sigma(V_i) = \$_j$, then an extra-cost of 3 will be incurred for this variable. This is because, V_i appears three times in the edge-segment P_3 and the \$'s do not appears in T_3. So we can change $\sigma(V_i)$ to be "t" or "f", by taking majority of the three text symbols to which V_i is aligned in the edge-segment. This change in mapping will actually decrease the cost. So, we can now assume that σ satisfies Property 1.

Next we convert σ to be block-respecting. In other words, for each edge $e_i = (v_j, v_k)$, the string $s_i = V_j V_k$ is aligned within the substring "$fttf$" of the ith block of T_3. Let us consider the first edge $e_i = (v_j, v_k)$ that violates this condition. By shortening $\sigma(D_{i-1})$ and lengthening $\sigma(D_i)$ appropriately, we can make e_i to be block-respecting. This idea is expressed in Figure .

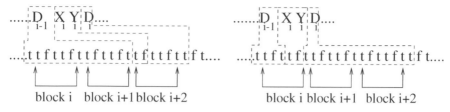

Fig. 2. Making the ith edge block respecting

By repeating this process, we can convert σ to be block-respecting without any additional cost.

Next we ensure that for each edge $e_i = (v_j, v_k)$, at least one of V_j or V_k is mapped to "t". If V_j and V_k are both mapped to "f", we pick one of them arbitrarily, say V_j, and map it to "t". This would increase the cost by 1, due to the mismatch in the vertex segment. We then change $\sigma(D_{i-1})$ appropriately, so that, $V_j V_k$ is aligned to "tf" in the text. Previously, $V_j V_k$ was mapped to "ff". Since T_3 does not contain the substring "ff", this would have caused a mismatch. By aligning $V_j V_k$ to "tf", we remove this mismatch and reduce the cost by 1. This compensates the increase in cost in the vertex segment. We need to handle the other two places in the edge-segment where V_j occurs, without increasing the cost. As V_j is mapped to "t" now, the strings that we need to match can only be "tt", "tf" or "ft". All these are substrings of "$fttf$". They can be accommodated by adjusting the mapping to the appropriate dummy variable so that there is no increase in cost. So the total change in cost is 0.

We can now assume that σ satisfies Property 1 and Property 2. We construct a set VC, by adding the vertex v_i to it iff $\sigma(V_i) = $ "t". This would be a vertex cover. $cost(\sigma) = 11n + |VC|$. Cost of σ^* is, at least, $11n + VC^*$. Otherwise, we can use the above mentioned procedure to get a vertex cover smaller than VC^*. This proves that, $|VC| - |VC^*| \leq cost(\sigma) - cost(\sigma^*)$. We have proved that our reduction is an L-reduction. $\square$

COROLLARY 1 *The 1,2-Matching problem cannot be approximated within a factor of 3313/3312.*

PROOF: It is known that the vertex cover problem is not approximable within a factor of 145/144, even on graphs with maximum degree at most three [ ]. For ease of exposition, we gave the L-reduction from 3-regular graphs. The same proof works even for graphs with maximum degree at most three. In the above L-reduction we ensured that the cost of the optimal solution for the output instance is no more than 23 times the optimal vertex cover of the input graph. We also gave a way to translate a given solution σ of the matching problem into a vertex cover VC, such that $|VC| - |VC^*| \leq cost(\sigma) - cost(\sigma^*)$, where VC^* and σ^* are the corresponding optimal solutions. Using these, we can get the required inapproximability bound. $\square$

We next adapt the above proof to show that even when the alphabet is restricted to be binary, the problem is MAXSNP-Complete.

THEOREM 2 *The 1,2-Matching problem is MAXSNP-Complete, even if the alphabet is binary.*

Proof: We adapt the proof of Theorem here. We use the same L-reduction with some changes. Our alphabet is now $\{t, f\}$. The pattern is $P = P_1 P_2 P_3$, where P_1 and P_3 are same as in the original L-reduction. P_2 is now a string "$ff \ldots f$" of length $1.5n$. The text $T = T_1 T_2 \hat{T}_2 T_3$ is now made of four segments. T_1 and T_3 are as in the original proof. T_2 is the string "$ff \ldots f$" of length $1.5n$. $\hat{T}_2 = $ "$tt \ldots t$", a string of length n.

By adapting the original proof, we can translate an optimal vertex cover VC^* into an assignment with an extra-cost of $|VC^*|$ (note that D_0 can be mapped appropriately to match $\hat{T}_2$ also). As the length of the text is now $12.5n$, we can show that the cost of the optimal assignment is, at most, $26|VC^*|$.

We next modify the algorithm that converts a given assignment σ to a vertex cover VC. The main issue is that $|\sigma(V_i)|$ could be more than 1, for some V_i. We cannot use the argument given in the previous proof to address this issue, as we no longer have n different special symbols. We make a series of transformations to σ, so that $\forall i, |\sigma(V_i)| = 1$, without increasing the cost of σ.

We shall first transform σ so that the first "f" of P_2 is matched within T_2. If σ does not have this property, P_2 is matched completely within $\hat{T}_2$ and T_3. Notice that any substring of $\hat{T}_2 T_3$ of length l has, at least, $2/3 * l$ "t"'s. So, P_2 will incur an extra-cost of at least, $|P_2| * 2/3 = n$. By mapping all V_i's to "t", we can get an equally good assignment (with extra-cost of n).

We can now assume that the first "f" of P_2 is matched within T_2. Now, if a suffix of P_2 is matched to a prefix of T_3, then the entire $\hat{T}_2$ will be matched within P_2, incurring an extra cost of n. So we can assume that P_2 is matched within T_2 and $\hat{T}_2$.

We shall now transform σ so that $|\sigma(V_i)| = 1$, for all V_i. Suppose $|\sigma(V_i)| > 1$, for some V_i. Let $\sigma(V_i) = ayb$, where $a, b \in \{t, f\}$ and $y \in (t + f)^*$. V_i occurs in three places in P_3, as the first variable or as the second (this would depend on how the edge was represented; as $(v_i, \_)$ or as $(\_, v_i)$). If the majority is the first place, we map V_i to "b", else we map it to "a". Such a change in the mapping has to be accounted for in four places, once in P_1 and thrice in P_3. First let us consider its occurrence in the vertex-segment P_1. We have shrunk $\sigma(V_i)$. The substring $V_{i+1} \ldots V_n$ of P_1 will slide to the left over T_1 and T_2 that are made of f's only. So, there is no change in cost for P_1. The sliding also causes one or more f's of P_2, which were originally matched with a prefix of $\hat{T}_2$, to now match with a suffix of T_2. As $\hat{T}_2$ is a string of t's and T_2 is a string of f's, the cost decreases by at least 1. To avoid "disturbing" the alignments in the edge-segments, we increase the length of $\sigma(D_0)$ by $|y| + 1$ and the sliding has no effect in P_3. Now let us consider the three occurrences of V_i in P_3. Let us assume that V_i occurs a majority number of times as the first variable (the other case is handled similarly). Consider each of the three places where V_i occurs in the edge-segment. Let e_r be the edge in consideration. Then, we set $\sigma(D_{r-1}) \leftarrow \sigma(D_{r-1})ay$. This does not change the cost. If V_i is the second variable of e_r, we set $\sigma(D_r) \leftarrow yb\sigma(D_r)$. This could increase the cost by one, as a and b

may be different. The latter case will occur at most once, as we took majority. So we have made $|\sigma(V_i)| = 1$, without increasing the cost.

Repeating the above procedure for each variable, we have managed to change σ so that each V_i is either mapped to "t" or to "f". σ now satisfies Property 1 of the original proof. We can continue as in the original proof to satisfy Property 2 and get the required result. ☐

COROLLARY 2 *The 1,2-Matching problem cannot be approximated within a factor of 3313/3312, even when the alphabet is binary.*

PROOF: A straightforward adaptation of Corollary   to the above theorem would yield a bound of 3745/3744. This can be improved to 3313/3312 if we could reduce the length of the text from $12.5n$ to $11n$. We do this by using the block "$tfttf$" instead of "$tfttft$" in T_3 and adding a single constant t at the end of the text. The proof of the above theorem has to be modified slightly to handle this. ☐

COROLLARY 3 *The α, β-Matching problem is MAXSNP-Complete and cannot be approximated within a factor of* $1 + \frac{1}{144(\beta+21\alpha)(\beta-\alpha)}$.

3   Restrictions to Weighted CS-Matching

We showed that the 1,2-Matching problem is MAXSNP-Complete. In this section, we give a 2-approximation algorithm for the problem. We also consider whether restrictions of the problem have better upper bounds. The first restriction we consider is when the number of variables is bounded by a constant. We show that this problem can be solved in polynomial time. In the second restriction, we require that the pattern be a string made of variables only. We show that this problem is NP-Complete and also MAXSNP-Complete. We give an approximation algorithm for this problem, when the alphabet is bounded. The following lemma is useful in analyzing the two restrictions.

LEMMA 1 *We can verify whether a given instance of the weighted CS-Matching problem has a feasible assignment or not in polynomial time.*

PROOF: Let the input pattern contain k variables, $X_1, X_2, \ldots, X_k$, each occurring $a_1, a_2, \ldots, a_k$ times respectively. Let the length of the text be n, the total length of constants in the pattern be c and set $n' = n - c$. It can be seen that a feasible assignment exists iff the equation

$$a_1 * x_1 + a_2 * x_2 + \cdots + a_k * x_k = n'$$

has positive integer solutions in the unknowns $x_1, \ldots x_k$.

We give a simple algorithm based on dynamic programming to solve the above equation. We compute a $n' \times n'$ boolean array A, where $A_{p,q}$ is true iff there exists an assignment of positive integers to the x_i's, such that $\sum x_i = p$ and $\sum a_i x_i = q$. We set $A_{k,l}$ to true, where $l = \sum a_i$. We set all other entries of the first k rows to false. To compute the rest of the array, we use the recurrence relation,

$$A_{p,q} = \bigvee_{i=1}^{k} A_{p-1, q-a_i}$$

Finally, there exists a positive integer solution to the equation iff for some p, $A_{p,n'}$ is true. □

The above algorithm can be modified to find a feasible assignment. Any feasible assignment has cost at most $2n$, where n is the length of the text. The optimal assignment has cost at least n. So we have a 2-approximation algorithm.

COROLLARY 4 *There is a 2-approximation algorithm for the 1,2-Matching problem.*

We next discuss a problem that occurs in the analysis of the restricted versions that we consider below.

Fixed Length Weighted CS-Matching Problem: As usual, the input consists of a text T of length n and a pattern P over k variables $V_1, V_2, \ldots, V_k$, where the variable V_i occurs a_i times in P. We are also given a sequence of k integers $l_1, l_2, \ldots, l_k$ satisfying the equation, $\sum a_i * l_i = n - c$, where c is the total length of the constants in P. Among all assignments σ that satisfy the condition, $\forall i \ |\sigma(V_i)| = l_i$, the goal is to find an assignment with the least cost.

The above problem can be solved in polynomial time. We can compute the required mapping for each variable V_i as follows. As the lengths of all variables are fixed, the substrings in the text that align with each occurrence of V_i are also fixed. Let these be $y_1, y_2, \ldots, y_{a_i}$, each of length l_i. We set $\sigma(V_i)$ to $a_1 a_2 \ldots a_{l_i}$, where a_j is a constant that occurs the maximum number of times in the jth position of the strings $y_1, y_2, \ldots, y_{a_i}$. It can be seen that this assignment has the least cost of all assignments that satisfy the specified lengths.

3.1   Restriction 1: Bounded Number of Variables

THEOREM 3 *If the number of variables is bounded by a constant K, an optimal assignment can be found in polynomial time.*

PROOF: Finding the assignment σ can be viewed as a two stage process. We first fix the length of the mapping for each variable and then fix the actual mapping. Once we fix the lengths for $\sigma(V_i)$'s, we get an instance of the *fixed length weighted CS-Matching* problem, which was shown to be solvable in polynomial time. As each variable can only have length between 1 and n, the number of ways in which we can fix the lengths is bounded by n^K. So, when K is a constant, the problem can be solved in polynomial time. □

3.2   Restriction 2: Variable-Only Patterns

THEOREM 4 *The CS-Matching problem is NP-Complete even when the pattern is restricted to be a string of variables only and the alphabet is restricted to be binary.*

PROOF: We give a reduction from 1in3SAT. In the 1in3SAT problem, given a 3SAT formula, we need to check if there is a truth assignment that satisfies exactly one literal in each clause. This problem is known to be NP-Complete [ ].

Let the input formula for the 1in3SAT problem be over n boolean variables $x_1, x_2, \ldots x_n$ and contain m clauses. Our reduction outputs an instance of the matching problem over the alphabet $\Sigma = \{\$, a\}$ and the variable set V. For each boolean variable x_i, we add two variables X_i and $\overline{X_i}$ to V. We also add another variable Z to V. We output the pattern

$$ZX_1\overline{X_1}ZX_2\overline{X_2}Z \ldots ZX_n\overline{X_n}Zs_1Zs_2Z \ldots ZS_mZ$$

where s_j is a string that encodes the jth clause. For example, if the jth clause is $(x_1 \vee \overline{x_2} \vee x_3)$ then s_j would be the string $X_1\overline{X_2}X_3$. The text is the string $\$(aaa\$)^n(aaaa\$)^m$. This defines the polynomial reduction. We next show that the formula has a 1in3 satisfying assignment iff the pattern matches the text.

First, if the input formula has a 1in3 satisfying assignment ϕ, we find a matching assignment σ as follows:

– Set $\sigma(Z) = \$$.
– If $\phi(X_i)$ is true, then set $\sigma(X_i) = $ "aa" and $\sigma(\overline{X_i}) = $ "a".
– If $\phi(X_i)$ is false, then set $\sigma(X_i) = $ "a" and $\sigma(\overline{X_i}) = $ "aa".

It can be seen that σ is a matching assignment.

Next, let us assume that the pattern matches the text via a matching assignment σ. Then $\sigma(Z)$ has to start and end with $\$$. As the number of Z's in the pattern is same as the number of $\$$'s in the text, Z has to be mapped to $\$$. So, for any pair of variables X_i and $\overline{X_i}$ either X_i is mapped to "aa" and $\overline{X_i}$ to "a" or vice versa. We then exhibit a 1in3 satisfying assignment for the input formula. We set x_i to true, if X_i is mapped to "aa", and to false, otherwise. As the pattern and the text are matched, any string s_j is aligned with the string "aaaa". Thus, of the three variables in s_j, exactly one is mapped to "aa" and the other two are mapped to "a". This implies that the truth assignment we constructed is a 1in3 satisfying assignment.   □

THEOREM 5 *The 1,2-Matching problem is MAXSNP-Complete even when the over Variable-Only patterns and binary alphabet.*

PROOF: We shall adapt the proof of Theorem   to achieve this result. The only changes we require in the original L-reduction are that, $P_2 = ZZ \ldots Z$ and $T_2 = ff \ldots f$, where both are of length $11n$. The same proof can be used to show that we can construct an assignment whose cost is, at most, 45 times $|VC^*|$. For the other direction of the L-reduction, we can use the same proof, if we can first ensure that Z is mapped to "f". In any feasible assignment, Z cannot be mapped to a string of length greater than 1, because Z occurs $11n$ times and the text is of length $22n$. Every substring of the text of length l has at least $l/3$ f's. So, if Z is mapped to "t", there is an extra-cost of $11n/3$. But, we can easily construct an assignment of extra-cost n. Hence we can assume that Z is mapped to "f" and proceed with the original proof.   □

THEOREM **6** *For Variable-Only patterns, when the alphabet size is bounded by a constant D, we can approximate the 1,2-Matching problem within a factor of $1 + \frac{D-1}{D}$.*

PROOF: We first use the algorithm given in Lemma   to check if there is some feasible assignment. The algorithm can be modified to return the lengths of the assignment for each variable. We now have an instance of the fixed length weighted CS-Matching problem. We use the algorithm given for the latter problem to find the best assignment with these lengths. Since we use the majority rule in fixing the assignment, we are guaranteed to have at least one match for every $(D-1)$ mismatches. As the pattern has variables only, the constructed assignment has cost at most $n(1 + \frac{D-1}{D})$. Hence we have the required approximation bound.   □

4   Conclusions and Open Problems

The problem of context sensitive string matching, which has practical applications, is known to be NP-Complete. We presented a weighted version of the problem and showed it to be MAXSNP-Complete, even over a binary alphabet. We also showed that the problem cannot be approximated within a factor of 3313/3312. We considered an interesting restriction where the pattern is made of variables only and showed that even under this restriction, the problem is NP-Complete and MAXSNP-Complete. We also gave an approximation algorithm for this case, when the alphabet size is bounded.

There are several open problems in this area. We gave an approximation algorithm in the case where the pattern is void of constants and the alphabet is bounded. Finding an approximation algorithm without either or both of these restrictions could be the next step. Designing improved approximation algorithms even with both the restrictions is another avenue for research. Obtaining better inapproximability bounds is another interesting area. Studying the problem where the pattern can be matched to a substring of the text has practical ramifications.

Acknowledgments

We thank Jin-Yi Cai and Christine Heitsch for useful discussions and comments.

References

1. P. Alimonti and V. Kann. Hardness of approximating problems on cubic graphs. In *Proc. 3rd Italian Conf. on Algorithms and Complexity, Lecture Notes in Computer Science, 1203*, pages 288–298. Springer-Verlag, 1997.
2. A. Amir, Y. Aumann, G. Landau, M. Lewenstein, and N. Lewenstein. Pattern matching with swaps. In *Proc. 38th IEEE Conf. on Foundations of Computer Science (FOCS)*, pages 144–153, 1997.

3. A. Amir, R. Cole, R. Hariharan, M. Lewenstein, and E. Porat. Overlap matching. In *Proceeding of the Twelfth Annual Symposium on Discrete algorithms (SODA)*, pages 279–288, 2001.
4. D. Angluin. Finding patterns common to a set of strings. In *Journal of Computer and Systems Sciences*, volume 21, pages 46–62, 1980.   , ,
5. A. Apostolico and Z. Galil (eds.). *Pattern Matching Algorithms*. Oxford Univ. Press, 1997.
6. M. Crochemore and W. Rytter. *Text Algorithms*. Oxford Univ. Press, 1994.
7. J. Dassow, Gh. Paun, and A. Salomaa. Grammars based on patterns. In *Intl. Journal on Foundations of Computer Science*, volume 4, pages 1–14, 1993.
8. M.J. Fischer and M.S. Paterson. String matching and other products. In *Complexity of Computation, SIAM-AMS Proceedings*, pages 7:113–125, 1974.
9. M. R. Garey and D. S. Johnson. *Computers and Intractability: A Guide to the Theory of NP-Completeness*. Freeman, San Francisco, 1979.
10. C. Heitsch. *Computational Complexity of Generalized Pattern Matching*. Ph.D thesis, Dept. of Math., Univ. of California at Berkeley, 2000.
11. T. Jiang, A. Salomaa, K. Salomaa, and S. Yu. Decision problems for patterns. In *Journal of Computer and Systems Sciences*, volume 50, pages 53–63, 1995.
12. M. Karpinski. Approximating bounded degree instances of NP-Hard problems. In *Electronic Colloquium on Computational Complexity, ECCC Report TR01-042*, 2001.
13. M. Lothaire. *Combinatorics on Words*. Addison-Wesley, 1983.
14. V. Mitrana, Gh. Paun, G. Rozenberg, and A. Salomaa. Patttern systems. In *Theoretical Computer Science*, volume 154, pages 183–201, 1996.
15. E. Ohlebusch and E. Ukkonen. On the equivalence problem for e-pattern languages. In *Theoretical Computer Science*, volume 186, pages 231–248, 1997.
16. C. H. Papadimitriou. *Computational Complexity*. Addison-Wesley, 1994.
17. A. I. Zimin. Blocking sets of terms. In *Math. Sbornik*, volume 119, pages 363–375, 1982.

Two-Pattern Strings*

František Franěk[1], Jiandong Jiang[1,2], Weilin Lu[1,2], and William F. Smyth[1,3]

[1] Algorithms Research Group, Department of Computing & Software
McMaster University, Hamilton, Ontario, Canada L8S 4K1
smyth@mcmaster.ca
www.cas.mcmaster.ca/cas/research/groups.shtml
[2] Toronto Laboratories, IBM Canada
8200 Warden Avenue, Markham, Ontario, Canada L6G 1C7
[3] School of Computing, Curtin University
GPO Box U-1987, Perth WA 6845, Australia

Abstract. This paper introduces a new class of strings on $\{a, b\}$, called **two-pattern strings**, that constitute a substantial generalization of Sturmian strings while at the same time sharing many of their nice properties. In particular, we show that, in common with Sturmian strings, only time linear in the string length is required to recognize a two-pattern string as well as to compute all of its repetitions. We also show that two-pattern strings occur in some sense frequently in the class of all strings on $\{a, b\}$.

1 Introduction

In this paper we outline the results of an investigation of the properties of a new class of strings on $\{a, b\}$, derived by the successive action of a sequence of morphisms on the single letter a. All of the strings so determined are finite, and we deal with them from a computational point of view: initially, we are interested in efficient algorithms to recognize such strings and to compute the repetitions in them; then we go on to estimate their frequency of occurrence among all strings on $\{a, b\}$.

A previous paper [ ] specified linear-time algorithms to recognize and compute repetitions in finite substrings of Sturmian strings; the class of strings discussed here significantly extends this work.

Let p and q denote two distinct nonempty strings on $\{a, b\}$ such that $|p| \leq \lambda$ and $|q| \leq \lambda$, where λ is a finite integer called the **scope**. We call p and q **patterns of scope** λ. For any pair of finite positive integers i and j such that $i < j$, consider the morphism σ that maps single letters into **blocks**:

$$a \rightarrow p^i q, \quad b \rightarrow p^j q. \tag{1}$$

We call σ an **expansion of scope** λ and observe that it is specified by a 4-tuple $[p, q, i, j]$. Observe also that an expansion can be applied to any (finite or

* Supported in part by grants from the Natural Sciences & Engineering Research Council of Canada.

infinite) string on $\{a, b\}$ to yield an expanded string

$$\boldsymbol{y} = \sigma(\boldsymbol{x}).$$

Given any two morphisms σ_1 and σ_2, the composition $\sigma_1 \circ \sigma_2$ is therefore well defined: $\boldsymbol{z} = \sigma_1(\boldsymbol{y}) = \sigma_1(\sigma_2(\boldsymbol{x})) = (\sigma_1 \circ \sigma_2)(\boldsymbol{x})$.

Definition 1. *Suppose a positive integer λ and a finite sequence*

$$\sigma_1, \sigma_2, \ldots, \sigma_k$$

of expansions of scope λ are given, where

$$\sigma_r = [\boldsymbol{p_r}, \boldsymbol{q_r}, i_r, j_r]$$

for every $r = 1, 2, \ldots, k$. Then the string

$$\boldsymbol{x} = (\sigma_1 \circ \sigma_2 \circ \cdots \circ \sigma_k)(a)$$

*is a **complete two-pattern string of scope** λ if and only if every pair $(\boldsymbol{p_r}, \boldsymbol{q_r})$ of patterns is **suitable** (defined in Section 2).*

The definition of a suitable pair of patterns is deferred till Section 2 because it is necessarily somewhat technical. However, the main idea of a suitable pair is simple: $\boldsymbol{p}$ and $\boldsymbol{q}$ should be dissimilar enough that they can be efficiently distinguished from each other by an algorithm that recognizes complete two-pattern strings.

We can easily provide examples of complete two-pattern strings. If we suppose that $\lambda = 3$ and $\sigma_1 = [ab, ba, 2, 3]$, $\sigma_2 = [abb, aa, 1, 4]$, the following strings are all complete two-pattern strings of scope 3:

$$\sigma_1(a) = (ab)^2 ba;$$
$$(\sigma_1 \circ \sigma_1)(a) = (ab)^2 ba(ab)^3 ba(ab)^2 ba(ab)^3 ba(ab)^3 ba(ab)^2 ba;$$
$$(\sigma_1 \circ \sigma_2)(a) = (ab)^2 ba(ab)^3 ba(ab)^3 ba(ab)^2 ba(ab)^2 ba;$$
$$(\sigma_2 \circ \sigma_1)(a) = (abb)aa(abb)^4 aa(abb)aa(abb)^4 aa(abb)^4 aa(abb)aa.$$

Observe further that when the scope $\lambda = 1$, the choice of $\boldsymbol{p}$ and $\boldsymbol{q}$ is restricted to

$$(\boldsymbol{p}, \boldsymbol{q}) = (a, b) \quad \text{or} \quad (\boldsymbol{p}, \boldsymbol{q}) = (b, a). \tag{2}$$

If the further restriction is imposed that $j = i+1$, then all the strings generated by any finite sequence of expansions are finite substrings of Sturmian strings; in fact, in the terminology of [ ], these strings are exactly the set of "block-complete" finite substrings of Sturmian strings. We note that every complete two-pattern string of scope λ is also a complete two-pattern string of scope $\lambda+1$; thus in particular every block-complete finite substring of a Sturmian string is a complete two-pattern string.

As noted above, our initial interest in complete two-pattern strings is computational, following similar studies of Fibonacci [, ,] and Sturmian [,] strings. We pose two sets of questions:

(Q1) What is the complexity of determining whether or not a given string $x = x[1..n]$ is a fragment of a complete two-pattern string? Can an efficient algorithm be found to make this determination for every x?

(Q2) Given a fragment x of a complete two-pattern string, can an algorithm be found that computes all the repetitions in x in linear time?

Since complete two-pattern strings constitute a much more general class of strings than block-complete finite substrings of Sturmian strings, the following questions also become of interest:

(Q3) What is the frequency of occurrence of fragments x of complete two-pattern strings among all strings on $\{a, b\}$ of length n? What is the asymptotic frequency of occurrence of complete two-pattern strings among all infinite strings on $\{a, b\}$?

In this paper we provide a partial answer to (Q1) by outlining an algorithm that in $\Theta(n)$ time determines whether or not a given string $x[1..n]$ is complete two-pattern. Similar to the recognition algorithm in [ ], this algorithm outputs the sequence of expansions ( ) by which a is transformed into x — or more precisely, the sequence of **reductions**

$$p^i q \rightarrow a, \quad p^j q \rightarrow b \tag{3}$$

by which x is reduced to a. This sequence provides a complete specification of x. Since by ( ) each reduction decreases string length by a factor that exceeds

$$i|p| + |q| \geq 2, \tag{4}$$

the recognition algorithm thus yields as a byproduct a potential data compression technique for complete two-pattern strings x.

The reduction sequence is then used to provide partial answers to (Q2) and (Q3). Before going on to discuss these questions in more detail, we pause to provide an introduction and context for them, as well as an outline of the main results.

In dealing with (Q1), we need to cope with the possibility that at any stage of the reduction of x, there may be more than one reduction satisfying ( ): it then becomes possible that one of these reductions is a part of a sequence that reduces x to a, while another one is not. As long as this possibility exists, any recognition algorithm would be obliged to include provision for backtracking, leading possibly to an execution time exponential in the number of reductions. Our main result in this connection is to show however that backtracking is not required, and that therefore the algorithm that recognizes complete two-pattern strings requires only $\Theta(n)$ time.

Of course this does not yet fully solve (Q1). We conjecture that, just as for Sturmian strings [ ], there exists a $\Theta(n)$-time algorithm to determine whether or not a string is a *fragment* of a complete two-pattern string of scope λ. If such an algorithm were found, it would greatly extend the class of strings that could be efficiently compressed using reductions, or whose repetitions could be efficiently computed.

The view may be taken that interest in (Q2) has been superseded by other work. It has recently become clear that, as a result of research extending over a period of a quarter-century, the repetitions in any string $x[1..n]$ on an **indexed** alphabet — that is, an alphabet of size $\alpha \in O(n)$ that maps onto the integers $1..\alpha$ — can be computed in $\Theta(n)$ time. The main steps in this development are as follows:

- an algorithm to compute the suffix tree of x in $\Theta(n)$ time [ ];
- an algorithm to compute the s-factorization of x, given the suffix tree of x, in $\Theta(n)$ time [,];
- the identification of "maximal periodicities" or "runs" as a suitable encoding of repetitions in strings, and the computation of the leftmost occurrence of every distinct run in x in $\Theta(n)$ time, based on the s-factorization [ ];
- the proof that the number of runs in any string is $O(n)$, and the extension of the algorithm [ ] to compute all occurrences of every run in x in $\Theta(n)$ time, still based on the s-factorization [ ].

Impressive as this intellectual edifice is, it nevertheless appears, at least in the context of strings on the alphabet $\{a, b\}$, to be rather indirect in its approach, perhaps involving more sophistication than is really required. Indeed, it is not clear that the $\Theta(n)$-time algorithm given in [ ] is preferable in practice to classical $O(n \log n)$-time algorithms for suffix-tree construction. Further, the very long and technical proof that number of runs is linear in string length shows that a constant of proportionality exists, but provides no information about its size; at the same time, computer experiments described in [ ] provide convincing evidence that the maximum number of runs in any string is at most n, and that this maximum occurs in strings on $\{a, b\}$! Thus, in a sense, the existing theory serves to remind us of how little, rather than how much, we know of periodicity in strings, perhaps especially those on $\{a, b\}$.

For (Q2) we adopt a more direct approach, an extension of the methodology used in [ ] for Sturmian strings. Making use of the reduction sequence computed by the recognition algorithm, we show how to compute all the runs in complete two-pattern strings $x[1..n]$ in $\Theta(n)$ time. Essentially, we show that if y is derived from x by a reduction ( ), then the nontrivial runs in x can be computed directly from certain special configurations occurring in y; thus, over the whole reduction sequence, the runs in x can be computed on a step-by-step basis, from one reduction to the next. It is the special configurations that are of interest here, since they provide insight into the way in which repetitions are formed.

Finally, we report on progress with (Q3), in estimating the frequency of occurrence of complete two-pattern strings among all strings on $\{a, b\}$. We claim that for λ sufficiently large with respect to n, complete two-pattern strings are dense in the set of all strings. We claim also that for some values of k and fixed λ, the number of distinct strings of length k (the **complexity**) can exceed $2k$ — can in fact even be exponential in k.

Sections - deal with questions (Q1)-(Q3) respectively.

2   Recognizing Two-Pattern Strings in Linear Time

Before proceeding with our development, we need to provide a definition of the term "suitable pair" mentioned in Section :

Definition 2. *A string q is said to be **p-regular** if and only if there are strings $u \neq \varepsilon$, v together with nonnegative integers $n_1, \ldots, n_k$, $k \geq 1$, and r such that*

- *p is neither a prefix nor a suffix of u;*
- *p is neither a prefix nor a suffix of v;*
- *there are at most two integer values m_1 and m_2 such that for each $i \in 1..k$, $n_i = m_1$ or $n_i = m_2$, i.e. $|\{n_i : i \in 1..k\}| \leq 2$;*
- *$q = (up^r vp^{n_1})(up^r vp^{n_2}) \cdots (up^r vp^{n_k})u$;*
- *if $r = 0$, then $v = \varepsilon$ (the empty string).*

The next definition formalizes the notion that the two strings p and q are fundamentally distinct and can be used as "building blocks" for complete two-pattern strings.

Definition 3. *An ordered pair of nonempty strings (p, q) is said to be a **suitable pair of patterns** if and only if*

- *p is **primitive**, i.e. has no nonempty border;*
- *p is neither a prefix nor a suffix of q;*
- *q is neither a prefix nor a suffix of p;*
- *q is not p-regular.*

Using these definitions, we can now show how to reduce a nontrivial complete two-pattern string. Let

$$x = abbabaaabbababbababbababbabaaabbababbabaaabbababbababbabababbaba$$
$$aabbababbababbababbbabaaabbababbababbababbabaaabbababbabaaabbab.$$

Consider the reduction $\rho_1 = [a, bbab, 1, 3]$ and observe that according to Definition , $(a, bbab)$ is a suitable pair. Then applying the reduction

$$a(bbab) \to a, \quad a^3(bbab) \to b,$$

we find that

$$x_1 = \rho_1(x) = abaaababaaabaaabaaabab.$$

Similarly for $\rho_2 = [a, b, 1, 3]$, we find

$$x_2 = \rho_2(x_1) = ababbba,$$

while for $\rho_3 = [ab, bba, 2, 3]$,

$$x_3 = \rho_3(x_2) = a.$$

Thus $x = x[1..124]$ is completely described by the three-term reduction sequence

$$[a, bbab, 1, 3], \ [a, b, 1, 3], \ [ab, bba, 2, 3],$$

and so is a complete two-pattern string of scope 4.

In the context of possible applications to data compression, it is worth remarking that in general the number of terms in a reduction sequence is by ( ) at most $\lceil \log_2 n \rceil$ and may be much smaller. In our example, $n = 124$ and sequence length $3 = \log_{4.39} 124$.

We now introduce formally an idea mentioned in the introduction: a canonical reduction that is identified by patterns that are somehow "shortest":

Definition 4. *A reduction* $\rho = [\boldsymbol{p}, \boldsymbol{q}, i, j]$ *of a binary string* $\boldsymbol{x}$ *using patterns of scope* λ *is* λ-**canonical** *if and only if for every reduction* $\rho_1 = [\boldsymbol{p_1}, \boldsymbol{q_1}, i_1, j_1]$ *of* $\boldsymbol{x}$ *using patterns of scope* λ*:*

(a) *either* $|\boldsymbol{p}| < |\boldsymbol{p_1}|$*, or* $|\boldsymbol{p}| = |\boldsymbol{p_1}|$ *and* $|\boldsymbol{q}| < |\boldsymbol{q_1}|$*, or* $|\boldsymbol{p}| = |\boldsymbol{p_1}|$ *and* $|\boldsymbol{q}| = |\boldsymbol{q_1}|$*.*
(b) $\boldsymbol{x} = \boldsymbol{p_1}^{i_1} \boldsymbol{q_1}$ *implies* $\boldsymbol{x} = \boldsymbol{p}^i \boldsymbol{q}$*.*

It is then possible to prove that it suffices to reduce $\boldsymbol{x}$ using a sequence of canonical reductions:

Theorem 1. $\boldsymbol{x}$ *is a complete two-pattern string of scope* λ *if and only if there is a sequence of* λ-*canonical reductions* $\{\rho_1, \rho_2, \cdots, \rho_n\}$ *reducing* $\boldsymbol{x}$ *to a string a.*
□

We omit the fairly predictable details of the algorithm REC that is based on this theorem. We state however the main result:

Theorem 2. *For any* $\lambda \geq 1$*, the recognition algorithm REC determines in* $O(2\lambda^8 |\boldsymbol{x}|)$ *steps whether or not* $\boldsymbol{x}$ *is a complete two-pattern string of scope* λ*, and if so, the algorithm outputs the* λ-*canonical reduction sequence of* $\boldsymbol{x}$*.* □

3   Computing the Repetitions in Linear Time

We describe here the main ideas that permit the repetitions in a complete two-pattern string to be computed in linear time. Just as for Sturmian strings [ ], it turns out that the repetitions that occur in an expansion $\boldsymbol{y} = \sigma(\boldsymbol{x})$ of a two-pattern string $\boldsymbol{x}$ are formed as a result of the application of σ to certain well-defined configurations in $\boldsymbol{x}$. Thus, with the help of the expansion (reduction) sequence that determines a complete two-pattern string, it is possible to track and output the repetitions as they are formed by each expansion in the sequence. As mentioned in the introduction, a crucial factor that ensures the efficiency of this process is the encoding of the repetitions as runs.

Definition 5. *A* **repetition** *in* $\boldsymbol{x} = \boldsymbol{x}[1..n]$ *[ ] is a triple* (i, p, r) *of positive integers, where* $i < n$*,* $r > 1$*,*

$$\boldsymbol{x}[i..i+rp-1] = \boldsymbol{x}[i..i+p-1]^r,$$

and $\boldsymbol{x}[i + rp..i+(r+1)p-1] \neq \boldsymbol{x}[i..i+p-1]$*. The* **period** *of the repetition is* p *and its* **generator** *is* $\boldsymbol{x}[i..i+p-1]$*.*

Crochemore [ ] showed that Fibonacci strings, a special case of two-pattern strings, contain $\Omega(n \log n)$ repetitions. To avoid $\Omega(n \log n)$ processing just for output, we therefore introduce:

Definition 6. *A **run** in $x[1..n]$ [ ] is a 4-tuple (i, p, r, t) where*

$$(i, p, r), \ (i+1, p, r), \ \ldots, \ (i+t, p, r)$$

*are all repetitions, while $(i-1, p, r)$ and $(i+t+1, p, r)$ are not. The **period** and **generator** are defined as for (i, p, r).*

It is easy to prove that for any constant κ, all the runs in x whose period $p \le \kappa$ can be output in at most $c_\kappa n$ steps, where c_κ is a constant whose value depends only on κ. In particular, this result is true for the choice $\kappa = 3\lambda$. For $p > 3\lambda$, we require the following lemma, the main result of this section:

Lemma 1. *For $|p| \le \lambda$, $|q| \le \lambda$, let $\sigma = [p, q, i, j]$ be an expansion of x, so that $y = \sigma(x)$. Then for every run R in y whose period $p > 3\lambda$, one of the following holds:*

- *R is an expansion under σ of a run in x,*
- *R is determined by a square u^2 in x that is derived from a substring of x of one of the following forms:*

$$aa, ab, ba, bb, avbva, bvavb, bvaavb, avbv, bvav,$$

for any nonempty substring v. □

The details of "deriving" u^2 and of using it to "determine" R are lengthy and complicated, to be found at web site

$$\texttt{http://www.cas.mcmaster.ca/~franek/}$$

The analysis found there enables us to claim that

Theorem 3. *There exists an algorithm RUN such that for every integer $\lambda \ge 1$, RUN computes all the runs in every complete two-pattern string $x = x[1..n]$ of scope λ, based on the reduction sequence of x, in at most $c_\lambda n$ steps, where c_λ is a constant whose value depends only on λ.* □

4   Frequency of Two-Pattern Strings

In this section we first present results showing that infinite two-pattern strings (that is, two-pattern strings formed from an infinite sequence of expansions) have complexity at least $k+1$, while for some values of k the complexity can even be exponential in k. Since Sturmian strings have complexity $k+1$, we can accordingly claim that two-pattern strings are in some sense more frequent among all strings on $\{a, b\}$ than Sturmian strings are. Nevertheless, for fixed λ the relative frequency of two-pattern strings approaches zero as string length approaches infinity.

Another point of view is also of interest. We find that if we consider only those values of n that are close to λ, then two-pattern strings occur frequently among all strings of length n. In other words, for sufficiently large λ, a large proportion of strings on $\{a, b\}$ turn out to be two-pattern strings.

We begin by recalling the notation $\mathrm{LCP}(\boldsymbol{u}, \boldsymbol{v})$ and $\mathrm{LCS}(\boldsymbol{u}, \boldsymbol{v})$ for arbitrary strings $\boldsymbol{u}$ and $\boldsymbol{v}$: longest common prefix and longest common suffix, respectively. It is then convenient to define, for any integer $m \geq 0$,

$$\Delta_m = \Delta_m(\boldsymbol{u}, \boldsymbol{v}) = m|\boldsymbol{p}| + |\mathrm{LCP}(\boldsymbol{u}, \boldsymbol{v})| + |\mathrm{LCS}(\boldsymbol{u}, \boldsymbol{v})|.$$

Using this notation, we state:

Theorem 4. *Let $\boldsymbol{x}$ be an infinite two-pattern string of scope λ reducible by $[\boldsymbol{p}, \boldsymbol{q}, i, j]$. Then the complexity $C_k = C_k(\boldsymbol{x})$ satisfies*

(a) $C_k \geq k+1$ *when* $|\boldsymbol{p}| \leq k \leq \Delta_i$;
(b) $C_k \geq 2k - \Delta_i$ *when* $\Delta_i + 1 \leq k \leq \Delta_{j-1} + 1$;
(c) $C_k \geq k+1+(j-i-1)|\boldsymbol{p}|$ *when* $k \geq \Delta_{j-1} + 2$;

where $\Delta_m = \Delta_m(\boldsymbol{p}, \boldsymbol{q})$. $\square$

This result provides lower bounds on the complexity C_k that are in fact sharp. We have also established upper bounds on C_k that are however not sharp. To show that the complexity can for some values of k be much larger than $k+1$, consider an infinite two-pattern string $\boldsymbol{x}$ with substring $\boldsymbol{pqp}$, where $\boldsymbol{p} = aaaabbbb$ and $\boldsymbol{q} = aababbab$ are a suitable pair of patterns. It is easy to check that in the substring $\boldsymbol{pqp}$ there are 2^4 substrings of length 4 — for $k = 4$ the complexity is exponential in k.

In order to state our final results, we introduce the constant [ ]

$$\phi = \lim_{n \to \infty} P(n)/2^n \approx 0.26778684,$$

where $P(n)$ is the number of primitive strings (with no nonempty border) of length n on $\{a, b\}$. The frequency of occurrence of two-pattern strings for λ large with respect to n can then be estimated in terms of ϕ:

Theorem 5. *For $n \geq 2$, let $f(n)$ denote the frequency of occurrence among all strings of length n of two-pattern strings $\boldsymbol{x}[1..n] = \boldsymbol{pq}$, where $\boldsymbol{p}, \boldsymbol{q}$ is a suitable pair of scope $\lambda \geq \lceil n/2 \rceil$. Then $f(n) \geq \phi/2$.* $\square$

Theorem 6. *For $n \geq 4$, let $f(n)$ denote the frequency of occurrence among all strings of length n of two-pattern strings $\boldsymbol{x}[1..n] = \boldsymbol{p}^i \boldsymbol{q}$, where $i \geq 1$ and $\boldsymbol{p}, \boldsymbol{q}$ is a suitable pair of scope $\lambda \geq n-2$. Then $f(n) \geq 15\phi/16$.* $\square$

References

1. M. Boshernitzan & Aviezri S. Fraenkel, A linear algorithm for nonhomogeneous spectra of numbers, J. Algorithms **5** (1984) 187–198.

2. Maxime Crochemore, An optimal algorithm for computing the repetitions in a word, IPL **12-5** (1981) 244–250.   ,

3. Martin Farach, Optimal suffix tree construction with large alphabets, Proc. 38th Annual IEEE Symp. FOCS (1997) 137–143.

4. Aviezri S. Fraenkel & R. Jamie Simpson, The exact number of squares in Fibonacci words, TCS **218-1** (1999) 83–94.

5. František Franěk, Ayşe Karaman & W. F. Smyth, Repetitions in Sturmian strings, TCS **249-2** (2000) 289–303.   ,   ,   ,   ,

6. Leo J. Guibas & Andrew M. Odlyzko, Periods in strings, J. Combinatorial Theory, Series **A 30** (1981) 19–42.

7. Costas S. Iliopoulos, Dennis Moore & W. F. Smyth, A characterization of the squares in a Fibonacci string, TCS **172** (1997) 281–291.

8. Roman Kolpakov & Gregory Kucherov, On maximal repetitions in words, J. Discrete Algorithms 1 (2000) 159–186.   ,

9. Abraham Lempel & Jacob Ziv, On the complexity of finite sequences, IEEE Trans. Information Theory **22** (1976) 75–81.

10. Michael G. Main, Detecting leftmost maximal periodicities, Discrete Applied Maths. **25** (1989) 145–153.   ,

11. Jacob Ziv & Abraham Lempel, A universal algorithm for sequential data compression, IEEE Trans. Information Theory **23** (1977) 337–343.

Edit Distance with Move Operations

Dana Shapira and James A. Storer

Computer Science Department, Brandeis University
Waltham, MA 02254
{shapird,storer}@cs.brandeis.edu

Abstract. The traditional edit-distance problem is to find the minimum
number of insert-character and delete-character (and sometimes change
character) operations required to transform one string into another. Here
we consider the more general problem of strings being represented by
a singly linked list (one character per node) and being able to apply
these operations to the pointer associated with a vertex as well as the
character associated with the vertex. That is, in O(1) time, not only can
characters be inserted or deleted, but also substrings can be moved or
deleted. We limit our attention to the ability to move substrings and leave
substring deletions for future research. Note that O(1) time substring
move operations imply O(1) substring exchange operations as well, a
form of transformation that has been of interest in molecular biology. We
show that this problem is NP-complete, show that a "recursive" sequence
of moves can be simulated with at most a constant factor increase by a
non-recursive sequence, and present a polynomial time greedy algorithm
for non-recursive moves with a worst-case log factor approximation to
optimal. The development of this greedy algorithm shows how to reduce
moves of substrings to moves of characters, and how to convert moves
with characters to only insert and deletes of characters.

1   Introduction

The traditional edit-distance problem is to find the minimum number of insert-
character and delete-character operations required to transform a string S of
length n to a string T of length m. Sometimes the costs of inserts and deletes
may differ, and change-character operations may have a different cost from a
delete plus an insert. Here we restrict our attention to just the insert-character
and delete-character operations where both have unit cost, although we believe
that much of what we present can be generalized to non-uniform costs.

It is well known how to solve the edit distance problem in $O(n \cdot m)$ using
dynamic-programming (see for example the book of Storer [ ] for a presentation
of the algorithm and references). If the whole matrix is kept for trace back to
find the optimal alignment, the space complexity is $O(n \cdot m)$, too. If only the
values of the edit distance is needed, only one row of the matrix is necessary,
and the space complexity is $O(m)$.

In addition to the insert and delete operations, we allow move operations
that transfer a sequence of characters from one location in S to another at a

A. Apostolico and M. Takeda (Eds.): CPM 2002, LNCS 2373, pp. 85– , 2002.
© Springer-Verlag Berlin Heidelberg 2002

constant cost. One way to model the move operation is by viewing strings as singly-linked lists (one character per vertex), and allow operations to apply to the pointer associated with a vertex as well as the character associated with the vertex. To define the problem properly, we can assume that special characters # and $ are first added to S to form the string $\#S\$$, and the problem is to transform $\#S\$$ to $\#T\$$ with the stipulation that # and $ cannot be involved in any operation, although #'s pointer might be ($'s pointer is always nil). That is, # defines the list head, the process must produce a list that goes from # to $, and the characters stored at the vertices traversed are the transformed string, which must be equal to T (all vertices unreachable from # after the transformation is complete are considered deleted). For simplicity, we assume that all operations (insert-character, insert-pointer, delete-character, and delete-pointer) have the same unit cost. In terms of what can be done in $O(1)$ time, what has been gained with the addition of the insert-pointer and delete pointer operations is the ability to:

1. Move a substring in $O(1)$ time.
2. Delete a substring in $O(1)$ time.

We limit our attention to the ability to move substrings and leave substring deletions for future research. Note that $O(1)$ time substring move operations imply $O(1)$ substring exchange operations as well, a form of transformation that has been of interest in molecular biology. Move operations can perform transformations in $O(1)$ time that could not be done in $O(1)$ time in the standard edit distance model. For example, let $S = a^n b^m c^n d^m e^n$ and $T = a^n d^m c^n b^m e^n$, where $m << n$ but m is not a constant. The usual edit distance between S and T is $O(m)$, as we would like to swap every b with every d, and visa verse. Using the new model the edit distance is reduced to $O(1)$, by changing $O(1)$ pointers.

Kececioglu and Sankoff [ ] and Bafna and Pevzner [ ] consider the reversal model, which takes a substring of unrestricted size and replaces it by its reverse in one operation. Move operations can be simulated with $O(1)$ reversal operations. For example, instead of moving a substring B to the right in S, let C be the substring of S between A and its destination position (so for some possibly empty strings A and D we wish to transform $S = ABCD$ to $ACBD$) and we can simply reverse BC and then reverse each of B and C separately. However, a reversal cannot in general be simulated by $O(1)$ moves (so the reversal model is more powerful than the move model).

Muthukrishnan and Sahinalp [ ] consider approximate nearest neighbors and sequence comparison with block operations, and without "recursion". With block operations, they include moves, copies, deletes, and reversals. The addition of the copy and reversal block operations changes the problem greatly. Reversals are more powerful than moves (i.e., reversals can simulate moves with at most a constant factor increase in cost but it is not necessarily true that moves can simulate reversals). Copies allow one to do in $O(1)$ cost something that is not possible in $O(1)$ time under normal assumptions about the manipulation of lists. They show NP-completeness and give a close to log factor approximation algorithm for related problems, but their construction is much more complex than

presented here and does not seem to directly apply to this more simple model
of just deletes and moves.

Bafna and Pevzner [ ] refer to moves as transpositions; this is motivated by
observing that if $S = uvwx$ is transformed by moving substring w to $T = uwvx$,
then the effect is to have exchanged the two substrings w and v. For the case
that S is a permutation of the integers 1 through n, they give a 1.5 approximation
algorithm for the minimum number of transpositions needed to transform S to a
different permutation T. Although similar in related to the problem considered
here, the restriction that all characters are distinct greatly changes the problem.

Lopresti and Tompkins [ ] consider a model in which two strings S and T
are compared by extracting collections of substrings and placing them into the
corresponding order. Tichy [ ] looks for a minimal covering set of T with respect to a source S such that every character that also appears in S is included
in exactly one substring move; unlike our model, one substring move can be used
to cover more than one substring in T. Thus, S is constructed by using the copy
operation of substrings of the minimal covering set. Hannenalli [ ] studies the
minimum number of rearrangements events required to transform one genome
into another; a particular kind of rearrangement called a translocation is considered, where a prefix or suffix of one chromosome is swapped with the prefix or
suffix of the other chromosome, and a polynomial algorithm is presented which
computes the shortest sequence of translocations transforming one genome into
another.

We now give a formal description of the three operations *insert*, *delete*,
and *move*. Let Σ denote a finite alphabet. For a character $\sigma \in \Sigma$, a string
$S = s_1 \cdots s_n$ and a position $1 \le p \le n$, the operation $insert(\sigma, p)$ inserts
the character σ to the pth position of S. After performing this operation, S
is of the form $s_1 \cdots s_{p-1} \sigma s_p \cdots s_n$. The operation $delete(p)$ deletes the character which occurs at the pth position of S, and returns the character which
was deleted, i.e., $s = s_1 \cdots s_{p-1} s_{p+1} \cdots s_n$ and it returns the character s_p.
Given two distinct positions $1 \le p_1 \ne p_2 \le n$ and a length $1 \le \ell \le n -$
$p_1 + 1$, $move(\ell, p_1, p_2)$ moves the string at position p_1 of length ℓ to position p_2. After performing the move operation, if $p_1 < p_2$ then S is in the form
of $s_1 \cdots s_{p_1-1} s_{p_1+\ell} \cdots s_{p_2-1} s_{p_1} \cdots s_{p_1+\ell-1} s_{p_2} \cdots s_n$, and if $p_2 < p_1$, S is in the
form of $s_1 \cdots s_{p_2-1} s_{p_1} \cdots s_{p_1+\ell-1} s_{p_2} \cdots s_{p_1-1} s_{p_1+\ell} \cdots s_n$. For simplicity, we may
write $move(str, p_1, p_2)$, where str is the string which is moved.

In the following section we show that computing the edit-distance between
two linked lists is NP-complete (substring deletions are not needed for this construction, move-string operations suffice to imply NP-completeness). In Section
3 we simplify the problem of finding a constant factor approximation algorithm
by showing that the elimination of recursive moves cannot change the edit distance by more than a constant factor. In Section 4 we present a greedy algorithm
that works by repeatedly replacing a given number of copies of a longest common substring of S and T by a new character, and then section 5 shows how a
reduction to the standard edit distance algorithm can be used. Sections 6 and 7

then show that this greedy algorithm gives a log factor worst-case approximation to optimal. Section 8 mentions some areas of future research.

2   NP-Completeness of Edit Distance with Moves

THEOREM: Given two strings S and T, an integer $m \in N$, using only the three unit-cost operations *insert*, *delete*, and *move*. It is NP-complete to determine if S can be converted to T with cost $\leq m$.

PROOF: Since a non-deterministic algorithm need only guess the operations and check in polynomial time that S is converted into T with cost $\leq m$, the problem is in NP.

We employ a transformation from the bin-packing problem, which is:

Given a bin capacity B, a finite set of integers $X = \{x_1, ..., x_n\}$ where $x_i \leq B$, and a positive integer k, the bin packing problem is to determine if there is a partition of X into disjoint sets $X_1, ..., X_k$, such that the sum of the items in each X_i is exactly B. The BIN-PACKING problem is NP Complete in the strong sense (e.g., see Garey and Johnson [ ]); that is, even if numbers in the statement of the problem instance are encoded in unary notion (a string of n 1's representing the number n), it is still a NP-complete problem.

Given an instance B, $X = \{x_1, ..., x_n\}$, and k of the bin-packing problem, let a, #, and \$ denote three distinct characters, and let:

$$S = \$^k \prod_{i=1}^{n} \# a^{x_i}$$

$$T = (a^B \$)^k \#^n$$

$$m = n$$

Since the bin-packing problem is NP-complete in the strong sense, we can assume that the lengths of S and T are polynomial in the statement of the bin-packing problem.

CLAIM: S can be converted to T with a cost $\leq m$ if and only if there is a partition of X into disjoint sets $X_1, ..., X_k$ such that the sum of the items in each X_i is B.

For the **if** portion of the proof, suppose that there is a partition of X into disjoint sets $X_1, ..., X_k$, such that the sum of the items in each X_i is B or less. Then S can be transformed to T by, for each item $x_i \in X_j$, $1 \leq j \leq k$, moving the corresponding a's to between the \$ of the corresponding bin. That is, perform $move(a^{x_i}, k + \sum_{\ell=1}^{i-1} (x_\ell + 1), (j-1)B)$, for a total of $\sum_{j=1}^{k} \sum_{x_i \in X_j} 1 = \sum_{i=1}^{n} 1 = n$ operations.

For the **only if** portion of the proof, the full draft of this paper uses a sequence of lemmas to show that if the edit distance with moves between S and T is $\leq n$, then there is a bin packing in k bins of size B of the items of X.   □

3   Recursive Moves

In this section we simplify the problem of finding an approximation algorithm by showing that the elimination of *recursive* moves cannot change the edit distance by more than a constant factor. For simplicity we refer only to move operations, which is justified in section 6.

A sequence $A = a_1, a_2, ..., a_r$ of legal move operations produces a division, $\widehat{A}$, of the string S into blocks of characters. More formally, a move operation in A of the form $move(i, j, \ell)$ defines a partition of $S = s_1 \cdots s_n$ into four blocks: if $i < j$ the blocks are: $[s_1...s_{i-1}], [s_i...s_{i+\ell-1}], [s_{i+\ell}...s_{j-1}]$ and $[s_j...s_n]$, otherwise, the blocks are: $[s_1...s_{j-1}], [s_j...s_i], [s_i...s_{i+\ell-1}]$ and $[s_{i+\ell}...s_n]$.

The next move operation of A refines this partition by adding at most 3 blocks (two blocks at the source location and one at its destination). Note that any sub sequence of A defines a partition of S, and if $A_1 \subseteq A_2$ are two sub sequences of A, then the partition of S, $\widehat{A_2}$, defined by A_2 is a refinement of the partition $\widehat{A_1}$, defined by S_1.

Definition: A sequence, of move operations is *recursive* if it contains a move operation which moves a string for which its characters did not occur continuously in S.

For example, if S is the string *abcde* and the character b was moved to obtain the string $T = acdbe$, then moving the substring *dbe* or *ac* are both considered as recursive moves.

The following example shows us that performing recursive moves can reduce the cost of the edit-distance. Let $S = xababycdcdz$ and $T = xcddcyabbaz$. If we do not allow recursive moves the minimum cost of converting S into T is 6 ($Move(ab, 2, 7)$, $Move(a, 3, 8)$, $Move(b, 4, 8)$, $Move(cd, 7, 2)$, $Move(c, 8, 3)$ and $Move(d, 9, 3)$). By allowing recursive moves we can reduce the cost to be 4 operations ($Move(b, 5, 4)$, $Move(d, 10, 9)$, $Move(abba, 2, 7)$ and $Move(cddc, 7, 2)$).

THEOREM: *Suppose there is a recursive sequence, A, of size n which converts S into T. Then a non recursive optimal algorithm uses no more than $3n$ moves.*

PROOF: By induction on n.

A worst case example: In the full draft of this paper we give an infinite class of strings for which non-recursive is a factor of 3 more costly than recursive.

4   The Greedy Algorithm

In this section we present a polynomial time approximation algorithm for the minimum move edit-distance. It is a greedy method that reduces the two strings S and T to two other strings, so that we can perform the traditional edit distance algorithm on the new strings. Define $LCS(S, T)$ as the longest common substring of the two strings S and T. For example: $LCS(abcd, edbc) = 2$, since bc is the longest common substring of S and T, and consists of 2 characters, but $LCS(abc, def) = 0$, since there is not any common character of these two strings.

The algorithm uses two procedures. The $ed(S, T)$ procedure computes the traditional edit distance of S and T by using the dynamic programming method. The $check\_move(S, T)$ procedure checks whether we can reduce the edit-distance by using move operations instead of inserting and deleting the same character. The algorithm is given in Figure 1.

> Stage 1: **while** $(|LCS(S, T)| > 1)$ {
> $P \leftarrow LCS(S, T)$
>   Let A be a new character, i.e., $A \notin \Sigma$.
>   Replace the same number of occurrences of P in S and in T by A.
> $\Sigma \leftarrow \Sigma \cup \{A\}$
>   }
> Stage 2: $d \leftarrow ed(S, T)$
> Stage 3: $d \leftarrow check\_move(S, T)$
> **return** d

Fig. 1. The Greedy algorithm

In this section we explain the algorithm and in the following section we discuss the *check_move* procedure it uses. Consider the following example: let $\Sigma = \{a, b, c, d, e\}$, $S = cdeab$ and $T = abcde$. After the first stage of the greedy algorithm, $S' = AB$ and $T' = BA$, where $A = cde$ and $B = ab$. In the second stage, by performing the traditional edit distance on S' and T', we find that $d \leq 2$. The third stage does better by using check_move to determine that S can be converted to T by performing $A \leftarrow delete(1)$ and $insert(A, 2)$ (which deletes and inserts the same character A), and therefore $d = 1$ since we can simply move the string cde to the end of S.

The greedy algorithm reduces the strings S and T to (possibly) shorter strings by replacing the $LCS(S, T)$ by a new single character. In a first attempt it seems as if we must replace every occurrence of the $LCS(S, T)$ by a new character, without bothering about the same number of replacements in S and T. Otherwise, if the number of occurrences of the $LCS(S, T)$ in S and T is not equal, the copies which were not replaced by the same new character, are not noticed as resemble ones, which might increase the edit-distance. In the following Lemma we show that this is not the case. We denote the edit distance returned by this version of the greedy algorithm by $greedy'$ (i.e., the version without the restriction of the equal number of replacements in both S and T). Let opt denote the edit-distance returned by an optimal algorithm which allows move operations.

LEMMA 1: *The unrestricted version of the greedy algorithm is not bounded.*

PROOF: For every n there exists an example such that $\frac{greedy'}{opt} > n$. Let $\Sigma = \{a, b\}$. Let $S = (ab)^{4n}$ and let $T = (ab)^{2n}(ba)^{2n}$.

The optimal edit distance is 2 ($insert(b, 4n+1)$, $b \leftarrow delete(8n)$). By this version of the greedy algorithm S is reduced to the string AA and T is reduced to the string $A(ba)^{2n}$, where $A = (ab)^{2n}$. The edit distance is $2n + 1$ which includes the operations: $A \leftarrow delete(2)$, $insert(b, i-1)$ and $insert(a, i)$ for i, $4n + 2 \leq i \leq 8n$. Now, $\frac{greedy'}{opt} = \frac{2n+1}{2} > n$. $\qquad\square$

The problem is that A occurs twice in T and only once in S. Therefore the algorithm could not identify any resemblance between A and $(ba)^{2n}$. We overcome this problem by replacing the same number of occurrences of the $LCS(S, T)$ in S and in T as done by the greedy algorithm with the replace operation of the while loop.

Note that the greedy algorithm is based on the traditional edit distance algorithm. Therefore, it does not perform any recursive move, as every block participates in no more than one operation. However we have shown that by not allowing recursive moves we do not increase the number of move operations by more than a constant factor.

We now examine the running time of the greedy algorithm. The first stage can be done using a suffix trie in $O(min(n, m) \cdot max(n, m))$ processing time, where n and m are the length of the strings S and T, respectively. We construct the suffix trie, in $O(n + m)$ processing time, for the string $S\$T\#$, where $\$$ and $\#$ are two new symbols. We then traverse this suffix trie in post-order and label each vertex as to whether it has descendants in only S, in only T, or in both S and T (once you know this information for the children it is easily computed for the parent), to find the non-leaf vertex of lowest virtual depth that has both. This is repeated at most $\frac{min(n,m)}{2}$ times, which happens when all the common substring consist of 2 characters, and all characters participate in these common substrings. The second stage can be done in $O(n \cdot m)$ processing time, using the dynamic programming method. In the following section we prove that the third stage can be done in $O(n + m)$ processing time. Therefore, the entire processing time of the greedy algorithm is $O(n \cdot m)$.

5   Identifying Move Operations

In the third stage of the greedy algorithm we are interested in identifying move operations which were done in the second stage. A move operation of a character σ is simply an insert and a delete operation of the same character σ. At a first sight we might think that we cannot separate these two stages. It seems as if in every stage of the dynamic algorithm, after computing the cheapest operation of the current character, we must check if we could reduce the cost by combining it with an opposite operation of the same character and changing it to a move operation. We show that it is enough to identify move operations after computing the edit-distance, and we do not have to take it into account in the inner stages.

We use the following notations: Let P denote a way to convert a string S into a string T by using inserts and deletes. Let us denote by I_σ^P / D_σ^P the number of insertions/deletions of the character $\sigma \in \Sigma$ which where done when converting S into T using the path P. The edit-distance between the string S and T which is done according to path P would be denoted by ed^P. The edit-distance between S and T, including move operations is denoted by edm^P, i.e., if P includes both $insert(\sigma)$ and $delete(\sigma)$ for any $\sigma \in \Sigma$, this would be calculated as one operation.

If we are interested in the minimum edit-distance we use ed for the traditional edit distance and edm for the edit-distance with move operations.

LEMMA 2: *For any two paths P and P' and $\sigma \in \Sigma$, $|I_\sigma^P - D_\sigma^P| = |I_\sigma^{P'} - D_\sigma^{P'}|$.*

PROOF: Denote by n_σ^S and n_σ^T the number of appearances of a character σ in the strings S and T, respectively. For any path P which converts S into T $|I_\sigma^P - D_\sigma^P| = |n_\sigma^S - n_\sigma^T|$.   □

CONCLUSION 1: *Any two paths P and P' differ only by move operations.*

Note that $\forall a, b \in \mathcal{N}$, $a + b = 2min(a, b) + |a - b|$. Assuming that the cost of insert, delete and move operations are equal, we obtain:

$$ed^P = \sum_{\sigma \in \Sigma} \left(I_\sigma^P + D_\sigma^P \right) = \sum_{\sigma \in \Sigma} \left(2 \cdot min(I_\sigma^P, D_\sigma^P) + |I_\sigma^P - D_\sigma^P| \right) \qquad (1)$$

$$edm^P = \sum_{\sigma \in \Sigma} \left(I_\sigma^P + D_\sigma^P - min(I_\sigma^P, D_\sigma^P) \right) \qquad (2)$$

$$= \sum_{\sigma \in \Sigma} \left(min(I_\sigma^P, D_\sigma^P) + |I_\sigma^P - D_\sigma^P| \right) \qquad (3)$$

LEMMA 3: *The minimal edit distance with move operations occurs in any optimal path of the traditional edit-distance.*

PROOF: Suppose P and P' are two paths converting S into T, and that $ed^P < ed^{P'}$. By using Lemma 2 and equation (1) we find that

$$\sum_{\sigma \in \Sigma} \left(2 \cdot min(I_\sigma^P, D_\sigma^P) \right) < \sum_{\sigma \in \Sigma} \left(2 \cdot min(I_\sigma^{P'}, D_\sigma^{P'}) \right)$$

So by using Lemma 2 again and equation (3) we find that $edm^P < edm^{P'}$.   □

Lemma 2 and Lemma 3 show that after computing the edit-distance, we can take **any** optimal path which transfers S into T, and reduce the cost by exchanging inserts and deletes of the same character with one move operation. This can be done in $O(n + |\Sigma|)$ time, with the help of a $|\Sigma|$ size array.

6   Reduction to Only Move Operations

Using Lemma 2 we already know that for any two paths P and P' and any character $\sigma \in \Sigma$, $|I_\sigma^P - D_\sigma^P| = |I_\sigma^{P'} - D_\sigma^{P'}|$. Recall that n_σ^S and n_σ^T denote the number of appearances of σ in the strings S and T, respectively.

Given two strings S and T, we preprocess these strings and construct two new strings S' and T' as follows:

1. Let # and \$ be two new characters, i.e. #, \$ $\notin \Sigma$.
2. Define $\Sigma^1 = \{\sigma \in \Sigma : n_\sigma^S < n_\sigma^T\}$ and $\Sigma^2 = \{\sigma \in \Sigma : n_\sigma^T < n_\sigma^S\}$.

3.

$$S' \leftarrow S \cdot \prod_{\sigma \in \Sigma^1} \left((\#\sigma)^{n_\sigma^T - n_\sigma^S} \right) \$^{\sum_{\sigma \in \Sigma^2} (n_\sigma^S - n_\sigma^T)}$$

$$T' \leftarrow T \cdot \#^{\sum_{\sigma \in \Sigma^1} (n_\sigma^T - n_\sigma^S)} \prod_{\sigma \in \Sigma^2} \left((\$\sigma)^{n_\sigma^S - n_\sigma^T} \right)$$

For example: If $S = abcab$ and $T = abcdc$ then after preprocessing these strings we get that $S' = abcab\#d\#c\$\$$ and $T' = abcdc\#\#\$a\$b$. This way $\forall \sigma \in \Sigma \cup \{\#, \$\}$ $n_\sigma^{S'} = n_\sigma^{T'}$, and we can deal only with move operations.

LEMMA 4: *If the cost assigned to each insert, delete and move operation are all equal then $edm(S, T) = edm(S', T')$.*

PROOF: Suppose A is a sequence of move, insert and delete operations which are needed in order to convert S into T by a minimum cost. The insert and delete operations are done only when there is a missing or an additional character, respectively. The following operations convert S' into T'. Every insert operation is changed into a move operation of the appropriate character within the new characters into the same position in S. Every delete operation is a move operation of the character which is deleted to an appropriate position following a $\$$ sign. Every move operation remains unchanged. Therefore, $edm(S', T') \leq edm(S, T)$. As there are no two consequent $\#$ signs in S', but they must occur continuously in T', and as the $\$$ signs occur continuously in S' and must be separated by one character in T', there are at least $\sum_{\sigma \in \Sigma} |n_\sigma^T - n_\sigma^S|$ operations needed in order to locate the $\sum_{\sigma \in \Sigma} |n_\sigma^T - n_\sigma^S|$ $\#$ and $\$$ symbols at their final position. These operations do not influence the original characters of S and T. The number $\sum_{\sigma \in \Sigma} |n_\sigma^T - n_\sigma^S|$ is the number of insert and delete operations done when converting S into T. If $edm(S', T') < edm(S, T)$ it means that some of the move operations done to convert S into T were not done when converting S' into T'. Therefore, there is a cheaper way to convert S into T, which contradicts the minimalism of $edm(S, T)$. $\qquad\square$

7   Bounds between Optimal and Greedy

Finding a bound on the number of greedy phrases, as a function of the optimal phrases, gives us a bound on the number of move operations (see Lemma 6). Obviously, we are not concerned with the greedy blocks which contain optimal ones. Therefore, let us look at an optimal block in S which contains N greedy blocks. The following Lemma gives us a bound on the number of greedy blocks of the corresponding optimal block in T, which contain "most" of these phrases.

LEMMA 5: *Let N be the number of optimal phrases, and let L be the number of characters in the longest optimal phrase. If B is an optimal block which contains m greedy blocks in S, then at least $O(m - \log L)$ of these blocks are part of at most $O(\log L)$ greedy blocks, in the correspondence occurrence of B in T.*

PROOF: Denote by B' the sub block of B containing exactly these m greedy blocks. The greedy algorithm creates a non increasing sequence of the lengths

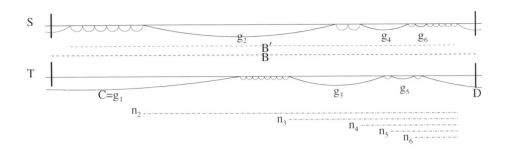

Fig. 2. Schematic view of Lemma 5's proof. The curve lines represent the greedy phrases of the optimal block given by the vertical lines

of the LCS's. As we know that B' is a string which occurs in both S and T, we examine the point in time when the greedy algorithm had chosen other phrases. Suppose C is the first greedy phrase chosen, which overlaps B'. As B' occurs in both S and T, and was not chosen by the greedy algorithm, it must be that C was chosen in T, and that $|C| \geq |B'|$. Obviously, C crosses an optimal block boundary. Without loss of generality, let us assume that it crosses the left boundary of B' in T. If C contains at least $O(m - \log L)$ greedy phases, we are done. If there exists a block in T which crosses B''s right boundary denote it by D. If C and D contain together at least $O(m - \log L)$ greedy phases, we are done. Let us define the sequence $G = \{g_1, g_2, ..., g_k\}$ to be a particular sequence of greedy blocks ordered by time, which occur in S or in T, with the following properties: The first greedy block in this sequence is C, i.e., $g_1 = C$, and for $i > 1$, g_i is a substring of B', which includes at least one more character of B' which does not already occur in the previous greedy blocks $\{g_1, ..., g_{i-1}\}$. Thus the g_i's form an increasing cover of B'. Note that the set G is finite since both S and T are finite strings, and that $|g_i| \geq |g_{i+1}|$.

OBSERVATION 1: If g_2 was chosen in T, then $C(= g_1)$, D and g_2, cover the m greedy blocks of S, in particular they include $O(m - \log L)$ greedy phases. (Since the string between C and D occurs in both S and T and is still free to be chosen).

OBSERVATION 2: If g_{i-1} and g_i $(i > 1)$, were both chosen in S/T, then these two greedy blocks end the G sequence, i.e. $i = k$. Since the adjacent block to g_i in S/T is the longest common substring in B' of S and T, and these two occurrences are still free to be chosen. In this case, about half of these k greedy blocks of T cover the $O(N - \frac{k}{2})$ greedy blocks of S.

We still have to prove that $k = O(\log n)$. That is, that the longest sequence of alternating g_i's is of order $\log n$. For $i \geq 2$, define n_i to be the number of characters in the suffix string of B' starting at the first position of g_i. Before choosing the block g_{i+1}, S and T share a common substring in B' of length $n_i - |g_i|$. We have chosen g_{i+1} since $|g_{i+1}| \geq n_i - |g_i|$. Using the fact that $|g_i| \geq |g_{i+1}|$, we find that $|g_i| \geq n_i - |g_i|$, thus

$$|g_i| \geq \frac{n_i}{2} \qquad (4)$$

Since the greedy blocks in S/T do not overlap, the difference between n_i and n_{i+2} ($i = 2, ..., k-2$) is at least $|g_i|$, i.e., $n_i - n_{i+2} \geq |g_i|$. Therefore, by using equation (4), $2|g_i| \geq n_i \geq |g_i| + n_{i+2}$, so we find that

$$|g_i| \geq n_{i+2} \qquad (5)$$

The longest sequence of $\{g_i\}$'s occurs when they are chosen alternately in S and in T and their lengths are as small as possible, i.e., when $|g_i| = \frac{n_i}{2}$. If $|g_i| = \frac{n_i}{2}$, then by using equation (5), after j stages: $n_i \geq 2n_{i+2} \geq 4n_{i+4} \geq \cdots \geq 2^j n_{i+2j}$. This alternate series terminates when the last block consists of exactly one character, i.e., when $n_{i+2j} = 1$. In particular, $L \geq |B| > n_2$, so we find that $L > 2^j \cdot 1$. Therefore, $\log L > j$. This means that there are at most $\log L$ greedy phrases in the G sequence, and that the $\{g_i\}$'s in T correspond to the m greedy blocks of B excluding those who belong to the $\{g_i\}$'s in S. $\qquad \square$

A greedy block is called a *primary* one, if it is a member of the sequence G. Intuitively, a primary block is one of the "big" greedy phrases either in S or in T. We have proved that the primary blocks alternate, and that the corresponding occurrence of the string of a primary block of S does not contain a primary block of T, and vise verse.

The following theorem gives us a bound on the number of greedy phrases.

THEOREM: *The number of phrases in a greedy parsing is at most a $\log(L)$ times the number of phrases in an optimal parsing.*

PROOF: Given an optimal and greedy parsing of the strings S and T, recall that the number of optimal phrases is denoted by N. We show that the greedy parsing does not consist of more then $O(N \log L)$ phrases. The reduction to only move operations which was presented in the previous section ensures that each greedy or optimal phrase in S occurs also in T, and vise verse.

Let us relate to the N optimal phrases of S. There are at most $N \log(L)$ primary phrases, of which about half of them are in S. We now count the number of the remaining phrases in S, by associating each one of them to a different primary phrase of S or to a different optimal boundary of S. Since there are not more than $N \log(L)$ primary phrases and not more than $N - 1$ optimal boundaries, this concludes our proof.

Consider a primary block, C, in T which covers ℓ greedy phrases of S, where $\ell >= 1$. The block C occurs in S, too. We refer to the correspondence string of the block C as $\mathcal{C}$. If C crosses an optimal boundary, this boundary can cross at most one of the ℓ original greedy phrases. We associate this particular phrase with this optimal boundary. Otherwise, suppose C is fully contained in an optimal block α in S, and therefore appears in the corresponding occurrence of α in T. If C in S is a non primary block of α, then the correspondence occurrence of the string $\mathcal{C}$ in T is contained in a primary block D of T, and we continue to the

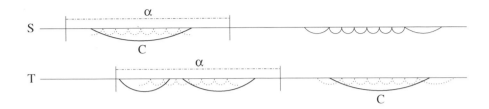

Fig. 3. Schematic view of the proof. The bold curve lines represent primary phrases of the greedy algorithm corresponding to different occurrences of the string C

following appearance of D in S. Finally we end this process when the string C occurs in a primary block of S.

Let us refer to the correspondent occurrence of this primary block in T. The greedy phrases of C were not merged by the greedy algorithm, since at least $\ell - 2$ of these original greedy phrases are covered by at most 2 greedy phrases. We associate these two uncovered phrases to this primary block of S. This process ends when there are not two adjacent original greedy phrases which remain attached in T. Otherwise these adjacent blocks can be merged into one larger block. Since every primary block is charged at most once, and each greedy block in T was separated (and therefore was counted), we have included all ℓ phrases in this count.   □

The following Lemma gives us a bound on the number of move operations when the bound on the number of phrases is known.

LEMMA 6: *Every parsing of N blocks, gives a lower bound of $\frac{N+1}{3}$ move operations, and an optimal upper bound of $N - 1$.*

PROOF: If these blocks occur in their reversed order, then every block except one must be moved, which gives us $N - 1$ operations. Every move operation creates at most 3 new boundaries (two in its source location, and one in its destination). Thus, N blocks, which have $N + 1$ boundaries, reflect at least $\frac{N+1}{3}$ move operations.   □

COROLLARY: *The number of move operations in a greedy parsing is at most a $\log L$ factor times the number of move operations in an optimal parsing.*

PROOF: Lemma 6 gives us at most a factor of three on the number of moves. □

To illustrate the different parsing of the greedy and optimal algorithms, suppose $S = abaxxxababaxxxxab$ and $T = baxxxxababaxxxaba$. Using the greedy algorithm we get that $S' = BAxab$ and $T' = baxAB$, where $A = xxxababaxxx$ and $B = aba$. The edit distance of S' and T' is 4 (since these blocks occur at their reverse order). The optimal parsing of S and T includes only two blocks in which require only one move operation,$move(baxxxab, 1)$.

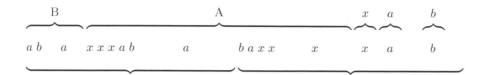

Fig. 4. The Different parsing of the Greedy and Optimal algorithms

8 Future Research

Here we have shown the edit distance problem with substring move operations to be NP-complete and have presented a greedy algorithm that is a worst-case log factor approximation to optimal. We have limited out attention to when all operations have unit cost, and hence an obvious area of future research would be to extend these ideas to non-uniform costs. Another area of interest is the incorporation of substring deletions, which are needed to capture the full power of the linked list model (to within a constant factor). Experiments with the greedy algorithm on real-life data (e.g., from molecular biology) are also of interest, although given the NP-completeness of an optimal computation, a framework for measuring performance of greedy for a particular application needs to be addressed.

Acknowledgments

We thank Maxime Crochemore for suggesting the edit distance with the exchange operation, which helped to motivate this work.

References

1. BAFNA V. AND PEVZNER P.A., Genome rearrangements and sorting by reversals, *34th IEEE Symposium on Foundations of Computer Science*, (1993) 148–157
2. BAFNA V. AND PEVZNER P.A., Sorting by transpositions, *34th SIAM J. Discrete Math., 11(2)*, (1998) 124–240
3. GAREY M.R. AND JOHNSON D.S., Computers and Intractability, A guide to the Theory of NP-Completeness, *Bell Laboratories Murry Hill, NJ*, (1979)
4. HAMMING R.W., Coding and information Theory, *Englewood Cliffs, NJ, Prentice Hall*, (1980)
5. HANNENHALLI S., Polynomial-time Algorithm for Computing Translocation Distance between Genomes *CPM*, (1996) 162–176
6. KECECIOGLU J. AND SANKOFF D., Exact and approximation algorithms for the inversion distance between two permutations. *Pro. of 4th Ann. Symp. on Combinatorial Pattern Matching, Lecture Notes in Computer Science 684*, (1993) 87–105

7. LIBEN-NOWELL D., On the Structure of Syntenic Distance, *CPM*, (1999) 50–65
8. LOPRESTI D. AND TOMKINS A., Block Edit Models for Approximate String Matching, *Theoretical Computer Science, 181*, (1997) 159–179

9. MUTHUKRISHNAN S. AND SAHINALP S.C., Approximate nearest neighbors and sequence comparison with block operations, *STOC'00, ACM Symposium on Theory of Computing*, (2000) 416–424

10. SMITH T.F. AND WATERMAN M.S., Identification of common molecular sequences, *Journal of Molecular Biology, 147*, (1981) 195–197

11. STORER J. A., *An Introduction to Data Structures and Algorithms*, Birkhauser - Springer, (2001)

12. TICHY W.F., The string to string correction problem with block moves, *ACM Transactions on Computer Systems, 2(4)*, (1984) 309–321

Towards Optimally Solving the Longest Common Subsequence Problem for Sequences with Nested Arc Annotations in Linear Time

Jochen Alber*, Jens Gramm**, Jiong Guo***, and Rolf Niedermeier

Wilhelm-Schickard-Institut für Informatik, Universität Tübingen
Sand 13, D-72076 Tübingen, Fed. Rep. of Germany
{alber,gramm,guo,niedermr}@informatik.uni-tuebingen.de

Abstract. We present exact algorithms for the NP-complete Longest Common Subsequence problem for sequences with nested arc annotations, a problem occurring in structure comparison of RNA. Given two sequences of length at most n and nested arc structure, our algorithm determines (if existent) in time $O(3.31^{k_1+k_2} \cdot n)$ an arc-preserving subsequence of both sequences, which can be obtained by deleting (together with corresponding arcs) k_1 letters from the first and k_2 letters from the second sequence. Thus, the problem is fixed-parameter tractable when parameterized by the number of deletions. This complements known approximation results which give a quadratic time factor-2-approximation for the general and polynomial time approximation schemes for restricted versions of the problem. In addition, we obtain further fixed-parameter tractability results for these restricted versions.

1 Introduction

Given two or more sequences over some fixed alphabet Σ, the Longest Common Subsequence problem (LCS) asks for a maximum length sequence that occurs as a subsequence in all of the given input sequences. This is considered to be a core problem of computer science with many applications, see, e.g., [, , , ,]. With the advent of computational biology, structure comparison of RNA and of protein sequences has become a central computational problem, bearing many challenging computer science questions. In this context, the Longest Arc Preserving Common Subsequence problem (LAPCS) recently has received considerable attention [, , , ,]. It is a sound and meaningful mathematical formalization of comparing the secondary structures of

---

* Supported by the Deutsche Forschungsgemeinschaft (DFG), project PEAL (parameterized complexity and exact algorithms), NI 369/1-1,1-2.
** Supported by the Deutsche Forschungsgemeinschaft (DFG), project OPAL (optimal solutions for hard problems in computational biology), NI 369/2-1.
*** Partially supported by the Deutsche Forschungsgemeinschaft (DFG), Zentrum für Bioinformatik Tübingen (ZBIT).

A. Apostolico and M. Takeda (Eds.): CPM 2002, LNCS 2373, pp. 99– , 2002.
© Springer-Verlag Berlin Heidelberg 2002

molecular sequences: For a sequence S, an *arc annotation* A of S is a set of un-ordered pairs of positions in S. Focusing on the case of two given arc-annotated input sequences S_1 and S_2, LAPCS asks for a maximum length arc-annotated sequence T that forms an arc-annotated subsequence of S_1 as well as S_2. More precisely, this means that if one deletes k_1 letters (also called *bases*) from S_1—when deleting a letter at position i, then *all* arcs with endpoint i are also deleted—and one deletes k_2 letters from S_2, then T and the resulting sequences are the same and also the arc annotations coincide. For related studies concerning algorithmic aspects of (protein) structure comparison using "contact maps," refer to [,]. Whereas LCS for two sequences without arc annotations is solvable in quadratic time (it becomes NP-complete when allowing for an arbitrary number of input sequences), LAPCS for two sequences is NP-complete [,]. According to Lin *et al.* [ ], however, LAPCS for *nested arc annotations* is "generally thought of as the most important variant of the LAPCS problem." Here, one requires that no two arcs share an endpoint and no two arcs cross each other. This problem is referred to by LAPCS(NESTED,NESTED). Answering an open question of Evans [ ], Lin *et al.* [ ] showed that LAPCS(NESTED,NESTED) is NP-complete. In addition, they gave polynomial time approximation schemes (PTAS) for (also NP-complete) special cases of LAPCS(NESTED,NESTED). Here, matches between two given input sequences are allowed only in a "local area" (of constant size) w.r.t. matching position numbers (refer to Section   for details). As to the general LAPCS(NESTED,NESTED) problem, Jiang *et al.* [ ] gave a quadratic time factor-2-approximation algorithm.

By way of contrast, our main result is an exact, fixed-parameter algorithm that solves the general LAPCS(NESTED,NESTED) problem in time $O(3.31^{k_1+k_2} \cdot n)$ where n is the maximum input sequence length. This gives an efficient algorithm in case of reasonably small values for k_1 and k_2 (the numbers of deletions allowed in S_1 and S_2, respectively), providing an *optimal* solution. Observe that the PTAS algorithms obtaining a good degree of approximation get prohibitive from a practical point of view due to enormous running times. And, the factor-2-approximation might be not good enough for biological and other applications. In summary, in terms of parameterized complexity [, ,], our result shows that LAPCS(NESTED,NESTED) is fixed-parameter tractable when parameterized by the number of deletions allowed, perhaps the most natural problem parameterization. Moreover, we obtain fixed-parameter tractability results for restricted versions of LAPCS(NESTED,NESTED) as studied by Lin *et al.* [ ]. In the restricted cases, we *additionally* prove fixed-parameter tractability when taking the length of the common subsequence as the problem parameter. Our algorithms employ the bounded search tree paradigm of parameterized complexity in a highly nontrivial way.

---

[1] Actually, we show fixed-parameter tractability for more general problems, allowing more complicated (e.g., crossing) arc annotations.

2   Preliminaries and Previous Work

Besides its central importance in computer science and applications, the classical, NP-complete LCS is particularly important from the viewpoint of parameterized complexity [, ,]. It is used as a key problem for determining the parameterized complexity of many problems from computational biology. In this setting, the input consists of several sequences over an alphabet Σ and a positive integer l, the question is whether there is a sequence from Σ^* of length at least l that is a subsequence of *all* input sequences. For example, with respect to parameter l the problem is known to be *fixed-parameter tractable*, that is, it can be solved in time $f(l) \cdot N^{O(1)}$, where N denotes the complete input size and f can be an arbitrarily fast growing function only depending on l. The decisive point here is that the combinatorial explosion seemingly unavoidable in optimally solving NP-hard problems can be completely restricted to the *parameter l*. Note, however, that for LCS this requires that $|\Sigma|$ is of constant size. If $|\Sigma|$ is unbounded, then there is concrete indication that LCS is *not* fixed-parameter tractable with respect to parameter l (by way of so-called W[2]-hardness which is well-known in parameterized complexity). We refer to [, ,] for any details.

Evans [,] initiated classical and parameterized complexity studies for the more general case that the input sequences additionally carry an *arc structure* each, which is motivated by structure comparison problems in computational molecular biology. For a sequence S of length $|S| = n$, an *arc annotation* (or *arc set*) A of S is a set of unordered pairs of numbers from $\{1, 2, \ldots, n\}$. Each pair (i, j) connects the two *bases* $S[i]$ and $S[j]$ at positions i and j in S by an arc. Since LAPCS is NP-complete even for two input sequences, here and in the literature attention is focused on this case. Let S_1 and S_2 be two sequences with arc sets A_1 and A_2, respectively, and let i_1, i_2, j_1, j_2 be positive integers. If $S_1[i_1] = S_2[j_1]$, we refer to this as *base match* and if $S_1[i_1] = S_2[j_1]$, $S_1[i_2] = S_2[j_2]$, $(i_1, i_2) \in A_1$, and $(j_1, j_2) \in A_2$, we refer to this as *arc match*. A common subsequence T of S_1 and S_2 induces a one-to-one mapping M from a subset of $\{1, 2, \ldots, n_1\}$ to a subset of $\{1, 2, \ldots, n_2\}$, where $n_1 = |S_1|$ and $n_2 = |S_2|$. Suppose that $M = \{\langle i_r, j_r \rangle \mid 1 \leq r \leq |T|\}$. Then T *induces* the base matches $\langle i_r, j_r \rangle$, $1 \leq r \leq |T|$. We say that T is an *arc-preserving subsequence (aps)* if the arcs induced by M are preserved, i.e., for all $\langle i_{r_1}, j_{r_1} \rangle, \langle i_{r_2}, j_{r_2} \rangle \in M$:

$$(i_{r_1}, i_{r_2}) \in A_1 \iff (j_{r_1}, j_{r_2}) \in A_2.$$

Depending on the arc structures allowed, various subproblems of LAPCS can be defined. We focus here on the NP-complete LAPCS(NESTED,NESTED) [, ,], where for both input sequences it is required that no two arcs share an endpoint and that no two arcs cross each other, i.e., for all (i_1, i_2), (i_3, i_4) with $i_1 < i_2$ and $i_3 < i_4$ part of the arc annotation of a sequence, it holds that $i_3 < i_1 < i_4$ iff $i_3 < i_2 < i_4$. We will also consider the less restrictive *crossing* arc structure, where the only requirement is that no two arcs share an endpoint. Lin *et al.* [ ] furthermore introduced the following special cases of LAPCS(NESTED,NESTED),

---

[2] Actually, in case of LCS running time $O(|\Sigma|^l \cdot N)$ is easily achieved.

motivated from biological applications [,]. For any positive integer c, in c-DIAGONAL LAPCS(NESTED,NESTED) base $S_1[i]$ ($S_2[j]$, respectively) is allowed to match only bases in the range $S_2[i - c, i + c]$ ($S_1[j - c, j + c]$, respectively). Herein, for $i_1 \leq i_2$, $S[i_1, i_2]$ denotes the substring of sequence S starting at position i_1 and ending at position i_2 and $S[i_1, +]$ denotes the suffix of S starting at position i_1. Similarly, in c-FRAGMENT LAPCS(NESTED,NESTED) all base matches induced by a *longest arc-preserving common subsequence (lapcs)* are required to be between "corresponding" fragments of size c of S_1 and S_2. Lin *et al.* [ ] showed that c-DIAGONAL LAPCS(NESTED,NESTED) is NP-complete for $c \geq 1$ and c-FRAGMENT LAPCS(NESTED,NESTED) is NP-complete for $c \geq 2$.

3   Search Tree Algorithm for LAPCS(NESTED,NESTED)

In this section, we describe and analyze Algorithm LAPCS which solves the LAPCS(NESTED, NESTED) problem in time $O(3.31^{k_1+k_2} \cdot n)$, where n is the maximum length of the input sequences. It is a search tree algorithm and, for sake of clarity, we choose the presentation in a recursive style: Based on the current instance, we make a case distinction, branch into one or more subcases of somehow simplified instances and invoke the algorithm recursively on each of these subcases. Note, however, that we require to traverse the resulting search tree in breadth-first manner, which will be important in the running time analysis. Before presenting the algorithm, we define the employed notation.

Recall that the considered sequences are seen as arc-annotated sequences; a comparison $S_1 = S_2$ includes the comparison of arc structures. Additionally, we use a modified comparison $S_1 \approx_{i,j} S_2$ that is satisfied when $S_1 = S_2$ after deleting at most i bases in S_1 and at most j bases in S_2. Note that we can check whether $S_1 \approx_{1,0} S_2$ or whether $S_1 \approx_{0,1} S_2$ in linear time. The subsequence obtained from an arc-annotated sequence S by deleting $S[i]$ is denoted by $S - S[i]$. When branching into the case of a simplified sequence $S - S[i]$ (or $S[2, n]$, resp.), the input for the recursive call is $S_{new} := S - S[i]$ (or $S_{new} := S[2, n]$, resp.)—hence, $|S_{new}| = |S| - 1$—and, therefore, $S_{new}[i] = S[i+1]$. For handling branches in which no solution is found, we use a modified addition operator "$\dotplus$" defined as follows: $a \dotplus b := a + b$ if $a \geq 0$ and $b \geq 0$, and $a \dotplus b := -1$, otherwise. We abbreviate $n_1 := |S_1|$ and $n_2 := |S_2|$.

The most involved case in the algorithm is Case (2.5), which will also determine our upper bound on the search tree size. The focus of our analysis will, in particular, be on Subcase (2.5.3). For sake of clarity, we, firstly, give an overview of the algorithm which omits the details of Case (2.5), and, then, present Case (2.5) in detail separately. Although the algorithm as given reports only the length of an lapcs, it can easily be extended to compute the lapcs itself within the same running time.

Algorithm LAPCS(S_1, S_2, k_1, k_2)
Input: Arc-annotated sequences S_1 and S_2, positive integers k_1 and k_2.
Return value: Integer denoting the length of an lapcs of S_1 and S_2 which can

be obtained by deleting at most k_1 symbols in S_1 and at most k_2 symbols in S_2. Return value -1 if no such subsequence exists.

(Case 0)   /* Recursion ends. */
  If $k_1 < 0$ or $k_2 < 0$ then return -1.   /* No solution found. */
  If $|S_1| = 0$ and $|S_2| = 0$, then return 0.   /* Success! Solution found. */
  If $|S_1| = 0$ and $|S_2| > 0$, then   /* One sequence done... */
      if $k_2 \geq |S_2|$, then return 0, else return -1. /* ...but not the other. */
  If $|S_1| > 0$ and $|S_2| = 0$, then   /* ditto */
      if $k_1 \geq |S_1|$, then return 0, else return -1.

(Case 1)   /* Non-matching bases. */
  If $S_1[1] \neq S_2[1]$, then return the maximum of the following values:
- $\text{LAPCS}(S_1[2, n_1], S_2, k_1 - 1, k_2)$ /* delete $S_1[1]$ */
- $\text{LAPCS}(S_1, S_2[2, n_2], k_1, k_2 - 1)$ /* delete $S_2[1]$ */.

(Case 2)   /* Matching bases */
  If $S_1[1] = S_2[1]$, then
  (2.1)   /* No arcs involved. */
  If both $S_1[1]$ and $S_2[1]$ are not endpoints of arcs, then return
  $1 \dotplus \text{LAPCS}(S_1[2, n_1], S_2[2, n_2], k_1, k_2)$.
  /* Since no arcs are involved, it is safe to match the bases. */
  (2.2)   /* Only one arc. */
  If $S_1[1]$ is left endpoint of an arc $(1, i)$ but $S_2[1]$ is not endpoint of an arc, then return the maximum of the following values:
- $\text{LAPCS}(S_1[2, n_1], S_2, k_1 - 1, k_2)$ /* delete $S_1[1]$ */,
- $\text{LAPCS}(S_1, S_2[2, n_2], k_1, k_2 - 1)$ /* delete $S_2[1]$ */, and
- $1 \dotplus \text{LAPCS}((S_1 - S_1[i])[2, +], S_2[2, n_2], k_1 - 1, k_2)$ /* match */.

  /* Since there is an arc in one sequence only, $S_1[1]$ and $S_2[1]$ can be matched only if $S_1[i]$ is deleted. */
  (2.3)   /* Only one arc. */
  If $S_2[1]$ is left endpoint of an arc $(1, j)$ but $S_1[1]$ is not endpoint of an arc, then proceed analogously as in (2.2).
  (2.4)   /* Non-matching arcs. */
  If $S_1[1]$ is left endpoint of an arc $(1, i)$, $S_2[1]$ is left endpoint of an arc $(1, j)$ and $S_1[i] \neq S_2[j]$, then return the maximum of the following values:
- $\text{LAPCS}(S_1[2, n_1], S_2, k_1 - 1, k_2)$ /* delete $S_1[1]$ */,
- $\text{LAPCS}(S_1, S_2[2, n_2], k_1, k_2 - 1)$ /* delete $S_2[1]$ */, and
- $1 \dotplus \text{LAPCS}((S_1 - S_1[i])[2, +], (S_2 - S_2[j])[2, +], k_1 - 1, k_2 - 1)$ /*match*/.

  /* Since the arcs cannot be matched, $S_1[1]$ and $S_2[1]$ can be matched only if $S_1[i]$ and $S_2[j]$ are deleted. */
  (2.5)   /* An arc match is possible. */
  If $S_1[1]$ is left endpoint of an arc $(1, i)$, $S_2[1]$ is left endpoint of an arc $(1, j)$, and $S_1[i] = S_2[j]$, then go through Cases (2.5.1), (2.5.2), and (2.5.3) which are presented below (one of them will apply and will return the length of the lapcs of S_1 and S_2, if such an lapcs can be obtained with k_1 deletions in S_1 and k_2 deletions in S_2, or will return -1, otherwise).

In Case (2.5), it is possible to match arcs $(1, i)$ in S_1 and $(1, j)$ in S_2 since $S_1[1] = S_2[1]$ and $S_1[i] = S_2[j]$. Our first observation is that, if $S_1[2, i-1] = S_2[2, j-1]$ (which will be handled in Case (2.5.1)) or if $S_1[i+1, n_1] = S_2[j+1, n_2]$ (which will be handled in Case (2.5.2)), it is safe to match arc $(1, i)$ with arc $(1, j)$: no longer apcs would be possible when not matching them. We match the equal parts of the sequences (either those inside arcs or those following the arcs) and call Algorithm LAPCS recursively only on the remaining subsequences. These cases only simplify the instance and do not require to branch into several subcases:

(2.5.1) /* *Sequences inside the arcs match.* */
  If $S_1[2, i-1] = S_2[2, j-1]$, then return
  $i \dotplus \text{LAPCS}(S_1[i+1, n_1], S_2[j+1, n_2], k_1, k_2)$.
(2.5.2) /* *Sequences following the arcs match.* */
  If $S_1[i+1, n_1] = S_2[j+1, n_2]$, then return
  $2 \dotplus (n_1 - i) \dotplus \text{LAPCS}(S_1[2, i-1], S_2[2, j-1], k_1, k_2)$.

If neither Case (2.5.1) nor Case (2.5.2) applies, this is handled by Case (2.5.3), which branches into four recursive calls: we have to consider breaking at least one of the arcs (handled by the first three recursive calls in (2.5.3)) or to match the arcs (handled by the fourth recursive call in (2.5.3)):

(2.5.3) Return the maximum of the following four values:

- $\text{LAPCS}(S_1[2, n_1], S_2, k_1 - 1, k_2)$ /* *delete* $S_1[1]$. */,
- $\text{LAPCS}(S_1, S_2[2, n_2], k_1, k_2 - 1)$ /* *delete* $S_2[1]$. */,
- $1 \dotplus \text{LAPCS}((S_1 - S_1[i])[2, +], (S_2 - S_2[j])[2, +], k_1 - 1, k_2 - 1)$
  /* *match* $S_1[1]$ *and* $S_2[1]$, *but do not match arcs* $(1, i)$ *and* $(1, j)$; *this implies the deletion of* $S_1[i]$, $S_2[j]$, *and the incident arcs.* */,
- l (computed as given below) /* *match the arcs.* */

Value l denotes the length of the lapcs of S_1 and S_2 in case of matching arc $(1, i)$ with arc $(1, j)$. It can be computed as the sum of the lengths l', denoting the length of an lapcs of $S_1[2, i-1]$ and $S_2[2, j-1]$, and l'', denoting the length of an lapcs of $S_1[i+1, n_1]$ and $S_2[j+1, n_2]$; each of l' and l'' can be computed by one recursive call. Remember that we already excluded $S_1[2, i-1] = S_2[2, j-1]$ (by Case (2.5.1)) and $S_1[i+1, n_1] = S_2[j+1, n_2]$ (by Case (2.5.2)). For the running time analysis, however, we will require that the deletion parameters k_1 and k_2 will be decreased by two in both recursive calls computing l' and l''. Therefore, we will further exclude those special cases in which l' or l'' can be found by exactly one deletion, either in S_1 or in S_2 (this can be checked in linear time); then, we need only one recursive call to compute l. Only if this is not possible, we will invoke the two calls for l' and l''. Therefore, l is computed as follows:

$$l := \begin{cases} j \dotplus \mathrm{LAPCS}(S_1[i+1,n_1], S_2[j+1,n_2], k_1-1, k_2) \\ \qquad \text{if } S_1[2,i-1] \approx_{1,0} S_2[2,j-1], \\ i \dotplus \mathrm{LAPCS}(S_1[i+1,n_1], S_2[j+1,n_2], k_1, k_2-1) \\ \qquad \text{if } S_1[2,i-1] \approx_{0,1} S_2[2,j-1], \\ 2 \dotplus (n_2-j) \dotplus \mathrm{LAPCS}(S_1[2,i-1], S_2[2,j-1], k_1-1, k_2) \\ \qquad \text{if } S_1[i+1,n_1] \approx_{1,0} S_2[j+1,n_2], \\ 2 \dotplus (n_1-i) \dotplus \mathrm{LAPCS}(S_1[2,i-1], S_2[2,j-1], k_1, k_2-1) \\ \qquad \text{if } S_1[i+1,n_1] \approx_{0,1} S_2[j+1,n_2], \\ 2 \dotplus l' \dotplus l'' \text{ (defined below) otherwise.} \end{cases}$$

Computing l', we credit the two deletions that will certainly be needed when computing l''. Depending on the length of $S_1[i+1,n_1]$ and $S_2[j+1,n_2]$, we have to decide which parameter to decrease: If $|S_1[i+1,n_1]| > |S_2[j+1,n_2]|$, we will certainly need at least two deletions in $S_1[i+1,n_1]$, and can start the recursive call with parameter $k_1 - 2$ (and, analogously, with $k_2 - 2$ if $|S_1[i+1,n_1]| < |S_2[j+1,n_2]|$ and both $k_1 - 1$ and $k_2 - 1$ if $S_1[i+1,n_1]$ and $S_2[j+1,n_2]$ are of same length):

$$l' := \begin{cases} \mathrm{LAPCS}(S_1[2,i-1], S_2[2,j-1], k_1-2, k_2) \text{ if } n_1-i > n_2-j, \\ \mathrm{LAPCS}(S_1[2,i-1], S_2[2,j-1], k_1, k_2-2) \text{ if } n_1-i < n_2-j, \\ \mathrm{LAPCS}(S_1[2,i-1], S_2[2,j-1], k_1-1, k_2-1) \text{ if } n_1-i = n_2-j. \end{cases}$$

Computing l'', we decrease k_1 and k_2 by the deletions already spent when computing l', where $k'_{1,1} := i-2-l'$ denotes the number of deletions spent in $S_1[1,i]$ and $k'_{2,1} := j-2-l'$ denotes the number of deletions spent in $S_2[1,j]$:

$$l'' := \mathrm{LAPCS}(S_1[i+1,n_1], S_2[j+1,n_2], k_1-k'_{1,1}, k_2-k'_{2,1}).$$

Correctness of Algorithm LAPCS. To show the correctness, we have to make sure that, if an lapcs with the specified properties exists, then the algorithm finds one; the reverse can be seen by checking, for every case of the above algorithm, that we only make matches when they extend the lapcs and that the bookkeeping of the "mismatch counters" k_1 and k_2 is correct. In the following, we omit the details for the easier cases of our search tree algorithm and, instead, focus on the most involved situation, Case (2.5).

In Case (2.5), $S_1[1] = S_2[1]$, there is an arc $(1,i)$ in S_1 and an arc $(1,j)$ in S_2, and $S_1[i] = S_2[j]$. In Cases (2.5.1) and (2.5.2), we handled the special situation that $S_1[1,i] = S_2[1,j]$ or that $S_1[i+1,n_1] = S_2[j+1,n_2]$. Observe that, if we decide to match the arcs (Case (2.5.3)), we can divide the current instance into two subinstances: bases from $S_1[2,i-1]$ can only be matched to bases from $S_2[2,j-1]$ and bases from $S_1[l+1,n_1]$ can only be matched to bases from $S_2[j+1,n_2]$. We will, in the following, denote the subinstance given by $S_1[2,i-1]$ and $S_2[2,j-1]$ as part 1 of the instance and the one given by $S_1[i+1,n_1]$ and $S_2[j+1,n_2]$ as part 2 of the instance. We have the choice of breaking at least one of the arcs $(1,i)$ and $(1,j)$ or to match them.

We distinguish two cases. Firstly, suppose that we want to break at least one arc. This can be achieved by either deleting $S_1[1]$ or $S_2[1]$. If we do not delete either of these bases, we obtain a base match. But, in addition, we must delete both $S_1[i]$ and $S_2[j]$, since otherwise we cannot maintain the arc-preserving property. Secondly, we can match the arcs $(1, i)$ and $(1, j)$. Then, we know, since neither Case (2.5.1) nor (2.5.2) applies, that an optimal solution will require at least one deletion in part 1 and will also require at least one deletion in part 2. We can further compute, in linear time, whether part 1 (or part 2, resp.) can be handled by exactly one deletion and start the algorithm recursively only on part 2 (part 1, resp.), decreasing one of k_1 or k_2 by the deletion already spent. In the remaining case, we start the algorithm recursively first on part 1 (to compute l') and, then, on part 2 (to compute l''). At this point we know, however, that an optimal solution will require at least two deletions in part 1 and will also require at least two deletions in part 2. Thus, when starting the algorithm on part 1, we can "spare" two of the $k_1 + k_2$ deletions for part 2, depending on part 2 (as outlined above). Having, thus, found an optimal solution of length l' for part 1, the number of allowed deletions remaining for part 2 is determined: we have, in part 1, already spent $k'_{1,1} := i - 2 - l'$ deletions in $S_1[2, i-1]$ and $k'_{2,1} := j - 2 - l'$ deletions in $S_2[2, j-1]$. Thus, there remain, for part 2, $k_1 - k'_{1,1}$ deletions for $S_1[i+1, n_1]$ and $k_2 - k'_{2,1}$ deletions for $S_2[j+1, n_2]$. This discussion showed that, in Case (2.5.3), our case distinction covers all subcases in which we can find an optimal solution and, hence, Case (2.5.3) is correct.

Running time of Algorithm LAPCS.

Lemma 1. *Given two arc-annotated sequences S_1 and S_2, suppose that we have to delete k'_1 symbols in S_1 and k'_2 symbols in S_2 in order to obtain an lapcs. Then, the search tree size (i.e., the number of the nodes in the search tree) for a call $LAPCS(S_1, S_2, k'_1, k'_2)$ is upperbounded by $3.31^{k'_1 + k'_2}$.*

Proof. Algorithm LAPCS constructs a search tree. In each of the given cases, we do a branching and we perform a recursive call of LAPCS with a smaller value of the sum of the parameters in each of the branches. We now discuss some cases which, in some sense, have a special branching structure. Firstly, Cases (2.1), (2.5.1), and (2.5.2) do not cause any branching of the recursion. Secondly, for Case (2.5.3), in total, we perform five recursive calls. For the following running time analysis, the last two recursive calls of this case (i.e., the ones needed to evaluate l' and l'') will be treated together. More precisely, we treat Case (2.5.3) as if it were a branching into four subcases, where, in each of the first three branches we have *one* recursive call and in the fourth branch we have *two* recursive calls.

In a search tree produced by Algorithm LAPCS, every search tree node corresponds to one of the cases mentioned in the algorithm. Let m be the number of nodes corresponding to Case (2.5.3) that appear in such a search tree. We prove the claim on the search tree size by induction on the number m.

---

[3] Note that there might be several lapcs for two given sequences S_1 and S_2. The length ℓ of such an lapcs, however, is uniquely defined. Since, clearly, $k'_1 = |S_1| - \ell$ and $k'_2 = |S_2| - \ell$, the values k'_1 and k'_2 also are uniquely defined for given S_1 and S_2.

For $m = 0$, we do not have to deal with Case (2.5.3). Hence, we can determine the search tree size by the so-called branching vectors (for details of this type of analysis we refer to [ ]): Suppose that in one search tree node with current sequences S_1, S_2 and parameters k_1', k_2', we have q branches. We analyze the size of the search tree in terms of the sum $k' := k_1' + k_2'$ of the two parameters. Suppose that in branch t, $1 \leq t \leq q$, we call LAPCS recursively with new parameter values $k_{1,t}'$ and $k_{2,t}'$, i.e., with a sum $k'(t) := k_{1,t}' + k_{2,t}'$. Letting $p_t := k' - k'(t)$, $1 \leq t \leq q$, the vector $(p_1, \dots, p_q)$ is called the *branching vector* for this branch. It corresponds to the recurrence

$$T_{k'} = T_{k'-p_1} + \cdots + T_{k'-p_q},$$

where T_i denotes the size of the search tree for parameter value i. The characteristic polynomial of this recurrence is

$$z^p - z^{p-p_1} - \cdots - z^{p-p_q}, \tag{1}$$

where $p = \max\{p_1, \dots, p_q\}$. If c is a root of ( ) with maximum absolute value, then $T_{k'}$ is $c^{k'}$ up to a polynomial factor and c is called the *branching number* that corresponds to the branching vector $(p_1, \dots, p_q)$. Moreover, if c is a single root, then even $T_{k'} = O(c^{k'})$. The branching vectors which appear in our search tree are $(1,1)$ (Case 1), $(1,1,1)$ (Cases 2.2, 2.3), $(1,1,2)$ (Case 2.4), $(1,1,2)$ (Case 2.5.3 with $m = 0$). The worst-case branching number c for these branching vectors is given for $(1,1,1)$ with $c = 3 \leq 3.31$.

Now suppose that the claim is true for all values $m' \leq m - 1$. In order to prove the claim for m, we have to, for a given search tree, analyze a search tree node corresponding to Case (2.5). Suppose that the current sequences in this node are S_1 and S_2 with lengths n_1 and n_2 and that the optimal parameter values are k_1' and k_2'. Our goal is to show that the branching of the recursion for Case (2.5.3) has branching vector $(1, 1, 2, 1)$ which corresponds to a branching number $c = 3.31$ (which can be easily verified using the corresponding recurrence). As discussed above, for the first three branches of Case (2.5.3), we only need one recursive call of the algorithm. The fourth branch is more involved. We will have a closer look at this fourth subcase of (2.5.3) in the following. Let us evaluate the search tree size for a call of this fourth subcase. It is clear that the optimal parameter values for the subsequences $S_1[2, i - 1]$ and $S_2[2, j - 1]$ are $k_{1,1}' = (i - 2) - l'$ and $k_{2,1}' = (j - 2) - l'$. Moreover, the optimal parameter values for the subsequences $S_1[i + 1, n_1]$ and $S_2[j + 1, n_2]$ are $k_{1,2}' = (n_1 - i) - l''$ and $k_{2,2}' = (n_2 - j) - l''$. Since, by Cases (2.5.1) and (2.5.2) and by the first four cases in the fourth branch of Case (2.5.3), the cases where $k_{1,1}' + k_{2,1}' \leq 1$ or $k_{1,2}' + k_{2,2}' \leq 1$ are already considered, we may assume that we have $k_{1,1}' + k_{2,1}', k_{1,2}' + k_{2,2}' \geq 2$. Hence, by induction hypothesis, the search tree size for the computation of l' is $3.31^{k_{1,1}' + k_{2,1}'}$, and the computation of l'' needs a search tree of size $3.31^{k_{1,2}' + k_{2,2}'}$.

---

[4] For the branching vectors that appear in our setting, c is always real and will always be a single root.

This means that the total search tree size for this fourth subcase is upper-bounded by

$$3.31^{k'_{1.1}+k'_{2.1}} + 3.31^{k'_{1.2}+k'_{2.2}}. \tag{2}$$

Note that, since k'_t is assumed to be the optimal value, we have

$$k'_t = n_t - l' - l'' - 2 \quad \text{for } t = 1, 2,$$

and, hence, an easy computation shows that

$$k'_{t,1} + k'_{t,2} = k'_t \quad \text{for } t = 1, 2.$$

From this we conclude that,

$$3.31^{k'_{1.1}+k'_{2.1}} + 3.31^{k'_{1.2}+k'_{2.2}} \le 3.31^{k'_1+k'_2-1}. \tag{3}$$

Inequality ( ) holds true since, by assumption, $k'_{1,1}+k'_{2,1}, k'_{1,2}+k'_{2,2} \ge 2$. Plugging Inequality ( ) into Expression ( ), we see that the search tree size for this fourth case of (2.5.3) is upperbounded by $3.31^{k'_1+k'_2-1}$. Besides, by induction hypothesis, the search trees for the first and the second branch of Case (2.5.3) also have size upperbounded by $3.31^{k'_1+k'_2-1}$ and the search tree for the third branch of Case (2.5.3) has size upperbounded by $3.31^{k'_1+k'_2-2}$.

Hence, the overall computations for Case (2.5.3) can be treated with branching vector $(1, 1, 2, 1)$. The corresponding branching number c of this branching vector is 3.31 (being the root of ( ) with maximal absolute value for the branching vector $(1, 1, 2, 1)$). This is the worst-case branching number among all branchings. Hence, the full search tree has size $3.31^{k'_1+k'_2}$. $\square$

Now, suppose that we run algorithm LAPCS with sequences S_1, S_2 and parameters k_1, k_2. As before let k'_1 and k'_2 be the number of deletions in S_1 and S_2 needed to find an lapcs. As pointed out at the beginning of this section, the search tree will be traversed in breadth-first manner. Hence, on the one hand, we may stop the computation if at some search tree node an lapcs is found (even though the current parameters at this node may be non-zero). On the other hand, if it is not possible to find an lapcs with k_1 and k_2 deletions, then the algorithm terminates automatically by Case (0). Observe that the time needed in each search tree node is upperbounded by $O(n)$ if both sequences S_1 and S_2 have length at most n. This gives a total running time of $O(3.31^{k_1+k_2} \cdot n)$ for the algorithm. The following theorem summarizes the results of this section.

Theorem 1 *The problem* LAPCS(NESTED, NESTED) *for two sequences S_1 and S_2 with $|S_1|, |S_2| \le n$ can be solved in time $O(3.31^{k_1+k_2} \cdot n)$ where k_1 and k_2 are the number of deletions needed in S_1 and S_2.* $\square$

4   Algorithms for Restricted Versions

In this section, we investigate the so-called c-FRAGMENT LAPCS(CROSSING, CROSSING) and the c-DIAGONAL LAPCS(CROSSING, CROSSING) problems.

In the setting of c-FRAGMENT, the sequences S_1 and S_2 are divided into fragments $S_1^{(1)}, \ldots, S_1^{(\lceil |S_1|/c \rceil)}$ and $S_2^{(1)}, \ldots, S_2^{(\lceil |S_2|/c \rceil)}$, respectively, each fragment of lengths exactly c (the last fragment may have length less than c) and we do not allow matches between distinct fragments, i.e., between bases in $S_1^{(t)}$ and $S_2^{(t')}$ with $t \neq t'$.

In the setting of c-DIAGONAL, a base $S_1[i]$ $(1 \leq i \leq |S_1|)$ is only allowed to match bases in the range $S_2[i - c, i + c]$.

We give algorithms for these problems when parameterized by the length l of the desired subsequence. The versions c-DIAGONAL and c-FRAGMENT of LAPCS (NESTED, NESTED) were already treated by Lin *et al.* [ ]. They gave PTAS's for these problems. The running times for the following algorithms are based on worst-case analysis. The algorithms are expected to perform much better in practice.

The problem c-FRAGMENT LAPCS(CROSSING, CROSSING). The algorithm presented here extends an idea which was used for the 1-FRAGMENT case by Lin *et al.* [ ]. We translate an instance of the c-FRAGMENT LAPCS problem into an instance of an INDEPENDENT SET problem on a graph $G = (V, E)$ of bounded degree. Since INDEPENDENT SET on graphs of bounded degree is fixed-parameter tractable, we also obtain a fixed-parameter algorithm for c-FRAGMENT LAPCS. The following lemma uses a straightforward search tree algorithm.

Lemma 2. *Let G be a graph of degree bounded by B. Then, an independent set of size k can be found in time $O((B + 1)^k B + |G|)$.*   □

The graph G which is obtained when translating an instance of the c-FRAGMENT LAPCS(CROSSING, CROSSING) problem has degree bounded by $B = c^2 + 2c - 1$, which gives us the following result.

Proposition 1 *The c-FRAGMENT LAPCS(CROSSING, CROSSING) problem parameterized by the length l of the desired subsequence can be solved in time $O((B + 1)^l B + c^3 n)$, where $B = c^2 + 2c - 1$.*

Proof. Let (S_1, A_1) and (S_2, A_2) be an instance of c-FRAGMENT LAPCS (CROSSING, CROSSING), where S_1 and S_2 are over a fixed alphabet Σ.

We construct a graph $G = (V, E)$ as follows. The set of vertices of G corresponds to all possible matches, i.e., we define

$$V := \{v_{i,j} \mid S_1[i] = S_2[j] \text{ and } \lceil i/c \rceil = \lceil j/c \rceil\}.$$

Note that for the c-FRAGMENT problem we are only allowed to match two symbols which are in the same fragment.

Since we want to translate the LAPCS instance into an instance of an INDEPENDENT SET problem on G, the edges of G will represent all conflicting

---

[5] We assume that both sequences have the same length, i.e., $n = |S_1| = |S_2|$. If the sequences do not have the same length, we extend the shorter sequence by adding a sequence of a letter not in the alphabet at its end.

matches. Since such a conflict may arise from three different situations, we let $E := E_1 \cup E_2 \cup E_3$, where

$$E_1 := \{\{v_{i_1,j_1}, v_{i_2,j_2}\} \mid (i_1 \neq i_2 \wedge j_1 = j_2) \vee (i_1 = i_2 \wedge j_1 \neq j_2)\} \qquad (4)$$

(the two matches represented by v_{i_1,j_1} and v_{i_2,j_2} share a common endpoint),

$$E_2 := \{\{v_{i_1,j_1}, v_{i_2,j_2}\} \mid ((i_1 < i_2) \wedge (j_1 > j_2)) \vee ((i_1 > i_2) \wedge (j_1 < j_2))\} \qquad (5)$$

(the two matches represented by v_{i_1,j_1} and v_{i_2,j_2} are not order-preserving), and

$$E_3 := \left\{ \{v_{i_1,j_1}, v_{i_2,j_2}\} \left| \begin{array}{c} ((i_1,i_2) \in A_1 \wedge (j_1,j_2) \notin A_2) \vee \\ ((i_1,i_2) \notin A_1 \wedge (j_1,j_2) \in A_2) \end{array} \right. \right\} \qquad (6)$$

(the two matches represented by v_{i_1,j_1} and v_{i_2,j_2} are not arc-preserving).

By construction, it is clear that all lapcs's of length l which match positions $Q_1 = \{i_1, \ldots, i_l\}$ in S_1 to positions $Q_2 = \{j_1, \ldots, j_l\}$ in S_2 one-to-one correspond to independent sets of the form $V' = \{v_{i_t,j_t} \mid t = 1, \ldots, l\}$ in the graph G. We then use the algorithm from Lemma   to determine an independent set in G of size l and, hence, an lapcs of length l for $(S_1, A_1), (S_2, A_2)$.

For the running time analysis of this algorithm, note that there can be up to c^2 vertices in G for each fragment. Hence, we can have a total of cn vertices.

Each vertex $v_{i,j}$ in G can have at most $c^2 + 2c - 1$ adjacent edges:

- If a base match $\langle i, j_1 \rangle$ shares with another base match $\langle i, j_2 \rangle$ the same base $S_1[i]$, then an edge must be imposed between the vertices v_{i,j_1} and v_{i,j_2}. There can be at most $c - 1$ such base matches, which share $S_1[i]$ with $\langle i, j \rangle$, and at most $c - 1$ base matches, which share $S_2[j]$ with $\langle i, j \rangle$. Thus, $v_{i,j}$ can have at most $2(c - 1)$ adjacent edges from the set E_1.
- If $S_1[i]$ is the first base in one fragment of S_1 and $S_2[j]$ is the last base in the same fragment of S_2, then the base match $\langle i, j \rangle$ can violate the order of the original sequences with at most $(c - 1)^2$ other base matches. Thus, at most $(c - 1)^2$ edges from E_2 will be imposed on vertex $v_{i,j}$.
- If $S_1[i]$ and $S_2[j]$ both are endpoints of arcs (i, i') and (j, j'), then all base matches involving $S_1[i']$ or $S_2[j']$ (but not both) with base match $\langle i, j \rangle$ cannot be arc-preserving. Since $S_1[i']$ and $S_2[j']$ can be in two different fragments and each of them has at most c matched bases, the edges from the set E_3 adjacent to vertex $v_{i,j}$ can amount to $2c$.

Thus, the resulting graph G has a vertex degree bounded by $B = c^2 + 2c - 1$. According to Lemma   , we can find an independent set in G of size l in time $O((B + 1)^l B + |G|)$. Moreover, since we have cn vertices, G can have at most $O(c^3 n)$ edges. The construction of G can be carried out in time $O(c^3 n)$. Hence, the c-FRAGMENT LAPCS(CROSSING, CROSSING) problem can be solved in time $O((B + 1)^l B + c^3 n)$, where $B = c^2 + 2c - 1$. $\qquad \square$

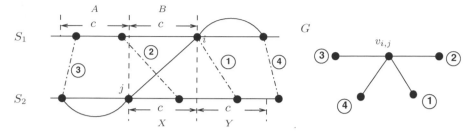

Fig. 1. c-**diagonal LAPCS.** Vertex $v_{i,j}$ is created for the base match $\langle i, j \rangle$. The lines 1, 2, 3, and 4 represent other four base matches, each of them corresponds to a vertex in G. Since the base match (1) shares base $S_1[i]$ with $\langle i, j \rangle$, an edge is imposed between vertices $v_{i,j}$ and vertex (1). Base matches (2) and $\langle i, j \rangle$ cross each other. Hence, an edge is also imposed between their corresponding vertices. It is clear that neither the pair of base matches (3) and $\langle i, j \rangle$ nor (4) and $\langle i, j \rangle$ is arc-preserving. Two edges are added to the graph G. Note that A, B, X, and Y are all substrings of length c

The problem c-DIAGONAL **LAPCS**(CROSSING, CROSSING). A similar approach as the one for c-FRAGMENT LAPCS(CROSSING, CROSSING) can be used to obtain a result for the c-DIAGONAL case. It can be shown that, in this case, that graph G for which we have to find an independent set has degree at most $B = 2c^2 + 7c + 2$.

Proposition 2 *The* c-DIAGONAL *LAPCS*(CROSSING, CROSSING) *problem parameterized by the length l of the desired subsequence can be solved in time* $O((B + 1)^l B + c^3 n)$, *where* $B = 2c^2 + 7c + 2$.

Proof. Again, we translate the problem into an INDEPENDENT SET problem on a graph $G = (V, E)$ with bounded degree and use Lemma   . For given arc-annotated sequences (S_1, A_1), (S_2, A_2), the set of vertices now becomes

$$V := \{v_{i,j} \mid S_1[i] = S_2[j] \text{ and } j \in [i - c, i + c]\},$$

since each position i in sequence S_1 can only be matched to positions $j \in [i - c, i + c]$ of S_2. The definition of the edge set E can be adapted from the case for c-FRAGMENT (cf. Equations ( ),( ), and ( )). We put an edge $\{v_{i_1,j_2}, v_{i_2,j_2}\}$ iff the corresponding matches $\langle i_1, j_1 \rangle$ and $\langle i_2, j_2 \rangle$ (1) share a common endpoint, (2) are not order-preserving, or (3) are not arc-preserving. Fig.   illustrates this construction. Obviously, $|V| \leq (2c + 1) \cdot n$. In the following, we argue that the degree of $G = (V, E)$ is upper-bounded by $B = 2c^2 + 7c + 2$:

- Because a base can be matched to at most $2c + 1$ bases in another sequence, a base match can have common bases with up to $2c + 2c = 4c$ other base matches. E.g., in Fig.   , base match $\langle i, j \rangle$ shares base $S_1[i]$ with the base match denoted by (1).

- We can observe in Fig.   that a vertex in G has a maximum number of edges from E_2, if the distance between the bases involved in its corresponding base match is equal to c. Consider, e.g., base match $\langle i, j \rangle$ in Fig.   . There, a base match crossing $\langle i, j \rangle$ must be from one of the following sets: $M_1 = \{ \langle i_1, j_1 \rangle \mid S_1[i_1]$ is in substring B, $S_2[j_1]$ is in substring $X \}$, $M_2 = \{ \langle i_2, j_2 \rangle \mid S_1[i_2]$ is in substring B, $S_2[j_2]$ is in substring Y, and $j_2 - i_2 \leq c \}$, or $M_3 = \{ \langle i_3, j_3 \rangle \mid S_1[i_3]$ is in substring A, $S_2[j_3]$ is in substring X, and $j_3 - i_3 \leq c \}$. The set M_1 can have at most c^2 elements. The number of elements of the other two sets can amount to $c^2 - c$. Therefore, each vertex in V can have at most $2c^2 - c$ edges which are imposed to guarantee the order-preserving property.
- If the two bases, which form a base match, both are endpoints of two arcs, like the base match $\langle i, j \rangle$ in Fig.   , then this base match cannot be in an arc-preserving subsequence with base matches, which involve only one of the other endpoints of the arcs. Two such base matches are marked in Fig.   with (3) and (4). Those base matches can amount to $4c + 2$.

Consequently, the graph G has degree bounded by $B = 2c^2 + 7c + 2$. With $(2c+1)n$ vertices, G has at most $O(c^3 n)$ edges. The construction of G can be done in time $O(c^3 n)$. Hence, the c-DIAGONAL LAPCS(CROSSING, CROSSING) problem can be solved in time $O((B + 1)^l B + c^2 n)$, where $B = 2c^2 + 7c + 2$.   □

The problems c-FRAGMENT (c-DIAGONAL) LAPCS(UNLIMITED,UNLIMITED). Note that the observation that the graph $G = (V, E)$ above has bounded degree heavily depends on the fact that the two underlying sequences have CROSSING arc structure. Hence, the same method does not directly apply for c-FRAGMENTED (c-DIAGONAL) LAPCS(UNLIMITED, UNLIMITED). However, if the "degree of a sequence" is bounded, we can upperbound the degree of G. The *degree* of an arc-annotated sequence (S, A) with UNLIMITED arc structure is the maximum number of arcs from A that are incident to a base in S. Clearly, the so-called *cutwidth* of an arc-annotated sequence (see [ ]) is an upper bound on the degree.

Proposition 3 *The c-FRAGMENT (and c-DIAGONAL, respectively) LAPCS (UN-LIMITED, UNLIMITED) problem with bounded degree b for its sequences, parameterized by the length l of the desired subsequence, can be solved in time $O((B + 1)^l B + c^3 n)$ with $B = c^2 + 2bc - 1$ (and in time $O((B' + 1)^l B' + c^3 n)$ with $B' = 2c^2 + (4b + c)c + 2b$, respectively).*

Proof. The c-FRAGMENT (c-DIAGONAL) LAPCS(UNLIMITED, UNLIMITED) problem can also be translated into an INDEPENDENT SET problem on a graph $G = (V, E)$ with bounded degree using a similar construction as shown above. The number of vertices in the resulting graph is equal to the one obtained in the proofs of Proposition   and   , but the bound on the degree changes. In the constructions in the proofs, we added three sets of edges to G. Since the first two sets have nothing to do with arcs, these edges remain in the graph for UNLIMITED arc structure. In the constructions above, $2c$ edges for c-FRAGMENT and $4c + 2$ edges for c-DIAGONAL are added into E_3 for a base match $\langle i,j \rangle$ with two arc endpoints, $(i,i_1) \in A_1$ and $(j,j_1) \in A_2$. These edges are between vertex $v_{i,j}$ and the vertices, which correspond to the base matches involving one of $S_1[i_1]$ and $S_2[j_1]$. In UNLIMITED arc structure with bounded degree b, a base $S_1[i]$ can be endpoint of at most b arcs, we denote them by $(i,i_1),(i,i_2),\ldots,(i,i_b)$. The third set of edges must be extended to include the edges between (i,j) and all vertices, which correspond to base matches involving one of $S_1[2],\ldots,S_1[b],S_2[2],\ldots,S_2[b]$. The amount of edges in this set can increase to $b(2c)$ for c-FRAGMENT and to $b(4c+2)$ for c-DIAGONAL LAPCS(UNLIMITED, UNLIMITED). The degree of the resulting graph for c-FRAGMENT is then bounded by $B = c^2 + 2bc - 1$, and the one for c-DIAGONAL by $B = 2c^2 + (4b+3)c + 2b$. □

5   Conclusion and Future Work

Adopting a parameterized point of view [, ,], we have shed new light on the algorithmic tractability of the NP-complete LONGEST COMMON SUBSEQUENCE problem with nested arc annotations, an important problem in biologically motivated structure comparison. Immediate questions arising from our work are to further improve the exponential terms of the running times of our exact algorithms and to determine the parameterized complexity of (non-restricted) LAPCS(NESTED,NESTED) when parameterized by subsequence length instead of number of deletions. Depending on what (relative) length of the longest common subsequence is to be expected, one or the other parameterization might be more appropriate. In the case of a "short" lcs, the subsequence length is more useful as a parameter, and in the case of "long" lcs, the number of deletions is the preferable parameter. Our complexity analyses are worst-case, however, and it is a topic of future investigations to study the practical usefulness of our algorithms by implementations and experiments. In this context, it is also meaningful to take a closer look at the (special case) problem "APS(NESTED,NESTED)," where one asks whether a given sequence forms an arc-preserving subsequence of another. Results in [ ] make us confident that this is doable in polynomial time, but, if so, to get the degree of the polynomial as small as possible remains a research issue of theoretical as well as practical importance, see [ ] for more details.

References

1. J. Alber, J. Gramm, and R. Niedermeier. Faster exact solutions for hard problems: a parameterized point of view. *Discrete Mathematics*, 229: 3–27, 2001.   ,   ,

2. H. L. Bodlaender, R. G. Downey, M. R. Fellows, M. T. Hallett, and H. T. Wareham. Parameterized complexity analysis in computational biology. *Computer Applications in the Biosciences*, 11: 49–57, 1995.
3. H. L. Bodlaender, R. G. Downey, M. R. Fellows, and H. T. Wareham. The parameterized complexity of sequence alignment and consensus. *Theoretical Computer Science*, 147:31–54, 1995.

4. P. Bonizzoni, G. Della Vedova, and G. Mauri. Experimenting an approximation algorithm for the LCS. *Discrete Applied Mathematics*, 110:13–24, 2001.
5. R. G. Downey and M. R. Fellows. *Parameterized Complexity*. Springer. 1999.   ,
   ,   ,
6. P. A. Evans. *Algorithms and Complexity for Annotated Sequence Analysis*. PhD thesis, University of Victoria, Canada. 1999.   ,   ,   ,
7. P. A. Evans. Finding common subsequences with arcs and pseudoknots. In *Proc. of 10th CPM*, number 1645 in LNCS, pages 270–280, 1999. Springer.   ,   ,
8. P. A. Evans and H. T. Wareham. Exact algorithms for computing pairwise alignments and 3-medians from structure-annotated sequences. In *Proc. of Pacific Symposium on Biocomputing*, pages 559–570, 2001.
9. M. R. Fellows. Parameterized complexity: the main ideas and some research frontiers. In *Proc. of 12th ISAAC*, number 2223 in LNCS, pages 291–307, 2001. Springer.   ,   ,
10. D. Goldman, S. Istrail, and C. H. Papadimitriou. Algorithmic aspects of protein structure similarity. In *Proc. of 40th IEEE FOCS*, pages 512–521, 1999.
11. J. Guo. *Exact Algorithms for the Longest Common Subsequence Problem for Arc-Annotated Sequences*. Diploma thesis, Universität Tübingen, Fed. Rep. of Germany. February 2002.
12. D. Gusfield. *Algorithms on Strings, Trees, and Sequences: Computer Science and Computational Biology*. Cambridge University Press. 1997.   ,
13. T. Jiang, G.-H. Lin, B. Ma, and K. Zhang. The longest common subsequence problem for arc-annotated sequences. In *Proc. of 11th CPM*, number 1848 in LNCS, pages 154–165, 2000. Springer. Full paper accepted by *Journal of Discrete Algorithms*.   ,   ,   ,
14. O. Kullmann. New methods for 3-SAT decision and worst-case analysis. *Theoretical Computer Science*, 223: 1–72, 1999.
15. G. Lancia, R. Carr, B. Walenz, and S. Istrail. 101 optimal PDB structure alignments: a branch-and-cut algorithm for the maximum contact map overlap problem. In *Proc. of 5th ACM RECOMB*, pages 193–202, 2001.
16. M. Li, B. Ma, and L. Wang. Near optimal multiple alignment within a band in polynomial time. In *Proc. of 32nd ACM STOC*, pages 425–434, 2000.
17. G.-H. Lin, Z.-Z. Chen, T. Jiang, and J. Wen. The longest common subsequence problem for sequences with nested arc annotations. In *Proc. of 28th ICALP*, number 2076 in LNCS, pages 444–455, 2001. Springer.   ,   ,   ,   ,
18. M. Paterson and V. Dancik. Longest common subsequences. In *Proc. of 19th MFCS*, number 841 in LNCS, pages 127–142, 1994. Springer.
19. D. Sankoff and J. Kruskal (eds.). *Time Warps, String Edits, and Macromolecules*. Addison-Wesley. 1983. Reprinted in 1999 by CSLI Publications.

Local Similarity Based Point-Pattern Matching[*]

Veli Mäkinen and Esko Ukkonen

Department of Computer Science, University of Helsinki
P.O Box 26 (Teollisuuskatu 23), FIN-00014 Helsinki, Finland
{vmakinen,ukkonen}@cs.Helsinki.FI

Abstract. We study local similarity based distance measures for point-patterns. Such measures can be used for matching point-patterns under non-uniform transformations — a problem that naturally arises in image comparison problems. A general framework for the matching problem is introduced. We show that some of the most obvious instances of this framework lead to NP–hard optimization problems and are not approximable within any constant factor. We also give a relaxation of the framework that is solvable in polynomial time and works well in practice in our experiments with two–dimensional protein electrophoresis gel images.

1 Introduction

The point-pattern matching problem is, given two finite sets A and B in d-dimensional Euclidean space, to find the best matching of A and B under some quality measure. Point-pattern matching problems arise in many applications. For example, in two–dimensional gel electrophoresis proteins are separated to form spots in a planar gel, and one wants to match the same proteins from different experiments; Figure   in Section   gives an example of a pair of gels to be matched. Such gels may contain partly different proteins, and protein spots may be distorted non-uniformly. Also the spot intensities and sizes vary from one gel to another (the amount of variance is actually what is interesting to the biologists using the gels). Therefore the problem is normally reduced to matching the midpoints of the detected spots.

In computational geometry, point-pattern matching problems under uniform transformations (like translation, rotation, and scaling) are widely studied (see [ ] for a survey). Allowing non-uniform transformations complicates the problem substantially, and a version of the problem has been shown NP-complete [,].

In practice, greedy methods are used to cope with non-uniform transformations: Some pairs of points ("landmarks"), that should match, are selected first and one point-pattern is interpolated with respect to the other so that the selected pairs become aligned. This procedure is repeated until all possible

---

[*] Supported by the Academy of Finland under grant 22584.

matching pairs are determined. Usually, the selected pairs are determined inter-
actively with the user. This kind of procedure was first proposed by Appel et.
al. [ ].

A key issue when coping with non-uniform transformations is that although
global similarity has not been preserved in the patterns to be compared one can
still assume that some local similarity is present. Otherwise the problem would
be infeasible. There are many proposals to formalize this criterion. Akutsu et.
al. [ ] and Jokisch and Müller [ ] used an absolute criterion for this: Find the
largest matching whose all pairs are consistent with a local similarity criterion.

In more detail, the criterion of [ ] is the following. Given positive parameter ϵ,
and two point-patterns $A = \{a_1, \ldots, a_m\}$ and $B = \{b_1, \ldots, b_n\}$ in d-dimensional
Euclidean space, find a maximum matching $M = \{(a_{i_1}, b_{j_1}), \ldots, (a_{i_l}, b_{j_l})\} \subseteq
A \times B$ satisfying for all h, k $(h \neq k)$

$$\frac{1}{1+\epsilon} < \frac{|b_{j_h} - b_{j_k}|}{|a_{i_h} - a_{i_k}|} < 1 + \epsilon.$$

The problem is NP-hard when $d \geq 2$. For $d = 1$, a dynamic programming
algorithm was given that solves this problem with time complexity $O(m^2 n^2)$.
As noted in [ ], this algorithm assumes that the points in the matching M
are in increasing order (i.e. $a_{i_1} < \ldots < a_{i_l}$ and $b_{j_1} < \ldots < b_{j_l}$). Without this
assumption, the problem is NP-hard also when $d = 1$ as shown by Jokisch and
Müller [ ].

In this paper we introduce a framework for defining local similarity preserving
point-pattern matching in Euclidean spaces. A matching of point-patterns A
and B preserves local similarity, if the points of A that are close to each other
are translated in about the same way in the matching to the points of B. This also
means that the neighborhoods of the points of A are preserved in the matching.

Our framework is based on the following technical idea: If the local similarity
is preserved then the translation which takes a point a of A to the corresponding
point of B should be the same as what one would expect based on the actual
translations of the neighbors of a. This suggests the concept of an expected
translation. We then define the cost of a matching as the difference between the
actual and the expected translations. One has to find matching for which this
difference is smallest possible.

We give different concrete instances of defining the expected translations
in the framework. Unfortunately it turns out that finding the corresponding
shortest distances and optimal matchings is typically NP–hard and not efficiently approximable. We also discuss different possibilities of allowing outlier points (i.e., mismatches) in the matchings, including a technique based on
many-to-one matchings. By relaxing the local similarity constraint we obtain a
polynomial–time solvable variant whose evaluation reduces to minimum weight
perfect matching in bipartite graphs. This method works well in practice: We report some performance results from our experiments with two–dimensional protein electrophoresis gels. The method finds fully automatically matchings that
are almost identical to our reference matchings obtained using an interactive
system.

2   Point-Patterns and Matchings

A *point-pattern* is any finite set in the Euclidean d–dimensional space. Our main application is in the two–dimensional space in which case the patterns are called 2D point-patterns. However, most of the results are valid in the general case.

Let A and B be two point-patterns. A *matching* of A and B is any set $M \subseteq A \times B$. With the obvious correspondence with the bipartite graph matching in mind, we call the members of M the *edges* of the matching. An edge (a, b) can be interpreted as a vector from a to b. Then it gives the translation that makes a to match b.

A matching M is *one–to–one* if whenever (a, b) and (a', b') are two different edges in M then $a \neq a'$ and $b \neq b'$, and *many–to–one* if (a, b) is an edge in M then at most one of a and b can belong to some other edge in M. A many–to–one matching is *perfect* if each point of A and B belong at least in one edge in M; in the standard graph theory terminology such an M is an *edge cover* of $A \cup B$. If $|A| = |B|$ then a one–to–one matching M of A and B such that $|M| = |A|$ is *perfect*.

3   Local Similarity Preserving Distance Functions

Our goal is to find the best perfect matching between point-patterns A and B such that the local neighborhoods of the points of A are preserved in the matching. That is, we want that points of A that are close to each other are translated to the points of B in about the same way. Hence, as the edges of the matching give these translations, we want that the edge vectors are locally similar.

Depending on the application, the neighborhood of a point that is expected to behave similarly in matching may vary significantly. Thus, it is of interest to study different ways of defining local similarity.

The actual distance between A and B will be defined as the smallest possible sum of differences between the actual translations (edges) applied on each element of A and the 'expected' translations that would be consistent with the neighboring translations.

A general framework for distances of this type will be defined as follows. Let $\mathcal{M}$ denote the set of all perfect matchings between A and B (which in this case have to be of equal size) and $\mathcal{M}^{\mathrm{many}}$ the set of all perfect many–to–one matchings between A and B. Let $f : \mathcal{M} \times \mathbf{R}^d \to \mathbf{R}^d$ be a function that given a perfect one-to-one matching $\mathcal{M}$ and a point a gives another point $f(M, a)$. Here the translation from a to $f(M, a)$ is the *expected translation* on a given by f while the *actual translation* on a is given by the edge(s) adjacent to a in the matching M. The purpose is that the expected translation is (almost) the same as the actual translations that M gives for the neighbors of a; later on we will give concrete examples of functions f. A corresponding function $f^{\mathrm{many}} : \mathcal{M}^{\mathrm{many}} \times \mathbf{R}^d \to \mathbf{R}^d$ for perfect many–to–one matching is defined similarly.

Now our local similarity distance will measure the difference between the expected and the actual translations as follows.

Definition 1 *Let* $M = \{(a_1, b_1), \ldots, (a_m, b_m)\}$ *be a perfect (one-to-one or many–to–one) matching between* A *and* B, *and let the* cost *of* M *with respect to* f *be*

$$w_f(A, B, M) = \sum_{k=1}^{|M|} |b_k - f(M, a_k)|.$$

The local similarity distance d_f *over perfect one-to-one matchings between* A *and* B *is defined as*

$$d_f(A, B) = \min\{w_f(A, B, M) \mid M \in \mathcal{M}\},$$

and the distance d_f^{many} *over perfect many–to–one matchings between* A *and* B *as*

$$d_f^{\text{many}}(A, B) = \min\{w_f(A, B, M) \mid M \in \mathcal{M}^{\text{many}}\}.$$

A matching M *that gives the value* $d_f(A, B)$ *or* d_f^{many} *is called the* best perfect matching *of* A *and* B *with respect to* f.

As we use only the perfect matchings in this definition, we assume in the sequel that the matchings considered are perfect if not explicitly stated otherwise.

There are many ways of choosing function f in Def. . It is desirable to have a function that captures local similarity and leads to a polynomial time computable distance. We will give below four different definitions for f, which all capture local similarity in some intuitive sense. The definitions range from very local to a full global condition in which all other points have an effect on the expected translation of a point. We are interested in the complexity of the distance evaluation problem in all these cases.

First think of the very local condition for local similarity in which the expected translation of a equals the actual translation of the nearest neighbor of a. This gives $f = f_{\text{NN}}$ as

$$f_{\text{NN}}(M, a) = a + (b' - a'), \tag{1}$$

where a' is the nearest neighbor of a in A, and $(a', b') \in M$. We call the corresponding distance function $d_{f_{\text{NN}}}$ the *nearest neighbor distance*.

An extension of the nearest neighbor distance would be the K-*nearest neighbors distance* $d_{f_{\text{KNN}}}$, where K is an integer, and the expected translations are given as

$$f_{\text{KNN}}(M, a) = a + \frac{\sum_{(a', b') \in M_K} b' - a'}{|M_K|}, \tag{2}$$

where $M_K \subset M$ consists of all $(a', b') \in M_K$ such that a' belongs to the K nearest neighbors of a in A.

Finally a global condition can be defined, where all actual translations have an influence on the expected translation. We make the strength of the influence to depend on the distance: We use a weighted average over the different actual translations, where the weights are inversely proportional to the distances from a:

$$f_{\text{I}}(M, a) = a + \frac{\sum_{(a', b') \in M, a' \neq a} \frac{b' - a'}{(|a - a'|)^r}}{\sum_{(a', b') \in M, a' \neq a} \frac{1}{(|a - a'|)^r}}, \tag{3}$$

where $r \geq 1$ is a constant. We call the corresponding distance d_{f_I} the *interpolation distance*.

To illustrate the three distance functions defined, one should easily see that if B is a translation of A then the distance between them equals 0 in all three cases.

4   Complexity of Local Similarity Preserving Distance

We will now prove the negative result that the three distances $d_{f_\text{NN}}, d_{f_\text{KNN}}, \text{and} d_{f_\text{I}}$ are NP-hard to compute. We will start with d_{f_NN}.

Theorem 2 *It is NP-hard to compute the nearest neighbor distances* d_{f_NN} *and* $d_{f_\text{NN}}^{\text{many}}$.

Proof. We use reduction from the 3-PARTITION problem [ ]. The reduction is similar to what Jokisch and Müller used for *distance-congruence* [ ]. The 3-PARTITION is defined as follows:

Definition 3 *Let P be a set of $3p$ elements, $S > 0$ a constant number, and $s : P \rightarrow \mathbb{N}$ a cost function with $S/4 < s(a) < S/2$ for all $a \in P$ such that $\sum_{a \in P} s(a) = pS$. The 3-PARTITION is the problem to find a partition $P_1, \ldots, P_p$ of P such that $\sum_{a \in P_i} s(a) = S$ for all $i \in \{1, \ldots, p\}$.*

The 3-PARTITION is strongly NP-complete [ ], and thus NP-complete even if the costs $s(a)$ are bounded by a polynomial in the length of the input. We use this property in the reduction.

We construct two point-patterns from an instance of the 3-PARTITION as shown in Figure . The two point-patterns, A and B, have both pS points. The points in A are grouped according to the costs of the elements of P: the group i corresponds to the ith element a_i of P and has $s(a_i)$ points. The points in B are grouped in equal size slots of size S corresponding to the partition of P.

With this reduction one can solve by computing the nearest neighbor distance between point-patterns A and B also the 3-PARTITION decision problem. To see this consider the following cases.

---

[1] One might notice the connection between the interpolation distance and the image warping problem. In image warping, we have an image that should be warped onto another surface. For some of the pixels we know where they should be mapped. We would like to interpolate these mapping vectors so that the whole image is continuously transformed onto the other surface. This kind of interpolation can be done using $f_\text{I}(M,p)$, where M represents the set of mapping vectors and p is a pixel whose mapping is to be calculated. The new location of p becomes $p' = p + f_\text{I}(M,p)$. Hence finding the interpolation distance can be seen as an inverse problem for the image warping. Such warping methods have been proposed in the literature, cf. [ ] for a survey. However, a common problem in these methods is overfitting; although smooth and continuous, the warp looks too variable for eye. Usually low-degree polynomials are incorporated to make the warp look more natural.

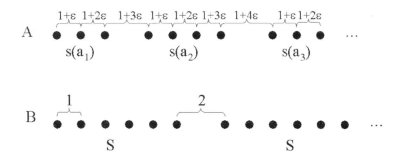

Fig. 1. Reduction from the 3-PARTITION to the nearest neighbor distance

1. Assume that there is a partition for the 3-PARTITION instance. Then there is a matching M_{yes} between A and B such that in each group of A all the points are mapped in the same order onto points in the same group of B; This matching is perfect and its cost is

$$w_{f_{\text{NN}}}(A, B, M_{\text{yes}}) = \sum_{a_i \in P} \left(2\epsilon + \sum_{j=2}^{s(a_i)-1} j\epsilon \right) \qquad (4)$$

   which has an upper bound $\sum_{a_i \in P}(2\epsilon + (\frac{S}{2})^2\epsilon) = 3p(2\epsilon + (\frac{S}{2})^2\epsilon)$.
2. Assume that there is no partition. Now we know that there is no perfect matching between A and B that would map all the points in each group of A into the same group of B; there must be a point $a \in A$ whose nearest neighbor a' is mapped in a perfect matching into a different group of B. This gives us the lower bound $1 - \frac{S}{2}\epsilon$ for the distance $d_{f_{\text{NN}}}(A, B)$, since a may be mapped onto the end point of one group of B and a' onto the beginning point of the next group. On the other hand, there are many-to-one matchings M that do not break the groups but these cannot define a partition for the 3-PARTITION problem. A proper many-to-one matching M must have at least one more edge than a perfect matching. Thus the distance $d_{f_{\text{NN}}}^{\text{many}}(A, B)$ is at least $w_{f_{NN}}(A, B, M_{\text{yes}}) + \epsilon$. Noting that a perfect matching is a special case of many-to-one matching, we altogether obtain the following lower bound: $d_{f_{\text{NN}}}^{\text{many}}(A, B) \geq \min(1 - \frac{S}{2}\epsilon, w_{f_{NN}}(A, B, M_{\text{yes}}) + \epsilon)$.

   By choosing small enough rational ϵ, the above lower bounds become larger than the distance ( ). Hence we can solve the decision problem 3-PARTITION by computing the nearest neighbor distance between A and B (either $d_{f_{\text{NN}}}$ or $d_{f_{\text{NN}}}^{\text{many}}$) and testing the value against ( ). The NP-hardness follows as A and B can be constructed in polynomial time from the instance of 3-PARTITION.

   The above reduction can be done in one–dimensional space as patterns A and B can be put on the same line. Hence the theorem is valid in Euclidean spaces of any dimension.   □

Theorem 4 *It is NP-hard to compute the K-nearest neighbors distance* $d_{f_{\text{KNN}}}$*.*

Proof. We construct point-patterns A and B from an instance of 3-PARTITION almost like in the previous proof, but we double S and all the values $s(a_i)$ until $s(a_i) > K$ for all $a_i \in P$. This guarantees that the K nearest neighbors will be found inside the same group in A. Note that the status of the 3-PARTITION decision problem remains the same in this transformation. Thus, we can safely assume that we have been given an instance in which $s(a_i) > K$ for all $a_i \in P$. The reduction is shown in Figure . The distances $x_1, x_2, \ldots$ between the groups in B will be defined below.

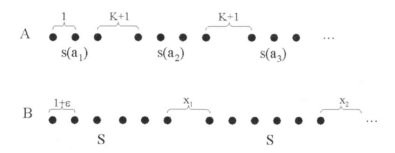

Fig. 2. Reduction from the 3-PARTITION to the K-nearest neighbors distance

Now, if there is a partition for the 3-PARTITION instance, then there is a perfect one-to-one matching M_{yes} between A and B with cost $w_{f_{\mathrm{KNN}}}(A, B, M_{\mathrm{yes}}) \leq pS\epsilon$. This is because the average vector over the K nearest neighbors for each point in A maps the point to at most distance ϵ from a point in B.

If there is no partition, then there must be at least one broken group in A whose points are mapped into different groups in B in any perfect one-to-one matching M_{no}. We will obtain a lower bound for $w_{f_{\mathrm{KNN}}}(A, B, M_{\mathrm{no}})$ by calculating the cost of mapping only one point. Here one must take the following fact into account: Since we are taking the average over the vectors of the K nearest neighbors, a point could be mapped in the "right" place even inside a broken group due to symmetry. To prevent the possibility for symmetry, we define the distances between groups in B as follows: $x_i = 2(K - \frac{1}{2})((\sum_{j=1}^{i-1} x_j) + i(S-1)(1+\epsilon))$ for $1 \leq i < p$.

Let a be a point of a broken group in A such that at least one of the K neighbors of a is mapped into a different group. We can now calculate the lower bound for $w_{f_{\mathrm{KNN}}}(A, B, M_{\mathrm{no}})$ by a case analysis. There are three cases: (1) All the neighbors of a that are mapped into a different group than a, are mapped to the left from where a is mapped. (2) All the neighbors of a that are mapped into a different group than a, are mapped to the right from where a is mapped. (3) The point a has neighbors such that some of them are mapped to the left and some to the right from the group which a is mapped into. A lower bound for the cost in case (1) is easy to achieve; clearly the smallest cost is achieved when a is mapped to the first element of the second group in B, one of the K neighbors is mapped to last element of the first group in B, and all the other $K-1$ neighbors are mapped to last element of the second group in B. A straightforward calculation gives that the cost of matching a is at least $(S-1)(1+\epsilon)$. The lower bound for the case (2) can be derived similarly, but evidently it is greater than $(S-1)(1+\epsilon)$. The case (3) goes also similarly; the smallest cost is achieved when a is mapped to the last element of some group in B, say B_j, one of the K neighbors is mapped to first element of the next group, B_{j+1}, and all the other $K-1$ neighbors are mapped to first element of the first group in B, B_1. The reduction is constructed such that not depending which group a is mapped into, the lower bound cost will be the same as before, $(S-1)(1+\epsilon)$. As M_{no} is arbitrary, it follows that $d_{f_{\text{KNN}}}(A,B) \geq (S-1)(1+\epsilon)$.

Now it is clear that choosing small enough rational value ϵ such as $\epsilon = \frac{1}{pS}$, we can solve the decision problem 3-PARTITION by computing the K-nearest neighbors distance $d_{f_{\text{KNN}}}$. Again, A and B can be put on the same line, hence the result is true in all Euclidean spaces.   □

Corollary 5 *It is NP-hard to compute the K-nearest neighbors distance $d_{f_{\text{KNN}}}^{\text{many}}$.*

Proof. The upper bound for the positive instances of the 3-PARTITION as well as the cases (1), (2), and (3) for the negative instances given in the proof of Theorem   hold also for $d_{f_{\text{KNN}}}^{\text{many}}$. We still show that in the negative case there can not be many-to-one matchings that do not break the groups and whose cost would be less than $pS\epsilon$.

As $|A| = |B|$, any perfect matching that is properly many-to-one must have at least one point of A, say a, that is mapped into two points in B, say b' and b''. The cost of these edges can be expressed as $|b' - (a + X')| + |b'' - (a + X'')|$, where X' and X'' are the average vectors that are calculated using the K neighbors of a. The average vectors depend only on a, not on its pairs in B, and thus $X' = X''$. Hence the cost of the two edges is $|b' - (a+X)| + |b'' - (a+X)| \geq |b-b'| \geq 1+\epsilon > 1$. This is also a lower bound for the total cost of any many-to-one matching. Thus, by choosing $\epsilon = \frac{1}{pS}$ we have $d_{f_{\text{KNN}}}^{\text{many}}(A,B) < 1$ for positive instances of the 3-PARTITION and $d_{f_{\text{KNN}}}^{\text{many}}(A,B) > 1$ for negative ones.   □

Theorem 6 *It is NP-hard to compute the interpolation distance $d_{f_{\text{I}}}$, when the constant r of the function f_{I} is > 1.*

Proof. Again, we use a reduction from the 3-PARTITION, illustrated in Figure   . The distances between the groups of A (values Y in the figure) and of B (values $x_1, x_2, \ldots$) are defined as follows: $x_i = 3S^2((\sum_{j=1}^{i-1} x_j) + i(S-1)(1+\epsilon))$ for $1 \leq i < p$, and $Y = x_{p+1}$. This guarantees that distance Y is longer than the diameter of B.

As before, we want to show that a partition for a positive 3-PARTITION instance corresponds to a one-to-one matching M_{yes} between A and B whose cost is minimal over all one-to-one matchings. We can estimate the cost $w_{f_{\text{I}}}(A, B, M_{\text{yes}})$ as follows. In M_{yes} all the points in the same group of A are mapped into the

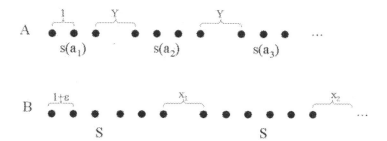

Fig. 3. Reduction from the 3-PARTITION to the interpolation distance

same group of B. The contribution $w((a,b))$ of one particular edge (a,b) of M_{yes} to the cost can be estimated upwards by assuming that the edge has at least one supporting edge in the same group (an edge (a',b') is called a supporting edge for (a,b) if $|(b-a)-(b'-a')| \leq \epsilon$ where ϵ is a parameter to be fixed below), and the points in the other groups try to take a as far away from b as possible. The goal is to select the parameter ϵ such that the effect of one supporting edge dominates over the total effect of the edges starting from the points in the other groups. Surely an upper bound for the worst effect is achieved when we assume that all the other points lie as near a as possible, i.e. at distance Y from a. Since we can place B after A on a same line, the points in the other groups can support a matching vector that would take a at most to distance $diam(A) + diam(B)$ from b, where $diam(S)$ means the diameter of point-pattern S. Now, we can estimate the cost contribution of the edge (a,b) in M_{yes} using ( ) and the fact that the points lie on a line:

$$w((a,b)) = \left| b - \left(a + \frac{\sum_{(a',b')\in M, a'\neq a} \frac{b'-a'}{(|a-a'|)^r}}{\sum_{(a',b')\in M, a'\neq a} \frac{1}{(|a-a'|)^r}} \right) \right|$$

$$< \left| b - \left(a + \frac{\frac{b-a-\epsilon}{1^r} + \frac{pS(b-a+diam(A)+diam(B))}{Y^r}}{\frac{1}{1^r} + \frac{pS}{Y^r}} \right) \right|$$

$$< \left| b - \left(a + \frac{\frac{b-a-\epsilon}{1} + \frac{pS(b-a+pS+pY)}{Y^r}}{1 + \frac{pS}{Y^r}} \right) \right|$$

$$< \left| b - \left(a + \frac{\frac{b-a-\epsilon}{1} + \frac{2pS(pS+pY)}{Y^r}}{1 + \frac{pS}{Y^r}} \right) \right|$$

$$\leq \left| b - a - \frac{b-a}{1 + \frac{pS}{Y^r}} \right| + \left| \frac{\epsilon}{1 + \frac{pS}{Y^r}} \right| + \left| \frac{2pS(pS+pY)}{Y^r(1 + \frac{pS}{Y^r})} \right|$$

$$< \left| b - a - \frac{b-a}{1 + \frac{pS}{Y^r}} \right| + \epsilon + \left| \frac{2pS(pS+pY)}{Y^r} \right|$$

$$< \epsilon + \epsilon + \epsilon,$$

for all $\epsilon > 0$, if we redefine Y such that $\frac{b-a}{1+\frac{pS}{Y^r}} > b - a - \epsilon$ and $\frac{2pS(pS+pY)}{Y^r} < \epsilon$.
It follows that the latter condition is stronger, so it is enough to choose Y such that $Y^r > 2pS(pS + pY)/\epsilon$. This is possible if and only if $r > 1$. The resulting upper bound is $d_{f_1}(A, B) < 3pS\epsilon$.

We have still to estimate the cost when there is no partition for the 3-PARTITION instance. We have constructed the point-patterns such that if there is a matching that breaks the groups in A, its cost is lower bounded by $(S-1)(1+\epsilon)$. This can be seen similarly as in the proof of Theorem .

Now it is clear that by choosing a small enough rational ϵ (e.g $\epsilon = \frac{1}{3pS}$), we can solve the 3-PARTITION by computing the interpolation distance d_{f_1}. Again, the result holds true in all Euclidean spaces.   □

Using similar arguments as in the proof of Corollary we also obtain the following.

Corollary 7 *It is NP-hard to compute the interpolation distance $d_{f_1}^{many}$ when $r > 1$.*   □

The inter-point distances in the reductions remain polynomially bounded with respect to the input values. Thus all the problems discussed above are strongly NP-hard.

As the upper bound for the distance in the positive instances of 3-PARTITION approaches 0 as $\epsilon \to 0$ while the lower bound in the negative case stays away from 0, it follows that also approximating the distances within a constant factor becomes NP–hard. However, this argument does not hold for the nearest neighbor distance $d_{f_{NN}}^{many}$. In this case our bound for the negative instances vanishes with ϵ.

Corollary 8 *It is NP-hard to approximate $d_{f_{NN}}$, $d_{f_{KNN}}$, $d_{f_{KNN}}^{many}$, d_{f_1}, and $d_{f_1}^{many}$ (the two last with $r > 1$) within any constant factor $c > 1$, i.e to give an answer that is $\le c \times OPT$, where OPT is the optimal solution for the corresponding minimization problem, and c is polynomial time computable on the input.*   □

4.1   Allowing Mismatches

We have considered so far only distances based on perfect matchings that cover A and B in full. In typical applications, however, one wants to leave some points out of the match if this makes it possible to find a very good one-to-one matching between some subsets of A and B. We call the outlier points the *mismatches* of the matching.

We will briefly consider two different models of allowing mismatches:

1. Fix a maximum allowed number κ for the mismatches in advance;
2. Assign some cost c for the mismatches and find a matching that minimizes the total cost consisting of the cost of the matched elements as well as of the mismatches.

Let us denote the set of all (not necessarily perfect) one-to-one matchings between A and B as $\mathcal{M}'$. The *mismatches induced by a matching* $M \in \mathcal{M}'$ consist of the elements of A and B that are not adjacent to any edge of M. We denote by $m(M)$ the total number of such elements.

There is no obvious way of defining the cost function c in the mismatch model (2). Therefore we just assume that there is available a generic cost function that gives the cost $c(A, B, M)$ of the mismatches induced by M.

Definition 9 *Let* $M = \{(a_1, b_1), \ldots, (a_m, b_m)\} \in \mathcal{M}'$ *be a (not necessarily perfect) one-to-one matching between* A *and* B, *and let the* cost *of* M *with respect to* f *be*

$$w_f(A, B, M) = \sum_{k=1}^{|M|} |b_k - f(M, a_k)|.$$

Let $\kappa \geq 0$ *be an integer. The* local similarity distance with at most κ mismatches d_f^κ *between* A *and* B *is defined as*

$$d_f^\kappa(A, B) = \min\{w_f(A, B, M) \mid M \in \mathcal{M}', m(M) \leq \kappa\},$$

and the local similarity distance with mismatch cost function c *between* A *and* B *as*

$$d_f^c(A, B) = \min\{w_f(A, B, M) + c(A, B, M) \mid M \in \mathcal{M}'\}.$$

The complexity results of Section   hold also for the above versions of the distances allowing mismatches. For an example, evaluating $d_{f_{NN}}^c$ is NP-hard since we can assign for each point a mismatch cost that is larger than the cost of M_{yes} in Theorem . Thus, if there is a partition for the 3-PARTITION instance, the optimal solution for $d_{f_{NN}}^c$ will be the same as for $d_{f_{NN}}$. If the function c depends on the problem instance, this result naturally does not hold, and a separate proof seems possible, depending on the details of c.

For $d_{f_{NN}}^\kappa$ we have a stronger result that holds for each fixed κ. We can add κ extra points in the reduction such that if any of them are matched, the cost will be larger than the cost of M_{yes}. Thus, if there is a partition for the 3-PARTITION instance, the optimal solution for $d_{f_{NN}}^\kappa$ will be the same as for $d_{f_{NN}}$.

5   Relaxed Problem and a Useful Heuristic

Our local similarity distances turned out NP–hard to evaluate. Therefore we will next look at relaxed formulations. In practice it might be possible to approximately evaluate the distance using an iterative approach. Such a method needs a good matching to start with. It turns out that sometimes we can ignore the local condition at first.

We define still another function $f = f_{id}$ as $f_{id}(M, a) = a$. This gives distance function $d_{f_{id}}$ which as compared to our earlier distance functions does not get anything for free but has to measure the translations of the elements of A in full.

Distance $d_{f_{id}}$ and its different mismatch variants can be evaluated fast using a reduction to a well–known graph matching problem, to be described next.

5.1   Reductions from $d_{f_{\mathrm{id}}}$ and Its Variants to Minimum Weight Perfect Matching

Notice that the distance $d_{f_{\mathrm{id}}}$ corresponds to a known graph problem; it is an Euclidean instance of *minimum weight perfect matching* in a bipartite graph, when identifying points with nodes and pairs with edges. Moreover, $d_{f_{\mathrm{id}}}^{\mathrm{many}}$ corresponds to *minimum weight edge cover* in a bipartite graph. Also, Eiter and Mannila [ ] have proposed a distance measure for point sets called *link distance*, which equals to $d_{f_{\mathrm{id}}}^{\mathrm{many}}$. They give in [ ] an efficient algorithm for computing $d_{f_{\mathrm{id}}}^{\mathrm{many}}$ since they show a linear reduction from minimum weight edge cover to minimum weight perfect matching.

The reduction used for $d_{f_{\mathrm{id}}}^{\mathrm{many}}$ can be modified to cover also $d_{f_{\mathrm{id}}}^{c}$ in a special case. In this case the cost function c has to be additive in the sense that $c(A, B, M)$ has a representation as a sum of individual mismatch costs. That is, each element $a \in A \cup B$ has a mismatch cost $c(a)$, which can depend on A and B but not on M, and $c(A, B, M) = \sum_{a \in \bar{M}} c(a)$ where $\bar{M}$ is the set of the mismatches induced by M.

We only sketch the modifications of the reduction that are needed in this case. On the top of page 122 in [ ], redefine the edge costs for edges of type $e = \{a_i, a_i'\}$ (resp. $e = \{b_j, b_j'\}$) as $w''(e) = c(a_i)$ ($w''(e) = c(b_j)$), where $c(a_i)$ ($c(b_j)$) is the mismatch cost assigned to point $a_i \in A$ ($b_j \in B$). The optimal perfect matching $M^{G''}$ in this new reduction graph G'' corresponds to an optimal solution M^c of $d_{f_{\mathrm{id}}}^{c}$ as follows: If $(a_i, b_j) \in M^{G''}$ then $(a_i, b_j) \in M^c$, if $(a_i, a_i') \in M^{G''}$ then a_i is a mismatch, and if $(b_j, b_j') \in M^{G''}$ then b_j is a mismatch.

For the distance d_f^{κ} a slightly different reduction is needed. Note first that we can assume that $\kappa - |m - n|$ is divisible by 2 (after removing κ mismatches we must be able to produce a perfect matching). The reduction is the following: Add $|m - n|$ points into the largest point-pattern, and then $(\kappa - |m - n|)/2$ points into both point-patterns. Denote this set of "mismatch sinks" by $S = S_A \cup S_B$, where S_A is the set of points added to B and S_B is the set of points added to A. Assign 0 cost for each pair of points of type (a, s_a) or (s_b, b), $a \in A, s_a \in S_A, b \in B, s_b \in S_b$. For all other pairs the cost is the Euclidean distance between the points. It is easy to see that by computing the perfect matching M^S between $A \cup S_B$ and $B \cup S_A$ one gets an optimal solution M^{κ} for $d_{f_{\mathrm{id}}}^{\kappa}$ as follows:

If $(a, b) \in M^S, a \in A, b \in B$ then include (a, b) into M^{κ}. The other κ pairs contain the mismatches.

Proposition 10 *The distances $d_{f_{\mathrm{id}}}, d_{f_{\mathrm{id}}}^{\kappa}$, $d_{f_{\mathrm{id}}}^{c}$ (with an additive c), and $d_{f_{\mathrm{id}}}^{\mathrm{many}}$ on point-patterns A and B of sizes m and n, respectively, can be computed asymptotically as fast as minimum weight perfect matching in a bipartite graph of $m + n$ nodes and mn edges.*

5.2   Heuristic Solution

As such, $d_{f_{\mathrm{id}}}$ does not measure local similarity, but if we replace the Euclidean distances between points with a local similarity measure between the neighborhoods of the points, we obtain a polynomial time solvable problem that takes local similarity into account.

Consider first that there are no mismatches. Assume that the data are such that when restricted to a small region in the data (e.g. a region with K nearest neighbors for a point), the mapping that is produced by e.g. optimal solution for the K-Nearest neighbor distance $d_{f_{KNN}}$, is close to translation. With this kind of data we can estimate the cost of matching an edge (a, b) by comparing the neighborhoods of a and b. Thus, we can assign a cost for each edge (a, b), $a \in A$, $b \in B$. With these costs, we can calculate the minimum weight perfect matching between A and B.

The cost for an edge (a, b) can be calculated as follows. Take K nearest neighbors for a in A, translate them with vector $b - a$, and match each translated point with its nearest neighbor in B. The result is a matching M^{ab} of size $K + 1$ (pair (a, b) is included). Assign cost $w((a, b)) = |b - f_{KNN}(M^{ab}, a)|$.

We can express this matching problem by slightly abusing our framework by defining

$$f_H(M, a) = -b + w((a, b)).$$

Now the corresponding distance $d_{f_H}(A, B)$ equals to the minimum weight perfect matching between A and B.

Note that when the minimum weight perfect matching M is calculated with these costs, the matching is not guaranteed to include any of M^{ab}; it may happen that a point is matched far away from where its neighbors are matched.

A heuristic solution for preventing the deviation between the matches of neighboring points in M would be the following. When calculating the cost for an edge (a, b), we combine the similarity of the neighborhoods with the distance between a and b. This should work in situations where a and b can be assumed to be relatively close to each others.

Adding the sensibility for mismatches in this model is non-trivial. First, the mismatches affect the calculation of the costs for the possible edges (a, b). To fix this, one could take the "best" portion of the K nearest neighbors that support the edge (a, b). The second problem is how to separate the mismatches from the matches. This can be done either by using $d_{f_H}^{\kappa}$ or $d_{f_H}^{many}$. They both can be solved in polynomial time using the reductions given in Sect. refsec:idreductions for $d_{f_{id}}$ (the reductions extend to the cases where non-negative costs are used as distances between points). In $d_{f_H}^{many}$, one can choose the minimum cost edge from each many-to-one subset of the matching to be a match, and the other points can be regarded as mismatches.

5.3   Experimental Results

We have implemented a heuristic using the ideas presented above and done some experiments on two-dimensional protein electrophoresis gels (see Figure   for an example).

The goal is to match two gel images so that the corresponding protein spots will be matched. Our abstraction of this problem is to identify the spots with

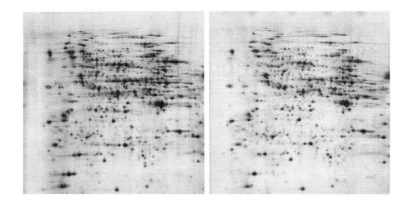

Fig. 4. Two-dimensional protein electrophoresis gel images

their midpoints and match these sets of midpoints. So the input for our method is two point-patterns, that both contain a lot of missing points and the transformation between the point-patterns is non-uniform.

To test our method, we had 15 pairs of point-patterns extracted from different gel images (about 1000 points in each point-pattern). We also had the "original" matchings for these point-patterns to compare, that were constructed with PDQuest^{TM} software, where the user can give landmarks (trusted matches) and the program tries to manage from the rest.

For computing the matchings, we used the implementation of Goldberg and Kennedy [ ] for minimum weight perfect matching in a bipartite graph. Due to the linear time reductions introduced before, also the versions where mismatches were allowed could be computed efficiently (typically in couple of seconds on a 700MHz Pentium III computer).

Because there were two kind of errors between these point-patterns, missing points and non-uniform motion, we decided to test first the sensitivity for pure non-uniformity. For this we removed all the points that were not matched in the original matching. We compared the original matching with the matching produced by our method. With most of the point-pattern pairs the result was fully consistent with the original matching and the largest error was 0.4% (meaning that 0.4% of the original matches were not detected).

Next we tested the real life situation, where both missing points (10-25%) and non-uniform motion were present. We used two mismatch models, $d_{f_H}^{\kappa}$ and $d_{f_H}^{\text{many}}$. In the first one, we needed some estimate for the number of mismatches κ. To test the upper limits of the method, we used the correct value for κ that was calculated from the original matching. The results are shown in Figure . Surprisingly, $d_{f_H}^{\text{many}}$ managed better than $d_{f_H}^{\kappa}$ although we knew the number of mismatches beforehand. However, $d_{f_H}^{\text{many}}$ naturally contained more false negatives than $d_{f_H}^{K}$ since also the mismatching points were matched. Altogether the results were quite good, on the average 93.7% of the original matches were found by $d_{f_H}^{\text{many}}$.

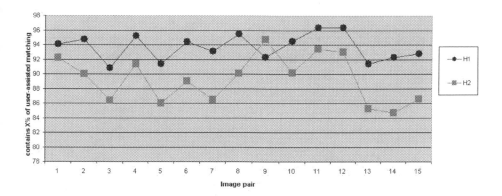

Fig. 5. Comparing the results of H1=$d_{f_H}^{many}$ and H2=$d_{f_H}^{\kappa}$ against the original user-assisted matching

We tried to improve the heuristic with basically the following iterative step: After a matching M is achieved (with $d_{f_H}^{\kappa}$ or $d_{f_H}^{many}$), assign a cost for each pair based on the Interpolation distance $(w((a,b)) = |b-(a+f_I(M,a))|)$. Remove $X\%$ of the pairs with the highest cost from M. Move the points in A based on the matching M using the Interpolation distance $(A' = \{a + f_I(M,a)|a \in A\}, A = A')$. Either recalculate matching and start the iteration again, or match each unmatched point in A to the nearest unmatched point in B if it is not more than Y times farther than the nearest matched point in B.

It turned out that the results did not change much after the first iteration (we used $X = 10\%$ and $Y = 2$). The results were not significantly better than with the basic approach, now the average was 94.9% when one iteration was used. However, the one iteration step was necessary for getting a useful matching, since the solution of $d_{f_H}^{many}$ contained a lot of false positives, which were now removed. An example of a matching produced by this iterative method is shown in Figure   .

As we can calculate the Interpolation distance for a given matching, we can now compare the quality of the matching proposed by our method with respect to the original matching. This comparison is shown in Figure   . As can be seen, the user-assisted matching has smaller cost in each image pair. However the difference is quite small, so the heuristic solution is quite close to the optimum (assuming that the optimum of d_{f_I} is close to the user-assisted matching).

6   Conclusions and Future Work

There are several possibilities for future studies, still. We have only considered a few example definitions for function f in our framework; other definitions could lead to more practical distance measures. Extremely interesting is the tradeoff

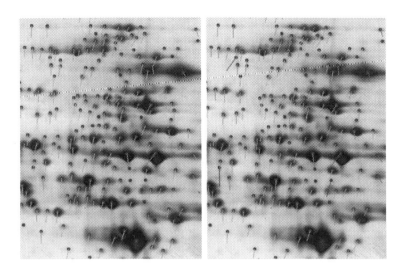

Fig. 6. Visualization of matchings. The original user-assisted matching is in the left, and the result of automatic matching is in the right. Both are portions of the matchings for the images in Fig. . As can be seen, the matching "flow" is relatively constant when restricted to a small neighborhood. This is essential for the heuristic to work. Experiments with gels with sparser selection of detected spots and greater non-uniformity have confirmed this fact

problem: How good local similarity function f one can use such that the problem can still be solved in polynomial time.

Another direction is to represent the problem as a *Metric Labeling Problem* [ ], where one seeks for a labeling of objects such that the labeling cost plus the difference between labelings of neighboring objects is simultaneously minimized. Here one could identify one point-pattern with objects and the other as labels. The labeling cost could be the same as what we used in the heuristic in Sect.   ; the prior estimate of local similarity between an object and a label. The difference between labelings of neighboring objects could simply be measured as the distance between labels (points in the other point-pattern). In addition, one would need to make the definitions symmetric (i.e. all labels must be used as well as all objects must be assigned a label). There are approximation algorithms for some metric labeling problems (e.g. [ ]) and even polynomial time algorithms for some special cases (e.g. [ ]). On the other hand, the requirement of symmetric definitions seems to make the problem harder; it is a special case of *Quadratic Assignment Problem*, which in general is NP-hard to approximate even in the metric case [   ].

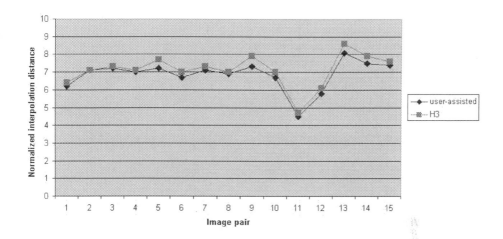

Fig. 7. Comparing the solution of the iterative heuristic (H3) to the solution of the user-assisted matching. The Normalized interpolation distance is d_{f_I}/m. Intuitively the value means how many units (e.g. pixels) on average a point must be moved once the effect of the overall matching is taken into account

Acknowledgements

We would like to thank Laura Salusjärvi from VTT Biotechnology for providing the data for the experiments.

References

1. T. Akutsu, K. Kanaya, A. Ohyama, and A. Fujiyama. Matching of Spots in 2D Electrophoresis Images. Point Matching Under Non-uniform Distortions. In *Proc. 10th Annual Symposium on Combinatorial Pattern Matching (CPM'99)*, LNCS 1645, pp. 212–222, 1999.
2. H. Alt and L. Guibas. Discrete Geometric Shapes: Matching, Interpolation, and Approximation, In J.-R. Sack, J. Urrutia, editors, *Handbook of Computational Geometry*, pp. 121–153. Elsevier Science Publishers B. V. North-Holland, Amsterdam, 1999.
3. R. D. Appel, J. R. Vargas, P. M. Palagi, D. Walther, and D. F. Hochstrasser. Melanie II - A Third Generation Software Package for Analysis of Two-dimensional Electrophoresis Images: II. Algorithms. *J. Electrophoresis* 18, pp. 2735–2748, 1997.
4. T. Eiter and H. Mannila. Distance Measures for Point Sets and Their Computation. *Acta Informatica*, 34(2):109–133, 1997.
5. M. R. Garey, D. S. Johnson. *Computers and Intractability: A guide to the theory of NP-completeness*. W. H. Freeman and Company, San Francisco, 1979.
6. A. V. Goldberg and R. Kennedy. An Efficient Cost Scaling Algorithm for the Assignment Problem. *Mathematical Programming*, Vol. 71, pp. 153–178, 1995.

7. D. S. Hochbaum. An Efficient Algorithm for Image Segmentation, Markov Random Fields and Related Problems. *Journal of the ACM*, Vol. 48, No. 4, pp. 686–701, 2001.
8. S. Jokisch and H. Müller. Inter-Point-Distance-Dependent Approximate Point Set Matching. *Research Report* No. 653, University of Dortmund, Department of Computer Science, July 1997.   ,   ,
9. J. Kleinberg and E. Tardos. Approximation Algorithms for Classification Problems with Pairwise Relationships: Metric Labeling and Markov Random Fields. In *Proc. 40th Annual IEEE Symposium on the Foundations of Computer Science (FOCS'99)*, pp. 14–23, 1999.
10. V. Mäkinen. Using Edit Distance in Point-Pattern Matching. In *Proc. 8th Workshop on String Processing and Information Retrieval (SPIRE 2001)*, IEEE CS Press, pp. 153–161, 2001.
11. M. Queyranne. Performance ratio of polynomial heuristics for triangle inequality quadratic assignment problems. *Oper. Res. Lett.*, 4, pp. 232–234, 1986.
12. D. Ruprecht and H. Müller. Image Warping with Scattered Data Interpolation. *IEEE Computer Graphics and Applications*, Vol. 3, pp. 37–43, 1995.

Identifying Occurrences of Maximal Pairs in Multiple Strings

Costas S. Iliopoulos[1] *, Christos Makris[2] **, Spiros Sioutas[2],
Athanasios Tsakalidis[3] ***, and Kostas Tsichlas[2]

[1] Department of Computer Science, King's College London
London WC2R 2LS, England
csi@dcs.kcl.ac.uk

[2] Computer Engineering & Informatics Dept. of Univesity of Patras
and Research Academic Computer Tecnology Institute (RACTI)
Rio, Greece, P.O. BOX 26500
{makri,sioutas,tsihlas}@ceid.upatras.gr

[3] Computer Engineering & Informatics Dept. of Univesity of Patras
and Research Academic Computer Tecnology Institute (RACTI)
Rio, Greece, P.O. BOX 26500
tsak@cti.gr

Abstract. A molecular sequence "model" is a (structured) sequence of
distinct or identical strings separated by gaps; here we design and analyze efficient algorithms for variations of the "Model Matching" and
"Model Identification" problems.

Keywords: String Algorithms, DNA Matching, Search Trees.

1 Introduction

Two important goals in computational molecular biology are finding regularities
in nucleic or protein sequences, and finding features that are common to a set
of such sequences. Both imply inferring patterns, unknown at first, from one or
more strings. In all cases, conservation of a pattern is not strict. Regularities in
a sequence may come under many guises. They may correspond to approximate
repetitions randomly dispersed along the sequence, or to repetitions that occur
in a periodic or approximately periodic fashion, or else to tandem arrays. The
length and number of repeated elements one wishes to be able to identify may
be highly variable. Patterns common to a set of sequences may likewise present
diverse forms. For various problems in molecular biology, in particular the study
of gene expression and regulation, it is important to be able to infer what has
been called "structured patterns". These correspond to an ordered collection
of p "boxes" (always of initially unknown content), p error rates (one for each

---

* Partially supported by Royal Society, Wellcome and NATO grants
** C. Makris was partially supported by a Royal Society grant
*** Partially supported by a Royal Society grant

A. Apostolico and M. Takeda (Eds.): CPM 2002, LNCS 2373, pp. 133–   , 2002.
© Springer-Verlag Berlin Heidelberg 2002

box) and $p - 1$ intervals of distances (one between each pair of successive boxes in the collection). When $p = 1$, we fall back into the classical pattern inference problem. Structured patterns allow to identify conserved elements recognized by different parts of a same protein or macromolecular complex, or by various complexes that then interact with one another.

Current combinatorial approaches on the other hand work with a model of how recognition happens that most certainly does not take into account all the elements that are involved in it. The models adopted by current combinatorial approaches are therefore probably a huge simplification of reality. Even without incorporating information that is not at present time, or may ever be obtainable, such models may be improved. This is worth the effort, because of two main advantages combinatorial approaches present. The first one comes from the fact that the algorithms are, by definition, exact. The second reason why it is worth trying to improve the models used by the combinatorial approaches is that, though limited, they mimic better how recognition really happens.

We will therefore examine next future models worth exploring and the diverse ways in which the models adopted by combinatorial approaches be ameliorated. Here we consider variants of the following *Model matching* problem: we are given a model M and a string S with length $n = |S|$ over an alphabet Σ and we want to find an occurrence of M in S. Considering the applications of these problems we may note that $\Sigma = \{A, C, G, T\}$, which constitute the bases of the DNA strings. The structure of model M is predetermined. In this way, the model consists of a sequence of boxes B_i (B_i are strings over Σ), which are divided by gaps. Each gap g_i must have a minimum min_i and a maximum value max_i. In section   we present $O(kn)$ algorithms for two variants of the "Model matching problem".

We also consider the *Model identification problem*: we are given a set of strings $S = \{S_1, S_2, \ldots, S_k\}, S_i \in \Sigma^*$ and we are asked to find a pair P of strings such that this pair occurs in S_i for each $1 \leq i \leq k$. In this setting, the pair P is a model M with two boxes which are identical and with various restrictions on gaps. Assume that $|S_1| + |S_2| + \cdots + |S_k| = n$. In section   we present two algorithms (with complexities $O(n+\alpha)$ and $O(n \cdot \log^2 n + k \cdot \log n \cdot \alpha)$) for variants of the above problem, where α is the size of the output.

2   A Simplified Model Matching Problem

In this section we present a solution for a simplified version of the Model Matching Problem defined above. In this variant, we assume that all model boxes are identical, i.e., $B = B_i = B_j, \forall i, j \leq k$. Let m be the length of each box and let k be the the number of boxes in M. Furthermore we have the restriction that the minimum and maximum bounds of a gap g_i coincide ($min_i = max_i$); each gap g_i is fixed. By context it will be clear whether by g_i we represent a gap or its length.

Thus, in this problem we are given a string $S \in \Sigma^*$ and a model M, like the one described above, and we are asked to find all occurences of M in S. We say that M occurs in position i of S if the following hold: $S[i, i + m] = B$,

$S[i + m + g_1, i + 2 \cdot m + g_1] = B$ and generally $S[i + j \cdot m + \sum_{l=1}^{j} g_l, i + (j+1) \cdot m + \sum_{l=1}^{j} g_l] = B, 1 \leq j \leq k - 1$

First we need to find all occurrences of the common box B in the sequence S. We store all such occurrences in a list L sorted by the position of the occurrences of the boxes. Note that we can generate gaps between consecutive occurrences of boxes on the fly by a single substraction of their occurrences and their sizes. We will use an inductive method to construct the final solution. The first step consists of the search of the occurrences of the boxes inside S. In the i-th step of the inductive algorithm we would like to extend our sequences of i boxes to sequences of $i + 1$ boxes maintaining the invariant that the newly formed gap between the i-th and the $(i+1)$-th box is equal to g_i. This is realized by sweeping in each inductive step the list L. However, we must store all occurrencs found until the i-th step, in a secondary list L'. Thus, in the $(i + 1)$-th step we sweep the list L', and for each occurrence of length i we try to extend it to length $i+1$. This is easily accomplished by checking for every occurrence $L_{i,j}$ of length i (i consecutive boxes define $i - 1$ consecutive gaps) in list L' whether there is a box B whose gap with the last box of $L_{i,j}$ has length g_i. If the length is g_i we keep the occurrence in the list L', otherwise we remove it.

Since the gaps have strict length it is easy to see that the maximum length of list L' will be $O(n)$. Thus, the space usage of the algorithm is linear. Since for each occurrence in list L' we check in $O(1)$ time (by maintaining pointers to the last box of each occurrence) whether it can be extended or not, we deduce that for each sweeping we need $O(n)$ time. Thus, the time complexity of the algorithm is $O(kn)$ since there will be at most k sweeps. Finally, if we allow for the boxes to be approximate we can easily use the above algorithm with the difference that the string searching algorithm for the occurrences of boxes must be as well approximate (see [ ], [ ] Section 5, [ ] Section 6 and [ ]). The choice of the string searching algorithm depends on the measure of approximation.

A more general version that the one described above is acquired by allowing the length of the gaps to be upper and lower bounded. In this way, the box B is a string of length m over the alphabet $\Sigma = \{A, C, G, T\}$ and min_i, max_i, where $1 \leq i \leq k - 1$, are integers with $min_i \leq max_i \leq n$ that bound the length of gap g_i. Here, we are given a sequence S with length n and a model M with k boxes and we are asked to find all occurrences of M in S. We say that M occurs in position i of S if the following hold: $S[i, i+m] = B$, $S[i+m+g_1, i+2 \cdot m+g_1] = B$ and $min_1 \leq g_1 \leq max_1$ and generally $S[i + j \cdot m + \sum_{l=1}^{j} g_l, i + (j+1) \cdot m + \sum_{l=1}^{j} g_l] = B$ and $min_j \leq g_j \leq max_j, 1 \leq j \leq k - 1$

The approach here is similar to the one described above. We extend the occurrence of i boxes to the occurrence of $i+1$ boxes satisfying the gap invariant. The problem here is that many boxes may satisfy the invariant and so we have many possible solutions. We need to keep track of all these solutions. Assuming that $min\{min_1, min_2, \ldots, min_{k-1}\} = 0$ and that $max\{max_1, max_2, \ldots, max_{k-1}\} = n - c$, where c is a constant, then it is obvious that the size of the list L' will be at most $O(n^2)$. Thus, the space complexity will be $O(n^2)$ while the time complexity will be $\Omega(n^2)$. However, if we assume that the maximum gap has constant length l, then we can guarantee that the size of the list L' will be $O(n)$. Thus, the procedure will be similar to the one described in section . The only difference will be that during the extension of an occurrence from length (with respect to the number of boxes) i to length $i+1$ we must check l positions in the string S. Thus, the space complexity will be $O(l \cdot n)$ while the time complexity will be $O(k \cdot l \cdot n)$. Since we assumed that $l = O(1)$ the space usage will be linear to n while the time complexity will be $O(k \cdot n)$. Finally, as in the previous problem we can also find approximate occurrences of the boxes and based on the respective list L solve the problem exactly as in the previous case.

3   The Model Identification Problem

In this section we are given a set of sequences and we want to identify a "simple" model that occurs at least once in each one of them. We are given a set of strings $S = \{S_1, S_2, \ldots, S_k\}, S_i \in \Sigma^*$ and we are asked to find a pair P of strings such that this pair occurs in S_i for each $1 \le i \le k$. In this setting, the pair P is a model M with two boxes which are identical and with various restrictions on gaps. Assume that $|S_1| + |S_2| + \cdots + |S_k| = n$. To control the output size we restrain the problem to finding maximal pairs of strings. A pair is said to be *left-maximal (right-maximal)* if the characters to the immediate left (right) of the two occurrences of the boxes are different and so we cannot extend the pair. A pair is *maximal* if it is both left-maximal and right-maximal. The *gap* in a pair is defined as the number of characters between the two occurrences of a box in the pair.

In Gusfield ([ ], Section 7.12.3) there is a simple algorithm that finds all maximal pairs in a string S of length n without a restriction on gaps. The basic tool behind his solution was the suffix tree. After contructing a suffix tree for the string S he uses a bottom-up approach (from leaves to roots) reporting in each level of the tree the maximal pairs. This can be accomplished by maintaining a *leaf-list* at each internal node v of the suffix tree that stores all occurrences of suffixes of S ending at a leaf in the subtree T_v of node v. Maximality of pairs is guaranteed by using a different leaf-list for every different symbol preceding a suffix ending at a leaf of T_v. Thus, each node is attached at most σ lists, where $\sigma = |\Sigma|$ is the number of different symbols of the alphabet Σ. The ascension of levels is realized by properly concatenating the leaf-lists of brother nodes into one leaf-list (always with respect to a symbol). The reporting step is realized by the cartesian product of a leaf-list with all the leaf-lists of its brothers that correspond to a different symbol. The time complexity is $O(n + \alpha)$, where α is the size of the output.

Brodal et al. ([ ]), based on this algorithm of Gusfield, devised an algorithm that finds all maximal pairs in an input string S of length n whose gaps are restricted. When the gap belongs in an upper and lower bounded interval then their algorithm works in $O(n \log n + \alpha)$. They also prove that if the upper bound on the gaps is removed then the maximal pairs are found in $O(n + \alpha)$. They basically use the algorithm of Gusfield but the leaf-lists are implemented by

AVL-trees (thus the occurrences are sorted) so that one can search for the start of the output in the list in logarithmic time. However, this mechanism works only for right-maximal pairs. In order to find only maximal pairs an additional AVL-tree is used to store intervals of occurrences in the leaf-lists that are preceded by the same character. Thus, they do not need to traverse occurrences that do not belong to the output.

The problem we consider here is a generalization of the previous problems into a set of strings S. We will consider two variations of this problem depending on the restrictions on the gaps. In the first version we assume that there is no restriction on the gaps. Thus, a pair in S_i may have gap g_i while the same pair (with respect to the boxes) in string S_j $(i \neq j)$ has gap g_j. In the second version we demand that the gap in a pair be the same in all strings, that is $g_j = g_i, \forall i, j \leq n$. However, the gap must be upper bounded (and lower bounded if we wish) by a constant. We extend the algorithm of the first version to acquire the algorithm for the second version. The main difficulty in this problem is that it is difficult to define the output complexity of the problem whereas in [ ] it was trivial and in [ ] it was pretty easy. In subsection   we describe the algorithm for the first version of the Model Identification problem while in subsection we describe the algorithm for the second version.

3.1   Maximal Pairs with Arbitrary Gaps

We use an approach similar to that of Gusfield. First we build a *generalized suffix tree* for the set of strings $S = \{S_1, S_2, \ldots, S_k\}, k \geq 2$ ([ ], Section 6.4). A Generalized Suffix Tree (GST) for a set of strings contains all suffixes of all strings in set S. The leaves of the GST contain the position in the string where the suffix occurs and in addition an identifier of the string in which the suffix belongs. Thus, the pair (i, j) stored in a leaf of a GST corresponds to the occurrence at position j of a suffix of a string S_i. Note that a leaf may store multiple pairs that correspond to same suffixes in different strings. The GST is a tree with out-degree of internal nodes at least 2 and at most $\sigma = |\Sigma|$. However, the algorithm that follows makes the silent assumption that the suffix tree is a binary tree. In order to binarize the GST we replace each node v with out-degree $|v| > 2$ by a binary tree with $|v|$ leaves and $|v| - 1$ internal nodes with $|v| - 2$ edges. Each edge is labelled with the empty string ϵ so that all new nodes have the same path-label as node v that they replace. The size of the initial GST was $O(n)$. The size of the binary GST is also $O(n)$ since for each edge in the initial GST we add two nodes in the binary GST. We assume that the size of the alphabet σ is $O(1)$.

As in Gusfield, we follow a bottom-up approach. Each internal node v is attached a set of leaf-lists. There is a leaf-list for every string in set S and each leaf-list stores all occurrences of suffixes of the respective strings that occur in a leaf in the subtree T_v. The leaf-list $L_{v,i}$ corresponds to occurrences of suffixes of S_i in the subtree T_v. Each leaf-list consists of sublists $L_{v,i}^{\sigma_j}$ that stores occurrences of suffixes of string S_i in T_v such that they are preceded by symbol σ_j in string S_i. The leaf-lists of the leaves consist of the set of suffixes they represent. First, we describe the bottom-up step and then the reporting step.

Assume that we currently check nodes v and w whose father is z. After the reporting step we need to construct the leaf-lists of their father z. This update step is accomplished by concatenation of the leaf-lists of v and w. Each sublist $L_{v,i}^{\sigma_j}$ is concatenated with the sublist $L_{w,i}^{\sigma_j}$ and this concatenation results in $L_{z,i}^{\sigma_j}$. The concatenation is a simple list operation that can be realized in $O(1)$ time since we need not worry about the order of the occurrences of the suffixes in these sublists. Since the binary GST contains $O(n)$ nodes, there are at most k leaf-lists and each leaf-list has at most $\sigma = O(1)$ sublists we conclude that for the whole binary GST this procedure is realized in $O(n)$ time (it is easy to see that in the worst-case for this setting $S_i = n/k$).

Before the construction of the leaf-lists of the father z we first have to perform the reporting procedure for all maximal pairs between nodes v and w. Note that the cartesian product between sublists of a leaf-list is not necessarily part of the input and thus we need to be very cautious with this procedure. Assume that we want to generate all candidate pairs from leaf-lists $L_{v,i}$ and $L_{w,i}$. To accomplish this we compute the cartesian product of sublist $L_{v,i}^{\sigma_j}$ with sublists $L_{w,i}^{\sigma_l}, 1 \leq l \leq \sigma$ and $l \neq j$ and then combine with all other cartesian products of all other leaf-lists. If there is a leaf-list for which all possible cartesian products of its sublists are empty then no output can be generated from these two nodes.

In order to obtain a good output-sensitive time complexity we store an auxiliary structure E_v with size $O(k)$. By $E_{v,i}$ we refer to the i-th entry that corresponds to the leaf-list of S_i. We would like this structure to answer in $O(1)$ time whether there is a leaf-list (note that there are k leaf-lists) whose all possible cartesian-products are empty. This is easy to find by traversing all sublists of the leaf-lists $L_{v,i}$ and $L_{w,i}$ and finding all of them in one of the two leaf-lists empty. If this is the case then we store in $E_{v,i}$, if $L_{v,i}$ is empty, the value zero, otherwise we store value one. By implementing E_v as a linear sorted list we can in $O(k)$ time find whether there is an empty leaf-list. In this case, no reporting is performed since there are not maximal pairs generated by nodes v and w that occur in all strings of the set S. Otherwise, we perform the reporting step. The maintenance of structures E_v is performed during the update step by spending $O(k)$ time. However, note that during this step we can store a flag that indicates whether there is an empty list and as a result we get the $O(1)$ bound. Since there are $O(n)$ nodes in the binary GST we conclude that $O(n)$ time is sufficient. The previous algorithm is summarized in the following theorem.

Theorem 1. *The above algorithm for the problem of computing all maximal pairs occurring in each string of a set of strings S without any restrictions on the gaps of the different pairs, uses linear space and its time complexity is $O(n + \alpha)$, where α is the size of the output.*

3.2   Maximal Pairs with Bounded Gaps

In the previous version of the problem we did not impose any restrictions on the gaps of the pairs. For example, an occurrence of a pair P in S_i may have gap with length 1 while an occurrence of the same pair in $S_j, i \neq j$ may have gap with length $|S_j| - c_1$, for an arbitrary constant c_1. In this version we demand that the gap in a pair, which occurs in all strings of the set S, be the same. Thus, if a pair P occurs in S_i with gap g_i then P must occur in all strings $S_j \in S, i \neq j$ with gap exactly g_i. We will see later in the description of the algorithm that it is imperative to add the additional restriction that the gaps must be upper bounded by a constant. The aim is to devise an algorithm with complexity $O(n \cdot polylog(n) + k \cdot \alpha \cdot \log n)$.

The base structure is the one described in the previous subsection. However, maintaining the output size in the time complexity is more complicated. Assume the leaf-lists $L_{v,i}$ and $L_{w,i}$ for brother nodes v and w. All maximal pairs produced by the $(\sigma - 1) \cdot \sigma$ cartesian products of their sublists are candidate maximal pairs and must be tested whether they occur in all other strings of set S with the same gap (from now and on pairs are equal with respect to the boxes and their gap). However, it is obvious that the $(\sigma - 1) \cdot \sigma$ cartesian products may lead to $O(n^2)$ candidate solutions. For the worst case, where $|S_1| = |S_2| = \cdots = |S_k| = n/k$ the number of pairs may be $O((\frac{n}{k})^2)$. All these candidate maximal pairs may not be in the output and thus we acquire an algorithm with time complexity $O(n^2)$. In order to alleviate this problem we add the following realistic restriction.

Invariant 1 *The maximum length of a gap may be at most b, where b is a constant, which is indepedent of the input size n.*

This invariant lowers considerably the combinatorial complexity of the cartesian products, which we call *limited cartesian products*. We previously mentioned that these cartesian products construct $O(n^2)$ candidate maximal pairs, but by using invariant we lower this bound to $O(bn)$. In this way, each occurrence in a sublist $L_{v,i}^{\sigma_j}$ may be combined with at most $2b$ occurrences of the sublists $L_{w,i}^{\sigma_l}, 1 \leq l \leq \sigma, l \neq j$. This procedure will be executed for each leaf-list corresponding to strings S_i and for all internal nodes of the GST. To bound the total complexity we use the *smaller-half trick*, which is given in the following lemma.

Lemma 1 *The sum over all nodes v of an arbitrary binary tree of size n of terms that are $O(n_1)$, where $n_1 \leq n_2$ are the weights (the number of leaves) of the subtrees rooted at the two children of v, is $O(n \log n)$.*

Thus, by applying lemma we guarantee that the total candidate solutions considered during the execution of the algorithm will have a total complexity of $O(n \log n)$. To apply this lemma in the limited cartesian products we must ensure that the pairs are constructed by traversing the smaller sublist between the sublists $L_{v,i}^{\sigma_j}$ and $L_{w,i}^{\sigma_l}$, where $l \neq j$. Assume, without loss of generality, that $|L_{v,i}^{\sigma_j}| \leq |L_{w,i}^{\sigma_l}|$. Then, for each occurrence $q \in L_{v,i}^{\sigma_j}$ we make at most $2b$ pairs choosing appropriate $q' \in L_{w,i}^{\sigma_l}$ such that the gap is bounded by b. To choose the appropriate q' we must execute a search operation with key the position of the occurrence. This in turn, requires that the sublists are sorted by position and they are structured in a search tree so that the search operations are efficient. First, the following lemma on the merge operation of two height-balanced trees from [ ] must be given.

Lemma 2 *Two height-balanced trees of size n and m, where $n \leq m$, can be merged in time $O(\log \binom{n+m}{n})$.*

Proof. Assume two height-balanced trees T and T' of size n and m respectively, where $n \leq m$. Starting, with the smallest element $e \in T$ we insert each element in T'. This is accomplished by using a finger (see [ ] for finger search trees) that points to the last inserted element. Thus, the total cost for the insertions is $O(\log n + \sum_{i=2}^{m} \log d_i)$, where d_i is the distance between the previously inserted element and the next element in T. Since $\sum_{i=2}^{m} d_i \leq n$ and the sum above maximizes when all d_i are equal ($d_i = n/m$) it follows that the total cost for insertions is $O(\log n + \sum_{i=2}^{m} \log d_i) = O(m \cdot \log(\frac{n}{m}))$. $\square$

By using lemma   we can implement efficiently a multiple search of a set of elements on the height-balanced tree. The lemma below describes this procedure.

Lemma 3 *Given a sorted list of elements $e_1, e_2, \ldots, e_n$ structured in a heigth-balanced tree T and a height-balanced tree T' of size m, where $m \geq n$, we can find $q_i = min\{x \in T' | x \geq e_i\}$ for all $1 \leq i \leq n$ in time $O(\log \binom{n+m}{n})$.*

Proof. The basic idea is to use the merge algorithm of lemma   while keeping the positions where insertions of elements $e_i \in T$ take place. This change in the merge algorithm does not affect the time complexity and as a result we can find all q_i in time $O(\log \binom{n+m}{n})$. $\square$

Finally, the lemma below taken from [ ] states that following merging patterns like the one described in the previous lemmas on the nodes of the generalized suffix tree we may achieve a time complexity of $O(n \log n)$.

Lemma 4 *Let T be an arbitrary binary tree with n leaves. The sum over all internal nodes $v \in T$ of terms $\binom{n_1+n_2}{n_1}$, where $n_1 \leq n_2$ are the weights of the subtrees rooted at the two children of v, is $O(n \log n)$.*

Proof. See [ ]. $\square$

Until this point we have shown in lemma   that if we implement the sublists of the binary generalized suffix tree by using height-balanced trees then we can guarantee a time complexity for the merge operations taking place in internal nodes between sublists of $O(n \log n)$. In addition, we saw that by using the smaller-half trick given in lemma   we guarantee that the total number of candidate solutions will be at most $O(n \log n)$ for the whole GST. This implies that the maximal pairs under the restriction that the gaps are bounded by a constant are at most $O(n \log n)$, which is a fact that is stated in [ ].

The only remaining pendency is how to query all other leaf-lists using the candidate solutions produced by one leaf-list. Assume sublists $L_{v,i}^{\sigma_j}$ and $L_{w,i}^{\sigma_l}$, where $l \neq j$, and their limited cartesian product C. All pairs $p_x \in C$ are candidate solutions and assume that g_x is the gap of pair p_x. We would like to have data structures in each sublist such that given a query lentgh of a gap g_x to answer efficiently the question of whether there is a pair with such gap in this sublist. If we solve this problem then the construction is completed.

In order to solve this problem we again make heavy use of invariant . Assuming that the gaps are arbitrary and assuming two sublists $L_{v,i}^{\sigma_j}$ and $L_{w,i}^{\sigma_l}$ with sizes $|L_{v,i}^{\sigma_j}|$ and $|L_{w,i}^{\sigma_l}|$ respectively, then the number of all possible pairs formed by their cartesian product is $O(|L_{v,i}^{\sigma_j}| \cdot |L_{w,i}^{\sigma_l}|)$. This combinatorial complexity of different pairs is prohibitive for our algorithm. However, using invariant , it is easy to see that the number of pairs reduces to $O(b \cdot min\{|L_{v,i}^{\sigma_j}|, |L_{w,i}^{\sigma_l}|\})$, which is linear to the size of the sublists. We store the gaps of all possible pairs between two sublists $L_{v,i}^{\sigma_j}$ and $L_{w,i}^{\sigma_l}$ at the leaves of a height-balanced tree, which we call henceforth *gap tree* $G_i^{j,l}$, in order of their length. Each leaf has a pointer to the sublists identifying the first occurrence of the pair. Since the number of these gaps is linear to the size of the sublists, so is the size of the height-balanced tree. In this way, each sublist $L_{v,i}^{\sigma_j}$ contains $\sigma - 1$ stuctures for the gaps of all possible pairs between this sublist and sublists $L_{v,i}^{\sigma_l}, l \neq j$. First, we must merge these structures (as well as their corresponding sublists) of node v and w and then call the reporting procedure. In the queries for a gap g_x we query each of the $\sigma - 1$ structures of the sublists of a leaf-list and if we find at least one such gap then we are certain that the pair with gap g_x occurs in the respective string.

The query time in these structures is logarithmic to the size of the corresponding sublists. Thus, the procedure of finding the appropriate pairs out of all candidate solutions is a procedure that demands $O(k \cdot \sigma^2 \cdot \frac{n}{k} \cdot \log^2 \frac{n}{k}) = O(n \log^2 n)$. We have not considered yet the construction algorithm of these structures and their time overhead. Assume the leaf-list $L_{v,1}$ and the list $L_{w,1}$, which correspond to occurrences of the string S_1 while nodes w and v are brothers. Assume, that from this pair of leaf-lists we construct the set of candidate solutions C. Now, assume all other pairs of leaf-lists $(L_{v,i}, L_{w,i})$, for all $1 < i \leq k$. Then for each candidate solution $p_x \in C$ we query with gap g_x all other $k - 1$ pairs of leaf-lists to find whether there are occurrences of pairs with this gap. If all pairs contain such a pair then we report it as a maximal pair that occurs in all strings in set S. If it is not contained in all pairs then we reject this candidate solution.

Assume sublists $L_{v,i}^{\sigma_j}$ and $L_{w,i}^{\sigma_l}$ where $l \neq j$. For each such pair between leaf-lists $L_{v,i}$ and $L_{w,i}$ we construct a height-balanced tree for the gaps of all possible pairs between those two sublists. Thus, each possible pair of sublists has attached a tree $G_i^{j,l}$. As a result there will be at most $(\sigma - 1) \cdot \sigma = O(1)$ such trees for every pair of leaf-lists. The query must be performed on every such structure and thus their construction precedes the query operation at this level. The construction of these trees is carried out in parallel with the merging of sublists. Thus, for sublists $L_{v,i}^{\sigma_j}$ and $L_{w,i}^{\sigma_l}$ we had before the merging $\sigma - 1$ gap trees $G_{v,i}^{j,y}, y \neq j$ (the gap tree of sublists of symbols σ_j and σ_y of the i-th string of node v) and $G_{w,i}^{l,y}, y \neq l$. After the merging of the two sublists we get $\sigma - 1$ new gap trees $G_{z,i}^{j,y}, y \neq j$. This is accomplished as following. We merge both sublists by using lemma . For each occurrence inserted in the largest tree we must also make changes to the $\sigma - 1$ gap trees. By attaching $\sigma - 1$ pointers to each occurrence in the sublists that point in the smallest formed gap in their respective gap trees we can in $O(1)$ time find where the new gaps must be inserted. Each occurrence in a sublist may create at most $O(2b)$ gaps in a single gap tree. Since these gaps may be far apart in the gap tree, we structure them in a double connected circular list where the head of the list will be the smallest gap. The occurrence will need only to maintain a pointer to the head of this list in each gap tree. Thus, after finding where to insert the occurrence in the larger sublist we pick its predecessor in the sublist (the previous occurrence) and we condider each double-connected circular list in the gap trees. It is easy to see that by invariant we have to make $O(b^2)$ work in each such gap tree and as a result the complexity in the merging of the gap trees due to a single insertion to the largest sublist is $O(\sigma \cdot b^2) = O(1)$.

Since, the gap trees have size proportional to the size of their respective sublists by applying lemma we are able to guarantee an $o(n^2)$ time complexity for the merging of gap trees. From the above, the total time complexity with respect to the merging of gap trees is $O(\sigma^2 \cdot b^2 \cdot n \cdot \log n) = O(n \log n)$. Thus, concluding the time complexities we get the following: a) $O(n \log^2 n)$ for testing candidate solutions, b)$O(n \log n)$ for merging sublists of occurrences and gap trees and c) $O(k \cdot \alpha \cdot \log n)$ for the complexity of generating the output of size α. The following theorem states the result.

Theorem 2. *The above algorithm for the problem of computing all maximal pairs occurring in each string of a set of strings S with equal gaps, uses linear space and its time complexity is $O(n \log^2 n + k \cdot \log n \cdot \alpha)$, where α is the size of the output.*

Acknowledgements

We would like to thank Marie-France Sagot for introducing the above problem to us.

References

1. A. Apostolico and Z. Galil. Pattern Matching Algorithms. Oxford University Press, ISBN 0-19-611367-5, 1997.
2. G. S. Brodal, R. B. Lyngs, C. N. Storm Pedersen and DKFZ Jens Stoye. Finding Maximal Pairs with Bounded Gaps. *Journal of Discrete Algorithms*, 0(0):1-27, 2000.

3. M. R. Brown and R. E. Tarjan. A Fast Merging Algorithm. *Journal of the ACM*, 26(2):211-226, 1979.
4. M. R. Brown and R. E. Tarjan. Design and Analysis of a Data Structure for Representing Sorted Lists. *SIAM Journal on Computing*, 9:594-614, 1980.
5. D. Gusfield. Algorithms on Strings, Trees, and Sequences. Cambridge University Press, ISBN 0-521-58519-8, 1999.
6. G. Navarro. A Guided Tour to Approximate String Matching. *ACM Computing Surveys*, 33(1):31-88, 2001.
7. G. A. Stephen. String Searching Algorithms. Lecture Notes Series on Computing - Vol. 3, World Scientific Publishing, ISBN 981-02-1829-X, 1994.
8. E. Ukkonen. Finding Approximate atterns in strings. *Journal of Algorithms*, 6:132-137, 1985.

Space-Economical Algorithms
for Finding Maximal Unique Matches

Wing-Kai Hon[1] and Kunihiko Sadakane[2]

[1] Department of Computer Science and Information Systems
The University of Hong Kong
wkhon@csis.hku.hk
[2] Graduate School of Information Sciences
Tohoku University, Japan
sada@dais.is.tohoku.ac.jp

Abstract. We show space-economical algorithms for finding *maximal unique matches* (MUM's) between two strings which are important in large scale genome sequence alignment problems. Our algorithms require only $O(n)$ bits ($O(n/\log n)$ words) where n is the total length of the strings. We propose three algorithms for different inputs: When the input is only the strings, their compressed suffix array, or their compressed suffix tree. Their time complexities are $O(n \log n)$, $O(n \log^\epsilon n)$ and $O(n)$ respectively, where ϵ is any constant between 0 and 1. We also show an algorithm to construct the compressed suffix tree from the compressed suffix array using $O(n \log^\epsilon n)$ time and $O(n)$ bits space.

1 Introduction

The suffix tree is a quite useful data structure for solving string problems. Many problems can be efficiently solved by using the suffix tree [ ]. However the problem of using the suffix tree is its size. It is said that the suffix tree occupies about $17n$ bytes for a string of length n. Although a space-efficient representation of the suffix tree [ ] has been proposed, it still occupies more than $10n$ bytes. Therefore it is difficult to apply the suffix tree to solve large scale problems. The problem is severe in treating genome scale strings. For example the whole sequence of human DNA has length about 3 billion base pairs. Then its suffix tree occupies at least 30 gigabytes, which is not realistic.

Therefore it is important to develop space-economical alternatives to the suffix tree. Recently many such data structures were proposed, for example space-efficient suffix trees [ ], the compressed suffix array [,], the FM-index [ ], data structures for bottom-up traversal of the suffix tree [ ], and data structures for longest common prefixes [ ]. However none of them achieves the same functionalities as the suffix tree.

In this paper we consider the problem of finding *maximal unique matches* (MUM's) between two strings A and B. An MUM is a substring that appears once in both A and B and is not contained in any longer such substring. The

A. Apostolico and M. Takeda (Eds.): CPM 2002, LNCS 2373, pp. 144–   , 2002.
© Springer-Verlag Berlin Heidelberg 2002

MUM's are used in the algorithm of Delcher et al. [ ] for aligning two long genome sequences. Details are described in Section .

Although this problem can be solved in linear time by using the suffix tree of the strings A and B, it is not space-efficient. We use data structures of linear size, that is, $O(n)$ bits where n is the total length of the two strings. These improve the space complexity of the algorithm using the suffix tree by a factor of $O(\log n)$ because the suffix tree requires $O(n)$ pointers, or equivalently $O(n \log n)$ bits.

Our data structures include the compressed suffix array (CSA), parentheses representation of a tree [ ] and the data structures for longest common prefixes (Hgt array) [ ]. Note that these data structures do not store suffix links of a suffix tree. Therefore we may not be able to solve the problem efficiently. However in this paper we found a good property of the data structure of the compressed suffix array that some suffix links can be simulated in constant time, which allows us to solve the problem efficiently.

We propose an algorithm to solve the problem of finding MUM's in $O(n)$ time using $O(n)$ bits space if we are given CSA and Hgt array for a string consisting of a concatenation of A and B, and the parentheses representation of the suffix tree for the string. This algorithm utilizes a property of the compressed suffix array to simulate suffix links in the suffix tree.

We also propose an algorithm to construct the Hgt array and the parentheses representation of the suffix tree from CSA in $O(n \log^\epsilon n)$ time and $O(n/\epsilon)$ bits space where ϵ is any constant between 0 and 1. Note that the CSA can be constructed in $O(n \log n)$ time and $O(n)$ bits space [ ]. Therefore we can solve the problem in $O(n \log n)$ time and $O(n)$ bits space if we are given the two strings.

Our data structures are space-efficient both theoretically and practically. For DNA sequences, our compressed suffix tree occupies only about $12n$ bits, which is much smaller than the existing suffix tree of size $10n$ bytes ($80n$ bits).

The rest of this paper is organized as follows. Section   describes the data structures of suffix trees, suffix arrays, and their compressed versions. Section   describes the genome sequence alignment problem, the algorithm of Delcher et al.[ ], and the properties of MUM. In Section   we propose new algorithms to find MUM's. Section   shows concluding remarks.

2   Preliminaries

2.1   Suffix Trees and Suffix Arrays

Let $T[1..n] = T[1]T[2] \cdots T[n]$ be a string of length n on an alphabet $\mathcal{A}$. Assume that the alphabet size $|\mathcal{A}|$ is constant. The j-th suffix of T is defined as $T[j..n] = T[j]T[j+1] \ldots T[n]$ and expressed by T_j. A substring $T[j..l]$ is called a prefix of T_j. The suffix array $SA[1..n]$ of T is an array of integers j that represent suffixes T_j. The integers are sorted in lexicographic order of the corresponding suffixes. The suffix tree of a string $T[1..n]$ is a compressed trie built on all suffixes of T. It has n leaves, each leaf corresponds to a suffix $T_{SA[i]}$. For details, see Gusfield [ ].

Let *leaf(i)* denote the leaf of the suffix tree that corresponds the suffix $T_{SA[i]}$. Let *lca(v, w)* be the lowest common ancestor of nodes v and w. Let *str(v)* denote the string obtained by the concatenation of edge labels from the root node to an internal node v of the suffix tree. For each internal node v of the suffix tree such that $str(v) = c\alpha$ where c is a character in $\mathcal{A}$ and α is a string, the suffix link *sl(v)* is defined as a node w such that $str(w) = \alpha$. Both the suffix tree and the suffix array occupy $O(n \log n)$ bits, which are not linear in n.

We construct the suffix tree of $T = A\$_1 B\$_2$, that is, the string T consists of a concatenation of two strings A and B, separated by unique terminators $\$_1$ and $\$_2$. Let n_1 and n_2 be the length of A and B, respectively.

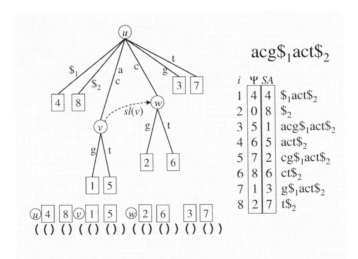

Fig. 1. The suffix tree and the suffix array for "acg$\$_1act\$_2$" and the parentheses representation

The topology of a suffix tree can be encoded in at most $4n$ bits [ ]. The tree is encoded into at most $2n$ nested open and close parentheses as follows. During a preorder traversal of the tree, write an open parenthesis when a node is visited, then traverse all subtrees of the node in alphabetic order of the edges from the node, and write a close parenthesis. Figure   shows the suffix tree, the suffix array, and the parentheses representation of a string "acg$\$_1act\$_2$." Leaf nodes are shown by boxes, and numbers in the boxes represent the elements of the suffix array. Internal nodes are shown by circles.

2.2   Compressed Suffix Arrays and the *Hgt* Array

The compressed suffix array is a compressed version of the suffix array. It occupies only $O(n/\epsilon)$ bits for a string of length n where $0 < \epsilon \le 1$ is any constant. Each entry $SA[i]$ of the suffix array can be computed in $O(\log^\epsilon n)$ time. We can also compute the inverse of the suffix array, $SA^{-1}[j]$, and the character $T[j]$, in $O(\log^\epsilon n)$ time [ ].

The compressed suffix array stores Ψ function instead of SA.

Definition 1.

$$\Psi[i] \equiv \begin{cases} i' \text{ such that } SA[i'] = SA[i] + 1 & (\text{if } SA[i] < n) \\ 0 & (\text{if } SA[i] = n) \end{cases}$$

The Ψ function occupies $O(n)$ bits and each entry can be computed in constant time [ ].

Because the parentheses representation of the tree topology of the suffix tree does not store the information of edge lengths, we need to store the following array.

Definition 2. $Hgt[i] = lcp(T_{SA[i]}, T_{SA[i+1]})$

The Hgt array occupies $n \log_2 n$ bits without compression. However it can be represented in linear size by using the suffix array or the compressed suffix array.

Theorem 1 (Sadakane [ ]). *Given i and $SA[i]$, the value $Hgt[i]$ can be computed in constant time using a data structure of size $2n + o(n)$ bits.*

We can compute $Hgt[i]$ in constant time if the suffix array is stored, or in $O(\log^\epsilon n)$ time if the compressed suffix array is stored.

3   Sequence Alignment and Maximal Unique Matches

Global alignment of sequences has been a long-standing topic in computational biology, and is becoming more and more important nowadays. With the advance in biotechnology, the number of organisms whose genomes are completely sequenced, increases rapidly over the years. Researchers can now obtain genomes of interest to compare with, in particular, those genomes which are closely related and shares large degree of homology. When such genomes are globally aligned, many interesting biological features may be identified. For example, an occurrence of a gene or other conserved regions may be suggested by exact matches, whereas SNPs (single nucleotide polymorphism), tandem repeats, large inserts or reversals can be found around the mismatches. On the other hand, with global alignment, we are able to make a good estimate on how close the two genomes are related, which can be used to construct evolutionary trees among a set of species.

Traditional global alignment algorithms takes $O(mn)$ time, for input sequences with length m and n respectively. This is, however, impractical for aligning two whole genomes where m and n are on the order of 10^6 or even more.

To circumvent the timing problem, Delcher et al.[ ] designed an efficient heuristic algorithm, which exploits a reasonable intuition that, if a long and identical sequence occurs exactly once in each genome (which they called an MUM), such a sequence is almost certain to be part of the global alignment. Therefore, when the two input genomes are fairly similar, long and identical substrings are likely to appear in both genomes. The whole genome alignment problem is now reduced to finding the MUM's, aligning the MUM's and finally aligning the local gaps between successive MUM's. In practice, the first two steps require $O(n)$ time and the last step is generally fast.

We focus on the first step because it is the most space-consuming task. We propose space-efficient algorithms for it.

Definition 3. *A pattern P is said to be a unique match (UM) of two strings A and B if P appears exactly once in both A and B. A pattern P is said to be a maximal unique match (MUM) if P is a UM and it is not contained in any longer such pattern.*

Note that unique match and maximal unique match are similar to maximal repeat and super-maximal repeat [ ] respectively.

Lemma 1. *An internal node of the suffix tree does not represent any MUM if it has an internal node as its child.*

Proof. Assume that an internal node v has a child internal node w. If descendants of w are from both A and B, v is not MUM because w is also a candidate of MUM and $str(v)$ is contained in $str(w)$. If all descendants of w are from either A or B, v is not MUM because $str(v)$ (and $str(w)$) appears at least twice in a string.

Therefore only internal nodes with two leaves are candidates of MUM. Note that not all such nodes are MUM's because some of them are not maximal.

The original algorithm using a suffix tree becomes as follows:

1. create a generalized suffix tree T for $A\$_1 B\$_2$.
2. mark each internal node v of T with exactly two child nodes, one is a leaf from A and the other is a leaf from B.
3. for each internal node v unmark $sl(v)$
4. report all marked nodes.

Note that in step 3 an internal node, pointed from another internal node v, is unmarked even if v is not marked. We can unmark the nodes in any order. This is crucial to the correctness of this algorithm and our algorithm. Note also that the total length of MUM's is not $O(n)$. Therefore we report them implicitly, by reporting pairs (p, l) representing a substring of length l that appears at position p in A.

Theorem 2. *The above algorithm reports all maximal unique matches of A and B in $O(n)$ time.*

Proof. First we show that all marked nodes in the suffix tree correspond to MUM's. Step 2 of the algorithm finds candidates of MUM. After Step 3 any marked node v corresponds to an MUM because of the following reason: Let v be a node of the suffix tree which is marked in step 2 but which does not correspond to any MUM. Then there exists a node w such that $str(w)$ contains $str(v)$. If $str(v)$ is a suffix of $str(w)$, there exists a node u such that $str(v)$ is a suffix of $str(u)$ and $|str(u)| = |str(v)| + 1$. This implies $sl(u) = v$. Therefore the node v is unmarked in step 3. If $str(v)$ is not a suffix of $str(w)$, there exists a node w' such that $str(v)$ is a prefix of $str(w')$. Then the node v has a descendant w'. This means that v has a node which is not a leaf. Therefore the node v has not been marked in step 2. Because each step takes $O(n)$ time, the algorithm runs in $O(n)$ time.

4   New Algorithms

4.1   In Case the Input Is the Compressed Suffix Array of the Strings

First we show a trivial algorithm for the case we are given the compressed suffix array of a concatenation of two strings A and B. We can find all MUM's as follows:

1. construct the Hgt array of $A\$_1 B\$_2$
2. for $i = 1, 2, \ldots, n$
   if $Hgt[i] < Hgt[i-1]$ and $Hgt[i] < Hgt[i+1]$
   and $leaf(i)$ and $leaf(i+1)$ come from different strings
   and $T[SA[i]-1]$ and $T[SA[i+1]-1]$ differ
   output $(SA[i], Hgt[i])$ that represents the substring $T[SA[i]..SA[i]+Hgt[i]-1]$

Note that the algorithm to construct the Hgt array is described in Section   .

Theorem 3. *The above algorithm reports all maximal unique matches of A and B in $O(n \log^\epsilon n)$ time and $O(n)$ bit space.*

Proof. If $Hgt[i] > Hgt[i-1]$ and $Hgt[i] < Hgt[i+1]$, $lca(leaf(i), leaf(i+1))$ is a node with two leaf children. Therefore the node represents a UM. If the left characters $T[SA[i]-1]$ and $T[SA[i+1]-1]$ differ, the node represents an MUM (see Gusfield [ ]). Because $Hgt[i]$ and $T[SA[i]-1]$ can be computed in $O(\log^\epsilon n)$ time, the algorithm runs in $O(n \log^\epsilon n)$ time.

4.2   In Case the Input Is a Compressed Suffix Tree of the Strings

We show that the problem can be solved in linear time and space when the input is a compressed suffix tree of the two strings. That is, the input is the CSA and the Hgt array of the concatenation of A and B, and the parentheses representation of the suffix tree. The above algorithm takes $O(n \log^\epsilon n)$ time because computing an element of the compressed suffix array takes $O(\log^\epsilon n)$ time. On the other hand, we can solve the problem in $O(n)$ time when the compressed suffix tree is given even if the compressed suffix array is still being used. Note that the compressed suffix tree does not have suffix links. However we can simulate suffix links by using the compressed suffix array.

Definition 4. *The compressed suffix tree of a string A consists of the compressed suffix array of A, the Hgt array of A, and the parentheses encoding of the topology of the suffix tree of A.*

Definition 5. *The parentheses encoding of a tree is defined by at most $2n$ nested open and close parentheses as follows. During a preorder traversal of the tree, write an open parenthesis when a node is visited, then traverse all subtrees of the node in alphabetic order of the edges from the node, and write a close parenthesis.*

A high-level description of the algorithm is the following:

1. compute a bit-vector $D[i]$ $(i = 1, 2, \ldots, n)$ such that $D[i] = 1$ if $leaf(i)$ is in A and $D[i] = 0$ if in B.
2. mark nodes which have two leaves from A and B.
3. unmark non-maximal nodes.
4. report all MUM's.

In step 1, to compute the bit-vector we compute $i = p, \Psi[p], \Psi^2[p], \ldots, \Psi[n-1]$ where p is the index such that $SA[p] = 1$, and set $D[i] = 1$ in first n_1 iteration and set $D[i] = 0$ in the following n_2 iteration. Because $\Psi^k[p] = p + k$, in first n_1 iteration the computed indices i correspond to the lexicographic order of suffixes of the string A. Therefore we set $D[i] = 1$. This takes $O(n)$ time because computing a Ψ value takes constant time.

In step 2, we scan the parentheses encoding of the suffix tree from left to right to find a pattern '(()())' which corresponds to an internal node of the suffix tree with two leaves. Let i be the lexicographic order of the left leaf, which can be computed during the scanning by counting the number of occurrences of '().' We mark nodes with two leaves in another bit-vector $V[i]$ $(i = 1, 2, \ldots, n)$. If $D[i]$ and $D[i+1]$ differ, set $V[i] = 1$.

Steps 3 and 4 become as follows:

3.1 define a temporary array W
3.2 for $i = 1, 2, \ldots, n$ $W[i] = V[i]$
3.3 for $i = 1, 2, \ldots, n$
    if $\Psi[i] + 1 = \Psi[i+1]$
    $W[\Psi[i]] = 0$
4.1 $i = SA^{-1}[1]$
4.2 for $j = 1, 2, \ldots, n$
    if $V[i] = 1$ and $W[i] = 1$
    output $(j, Hgt[i])$ that represents the substring $T[j..j + Hgt[i] - 1]$
    $i = \Psi[i]$

Theorem 4. *The above algorithm reports all maximal unique matches of A and B in $O(n)$ time and $O(n)$ bit space.*

Proof. First we show the correctness of the algorithm, that is, we show that the string $str(w)$ where $w = lca(leaf(i), leaf(i+1))$ is an MUM if and only if $V[i] = W[i] = 1$ after step 3.

Assume that $str(w)$ is a UM but not an MUM. If $str(w) \cdot \alpha$ is also a UM, the node w is not marked ($V[i] = 0$) because w has a non-leaf child. If $\alpha \cdot str(w)$ is also a UM, there exists i' such that $V[i'] = 1$, $i = \Psi[i']$. Furthermore, because $str(w)$ is unique in the strings $\alpha \cdot str(w)$ and $str(w)$ share the string $str(w)$. That is, $\Psi[i'] + 1 = \Psi[i' + 1]$. Therefore $W[i] = 0$ in step 3.

Assume that $str(w)$ is an MUM. We show $W[i] = 1$ by contradiction. Assume that $W[i] = 0$ after step 3. Then there exists i' such that $V[i'] = 1$ and $\Psi[i'] + 1 = \Psi[i' + 1]$. Then $str(w)$ is not an MUM because $str(lca(leaf(i'), leaf(i' + 1)))$ contains w.

Next we show the time and space complexity. Step 1 and 2 obviously are done in $O(n)$ time because computing a value of Ψ takes constant time. Step 3 also takes $O(n)$ time because we know $SA[i] = j$ in each iteration and therefore we can compute $Hgt[i]$ in constant time. Concerning the space complexity of the data structure, We need to store the Ψ function of the compressed suffix array ($O(n)$ bits), a compressed representation of the Hgt array ($2n + o(n)$ bits) and the parentheses representation of the suffix tree topology ($4n + o(n)$ bits). Therefore the total is also $O(n)$ bits.

The above algorithm requires only $12n$ bits space in practice: $3n$ for the Ψ function, $2n$ for the Hgt array, $4n$ for the parentheses sequence, and $3n$ bits for the three bit-vectors D, V and W.

4.3   In Case the Input Is only the Strings

Even if we are given only two strings, we can solve the problem using only $O(n)$ bits space. We show an algorithm to construct the compressed suffix tree of a string of length n from the string and its compressed suffix array in $O(n \log^\epsilon n)$ time and $O(n)$ space.

The Hgt array can be constructed from the text and its suffix array and the inverse of the suffix array in $O(n)$ time [ ]. This algorithm is adapted to using the compressed suffix array. Because we can compute $SA[i]$, $SA^{-1}[j]$ and $T[SA[i]]$ in $O(\log^\epsilon n)$ time, the algorithm runs in $O(n \log^\epsilon n)$ time using the compressed suffix array.

Next we construct the parentheses representation of the suffix tree topology. We can construct it by using only a compressed representation of the Hgt array, the compressed suffix array and a stack. We simulate a bottom-up traversal of the suffix tree by storing values $Hgt[i]$ ($i = 1, 2, \ldots, n$) in a stack [ ]. It takes $O(n \log^\epsilon n)$ time. Because the numbers in the stack are always monotone, we can encode them by the difference from the adjacent number. Then the space required to store the numbers is bounded by $O(n)$ bits if the differences are encoded by δ-code [ ] etc.

The parentheses encoding $C[v]$ of a tree rooted by a node v becomes $(C[w_1] C[w_2] \ldots C[w_k])$ where $w_1, w_2, \ldots, w_k$ are children of v. During the bottom-up traversal we keep pointers to the parentheses encoding of subtrees. If the parentheses encoding $C[w_i]$ of a subtree rooted by w_i occupies only $O(\log n)$ bits, we copy it to the end of $C[w_{i-1}]$. It takes constant time on RAM model. If $C[w_i]$

occupies more than $O(\log n)$ bits, we make a pointer from $C[w_{i-1}]$ to $C[w_i]$. It also takes constant time. Because we use pointers for only parentheses encodings that occupies more than $O(\log n)$ bits, the number of pointers is $O(n/\log n)$. Therefore the pointers occupies $O(n)$ bits.

5   Concluding Remarks

We have proposed linear space algorithms for finding MUM's between two strings, which is important to compute the alignment of two long genome sequences. We used the compressed suffix array and other space-economical data structures for suffix trees. Although it seems to be difficult to compute MUM's in linear time if the compressed suffix array is used, we found an algorithm for it by using a property of the compressed suffix array. This result will be extended to a space-economical representation for suffix links in a suffix tree. This remains as a future work.

References

1. A. L. Delcher, S. Kasif, R. D. Fleischmann, J. Peterson, O. White, and S. L. Salzberg. Alignment of Whole Genomes. *Nucleic Acids Research*, 27:2369–2376, 1999.   ,
2. P. Elias. Universal codeword sets and representation of the integers. *IEEE Trans. Inform. Theory*, IT-21(2):194–203, March 1975.
3. P. Ferragina and G. Manzini. Opportunistic Data Structures with Applications. In *41st IEEE Symp. on Foundations of Computer Science*, pages 390–398, 2000.
4. R. Grossi and J. S. Vitter. Compressed Suffix Arrays and Suffix Trees with Applications to Text Indexing and String Matching. In *32nd ACM Symposium on Theory of Computing*, pages 397–406, 2000.   ,
5. D. Gusfield. *Algorithms on Strings, Trees, and Sequences*. Cambridge University Press, 1997.   ,   ,   ,
6. T. Kasai, G. Lee, H. Arimura, S. Arikawa, and K. Park. Linear-time Longest-Common-Prefix Computation in Suffix Arrays and Its Applications. In *Proc. the 12th Annual Symposium on Combinatorial Pattern Matching (CPM'01)*, LNCS 2089, pages 181–192, 2001.   ,
7. S. Kurtz. Reducing the Space Requirement of Suffix Trees. *Software – Practice and Experience*, 29(13):1149–1171, 1999.
8. T. W. Lam, K. Sadakane, W. K Sung, and S. M Yiu. working draft.
9. J. I. Munro and V. Raman. Succinct Representation of Balanced Parentheses and Static Trees. *SIAM Journal on Computing*, 31(3):762–776, 2001.   ,
10. J. I. Munro, V. Raman, and S. Srinivasa Rao. Space Efficient Suffix Trees. *Journal of Algorithms*, 39(2):205–222, May 2001.
11. K. Sadakane. Compressed Text Databases with Efficient Query Algorithms based on the Compressed Suffix Array. In *Proceedings of ISAAC'00*, number 1969 in LNCS, pages 410–421, 2000.   ,
12. K. Sadakane. Succinct Representations of *lcp* Information and Improvements in the Compressed Suffix Arrays. In *Proc. ACM-SIAM SODA 2002*, pages 225–232, 2002.   ,   ,

The Minimum DAWG for All Suffixes
of a String and Its Applications

Shunsuke Inenaga[1], Masayuki Takeda[1,2], Ayumi Shinohara[1,2],
Hiromasa Hoshino[1], and Setsuo Arikawa[1]

[1] Department of Informatics, Kyushu University
33 Fukuoka 812-8581, Japan
{s-ine,takeda,ayumi,hoshino,arikawa}@i.kyushu-u.ac.jp
[2] PRESTO, Japan Science and Technology Corporation (JST)

Abstract. For a string w over an alphabet Σ, we consider a composite data structure called the *all-suffixes directed acyclic word graph* (*ASDAWG*). $ASDAWG(w)$ has $|w| + 1$ initial nodes, and the dag induced by all reachable nodes from the k-th initial node conforms with $DAWG(w[k :])$, where $w[k :]$ denotes the k-th suffix of w. We prove that the size of the *minimum ASDAWG(w)* ($MASDAWG(w)$) is $\Theta(|w|)$ for $|\Sigma| = 1$, and is $\Theta(|w|^2)$ for $|\Sigma| \geq 2$. Moreover, we introduce an *on-line* algorithm which directly constructs $MASDAWG(w)$ for given w, whose running time is linear with respect to its size. We also demonstrate some application problems, *beginning-sensitive pattern matching*, *region-sensitive pattern matching*, and *VLDC-pattern matching*, for which AS-DAWGs are useful.

1 Introduction

In the field of information retrieval, pattern matching on strings is one of the most fundamental and important problems. A variety of patterns have been considered so far, according to various kinds of purposes and aims. The most basic one is a *substring* pattern. Let Σ be a finite alphabet. We call an element in Σ a *character*, and one in Σ^* a *string*. We say a pattern string p is a substring of a text string w if $w = upv$ for some strings $u, v \in \Sigma^*$. When a text w is fixed and a pattern p is flexible, once constructing a suitable data structure for w, we can solve the substring matching problem in $O(|p|)$ time, where $|p|$ denotes the length of p. In order to solve the problem efficiently, much attention has extensively been paid to inventing efficient data structures, such as suffix trees [, ,], directed acyclic word graphs (DAWGs) [,], compact directed acyclic word graphs (CDAWGs) [, ,], suffix arrays [ ], compact suffix arrays [ ], suffix cacti [ ], compressed suffix arrays [,], and so on.

Meanwhile, the problem finding a *subsequence* pattern has also been widely studied. We say a pattern p is a subsequence of a text w if p can be obtained by removing zero or more characters from w. By means of the directed acyclic subsequence graph (DASG) for w, we can examine whether or not p is a subsequence of w in $O(|p|)$ time [,]. An *episode* pattern is a "length-bounded"

A. Apostolico and M. Takeda (Eds.): CPM 2002, LNCS 2373, pp. 153–   , 2002.

version of a subsequence pattern [ ]. An episode pattern is given in the form
a pair of a string p and an integer k, as $\langle p, k \rangle$. If p is a subsequence of x such th
x is a substring of w with $|x| \leq k$, we say that the episode pattern $\langle p, k \rangle$ match
w. The episode directed acyclic subsequence graphs (EDASGs) were introduc
in [ ], for a practical solution to the problem.

Now we propose a new kind of pattern matching problem: *Given a text str*
$w = w_1 w_2 \cdots w_n$ *($w_i \in \Sigma$), a string p and an integer k, examine whether or n*
p *is a substring of $w[k :]$ where $w[k :] = w_k \ldots w_n$.* (NOTE: if $k > |w|$, t
answer is always NO.) We name the pattern $\langle p, k \rangle$ a *beginning-sensitive patte*
a *BS-pattern* for short. For any string $w \in \Sigma^*$ $DAWG(w)$ denotes the DAW
of w. Using the DAWGs for all suffixes of w, this problem is solvable in $O(|$
time. This simple collection of the DAWGs is called the *naive all-suffixes direc*
acyclic word graph for w, written as the naive $ASDAWG(w)$. Since the size
$DAWG(w)$ is $O(|w|)$, that of the naive $ASDAWG(w)$ is $O(|w|^2)$.

In this paper we introduce a new composite data structure, named the *mi*
mum $ASDAWG(w)$ and denoted by $MASDAWG(w)$. $MASDAWG(w)$ is the m
imization of the naive $ASDAWG(w)$. We show that the size of $MASDAWG($
is $\Theta(|w|)$ if $|\Sigma| = 1$, and $\Theta(|w|^2)$ if $|\Sigma| \geq 2$. Also, we produce an *on-line*
gorithm that *directly* constructs $MASDAWG(w)$ in time linear in the size
$MASDAWG(w)$.

We show further two applications of $MASDAWG(w)$, one of which is as f
lows. Let $\Pi = (\Sigma \cup \{\star\})^*$, where $\star$ is a *wildcard* that matches any string.
pattern $q \in \Pi$ such as $q = a\star ba\star c$ is called a *variable-length-don't-care's p*
tern (*VLDC-pattern*), where $a, b \in \Sigma$. The *language $L(q)$* of a pattern $q \in$
is the set of strings obtained by replacing $\star$'s in q with strings. For examp
$L(a\star ba\star c) = \{aubavc \mid u, v \in \Sigma^*\}$. This language corresponds to a class of
pattern languages proposed by Angluin [ ]. We declare that the smallest auton
ton to recognize all possible VLDC-patterns matching a text w is a variant
$MASDAWG(w)$.

Finding a good rule to separate given two sets of strings, often referred to
positive examples and *negative examples*, is a critical task in knowledge discov
and data mining. In [ ], an efficient method, with which a subsequence patt
is considered as a rule for the separation, was given, and in [ ] one using
episode pattern was proposed. $MASDAWG(w)$ is believed certainly to be a go
"weapon" to develop a practical algorithm to find the best VLDC-patterns
distinguish given two sets of strings efficiently. In fact, our experimental res
has shown that the average size of the MASDAWGs for random texts of lengt
to 500 over a binary alphabet is proportional to $|w|^{1.24}$, in spite of the theoreti
space complexity, $\Theta(|w|^2)$.

2   All-Suffixes Directed Acyclic Word Graphs

Strings x, y, and z are said to be a *prefix*, *substring*, and *suffix* of string u
xyz, respectively. The sets of prefixes, substrings, and suffixes of a string w
denoted by $Prefix(w)$, $Sub(w)$, and $Suffix(w)$, respectively. The empty strin

enoted by ε, that is, $|\varepsilon| = 0$. Let $\Sigma^+ = \Sigma^* - \{\varepsilon\}$. The substring of a string
$\prime$ that begins at position i and ends at position j is denoted by $w[i:j]$ for
$\leq i \leq j \leq |w|$. For convenience, let $w[i:j] = \varepsilon$ for $j < i$. Let $w[i:] = w[i:|w|]$
$\prime$r $1 \leq i \leq |w| + 1$. Assume S is a subset of Σ^*. For any string $u \in \Sigma^*$,
$^{-1}S = \{x \mid ux \in S\}$.

Let $w \in \Sigma^*$. We define an equivalence relation $\equiv_w$ on Σ^* by

$$x \equiv_w y \Leftrightarrow x^{-1}Suffix(w) = y^{-1}Suffix(w).$$

$\prime$et $[x]_w$ denote the equivalence class of a string $x \in \Sigma^*$ under $\equiv_w$. The longest
ement in the equivalence class $[x]_w$ for $x \in Sub(w)$ is called its *representative*.

)efinition 1 (Directed Acyclic Word Graph (DAWG)). *$DAWG(w)$ is*
ιe dag (V, E) such that

$$V = \{[x]_w \mid x \in Sub(w)\},$$
$$E = \{([x]_w, a, [xa]_w) \mid x, xa \in Sub(w) \ and \ a \in \Sigma\}.$$

we designate the node $[\varepsilon]_w$ of $DAWG(w)$ as the initial state and the nodes
$\prime]_w$ with $x \in Suffix(w)$ as the final states, then the resulting automaton is the
nallest automaton that accepts the set $Suffix(w)$ [ ].

)efinition 2 (All-Suffixes DAWG (ASDAWG)). *$ASDAWG(w)$ is a kind*
$\prime$ deterministic automaton with $|w| + 1$ initial nodes, designated by integers
$1, \ldots, |w|$, in which the subgraph consisting of the nodes reachable from the
$\cdot$th initial node and of their out-going edges is $DAWG(w[k+1:])$.

The simple collection of $DAWG(w[1:])$, $DAWG(w[2:]), \ldots, DAWG(w[n])$,
$AWG(w[n+1:])$ $(n = |w|)$ is an example of $ASDAWG(w)$, referred to as
ιe *naive* $ASDAWG(w)$. The number of nodes of the naive $ASDAWG(w)$ is
$\langle|w|^2)$. By minimizing the naive $ASDAWG(w)$, we can obtain the *minimum*
$SDAWG(w)$, which is denoted by $MASDAWG(w)$. The naive $ASDAWG(abba)$
ιd $MASDAWG(abba)$ are shown in Fig. . The minimization is performed based
ι the equivalence relation defined as follows. Let an ordered pair $\langle u, [x]_u \rangle$ denote
node $[x]_u$ of $DAWG(u)$. Each node of the naive $ASDAWG(w)$ can be repre-
nted by a pair $\langle u, [x]_u \rangle$ with $u \in Suffix(w)$ and $x \in Sub(u)$. The equivalence
lation, denoted by $\sim_w$, is defined by

$$\langle u, [x]_u \rangle \sim_w \langle v, [y]_v \rangle \Leftrightarrow x^{-1}Suffix(u) = y^{-1}Suffix(v) .$$

node of $MASDAWG(w)$ corresponds to an equivalence class under $\sim_w$. We
rite $\langle u, [x]_u \rangle$ simply as $\langle u, [x] \rangle$ in case no confusion occurs.

roposition 1. *Let $u \in Suffix(w)$. Let x be a nonempty substring of u. We*
ctorize u as $u = hxt$ and assume h is the shortest such string. Then, $\langle hxt, [x] \rangle$
equivalent to $\langle sxt, [x] \rangle$ for every suffix s of h. (NOTE: The string x is not
cessarily the representative of $[x]_u$.)

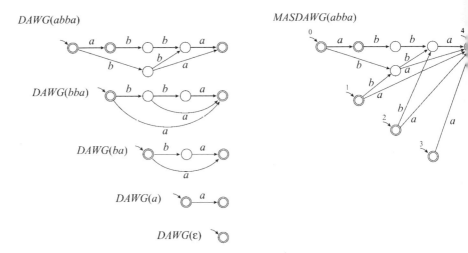

Fig. 1. $DAWG(x)$ for each string $x \in Suffix(w)$ is shown on the left, where $w = abba$. The collection of them is the naive $ASDAWG(w)$. On the right $MASDAWG(w)$ is displayed. While there are in total 16 nodes and 16 edges in the former, there are 9 nodes and 12 edges in the latter. For example, node $\langle abba, [b] \rangle$ and $\langle bba, [b] \rangle$ are equivalent due to Case 1 of Lemma and merged into one. Also, $\langle abba, [abb] \rangle$, $\langle bba, [bb] \rangle$, and $\langle ba, [b] \rangle$ are merged into one node, where the first two are equivalent due to Case 2 and the last two are equivalent due Case 3. The upper four sink nodes are equivalent due to Case 2 and the lower one is equivalent to them (see Lemma), and therefore the five are merged into the same sink node

Let $h_0, h_1, \ldots, h_r$ be the suffixes of the string h arranged in the decreasing order of their length. The above proposition implies an existence of the chain equivalent nodes

$$\langle h_0 xt, [x] \rangle, \langle h_1 xt, [x] \rangle, \ldots, \langle h_r xt, [x] \rangle.$$

In case more than one string belong to $[x]_u$, the chain length r is maximized choosing the shortest one as x. The chain, however, does not necessarily break at the node $\langle h_r xt, [x] \rangle$. The shortest string in $[x]_u$ is not necessarily the shortest in $[x]_{h_r xt}$: Shorter one may exist. Thus we need a more precise discussion.

Lemma 1. *Let $h \in \Sigma^+$ and $u, hu \in Suffix(w)$. If a node of $DAWG(u)$ equivalent to some node of $DAWG(hu)$, then it is also equivalent to some node of $DAWG(au)$ where a is the last character of the string h.*

Proof. Let $h = ta$ ($t \in \Sigma^*$). Assume $t \neq \varepsilon$. Let $x \in Sub(u)$ with $x \neq \varepsilon$, and $y \in Sub(tau)$ with $y \neq \varepsilon$. Assume $x^{-1}Suffix(u) = y^{-1}Suffix(tau)$. We have two cases to consider.

  – $x \equiv_u y$. In this case, every occurrence of the string y within tau must included within the u part. Thus, we have $x^{-1}Suffix(u) = y^{-1}Suffix(au)$.

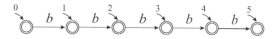

Fig. 2. $MASDAWG(w)$ for $w = b^5$. For every $i = 0, 1, \ldots, 4$, the initial node $[\varepsilon]_{b^i}$ of $DAWG(b^i)$ is equivalent to the node $[b]_{b^{i+1}}$ of $DAWG(b^{i+1})$

– $x \neq_u y$. In this case, (1) y is written as $y = sx$ where s is a nonempty string, and (2) there is an occurrence of y within tau that covers the boundary between a and u but the x part of the occurrence of $y = sx$ is contained in the u part of the string tau. In this case, by truncating an appropriate length prefix of s we can obtain a string z as a suffix of $y = sx$ such that $x^{-1}Suffix(u) = z^{-1}Suffix(au)$.

The proof is now complete.   □

The above lemma guarantees that the DAWGs sharing one node of $MASDAWG(w)$ are 'consecutive'. We therefore concentrate on the relation between two consecutive DAWGs. First, we consider the equivalence of the initial node.

Lemma 2. *Suppose $b \in \Sigma$ and $u, bu \in Suffix(w)$. Let $y \in Sub(bu)$ and assume y is the representative of $[y]_{bu}$. Then, the nodes $\langle u, [\varepsilon] \rangle$ and $\langle bu, [y] \rangle$ are equivalent under $\sim_w$ if and only if $y = b$ and u is of the form b^ℓ with $\ell \geq 0$.*

See, for example, $MASDAWG(bbbbb)$ shown in Fig. .
    As an extreme case of Lemma   where $\ell = 0$, the node $[\varepsilon]_\varepsilon$ of $DAWG(\varepsilon)$ is always equivalent to the sink node $[b]_b$ of the previous $DAWG(b)$.
    Next, we consider the equivalence of nodes other than the initial node.

Lemma 3. *Suppose $b \in \Sigma$ and $u, bu \in Suffix(w)$. Let $x \in Sub(u)$ with $x \neq \varepsilon$. Let $y \in Sub(bu)$ with $y \neq \varepsilon$. Assume x and y are the representatives of $[x]_u$ and $[y]_{bu}$, respectively. The equivalence $\langle u, [x] \rangle \sim_w \langle bu, [y] \rangle$ implies that if $y \in Prefix(bu)$ then $y = bx$ and $x \in Prefix(u)$, and otherwise $y = x$. Moreover, $\langle u, [x] \rangle \sim_w \langle bu, [y] \rangle$ holds if and only if either*

(Case 1) $x \notin Prefix(bu)$ and $y = x$;
(Case 2) $x \in Prefix(u)$, $x \equiv_{bu} y$, and $y = bx$; or
(Case 3) $x = b^i$, $y = b^{i+1}$, and u is of the form $b^\ell s$ such that $i \leq \ell$, and $s \in \Sigma^*$ *does not begin with b nor contain an occurrence of b^i.*

Proof. Suppose $x^{-1}Suffix(u) = y^{-1}Suffix(bu)$. Let $u[i+1 :]$ $(0 < i \leq |u|)$ be the longest member of this set.

1. When $y \in Prefix(bu)$. Then, $i = |y| - 1$ and $y = by'$ with $y' = u[1 : i]$. Since $u[i+1 :] \in x^{-1}Suffix(u)$, we have $u = hxu[i+1 :]$ for some $h \in \Sigma^*$. Namely, x is a suffix of $y' = u[1 : i]$.

(a) When $y' \notin Prefix(bu)$. We have $y \equiv_{bu} y'$ and

$$(y')^{-1}Suffix(u) \subseteq (y')^{-1}Suffix(bu) = y^{-1}Suffix(bu) \subseteq x^{-1}Suffix(u).$$

It derives from the assumption that $y^{-1}Suffix(bu) = x^{-1}Suffix(u)$. Thus, $(y')^{-1}Suffix(u) = x^{-1}Suffix(u)$, i.e., $x \equiv_w y'$. Since $y' \in Prefix(u)$, y' must be the representative of $[y']_u = [x]_u$. Consequently, we have $x = y'$.

(b) When $y' \in Prefix(bu)$. String y' is a prefix of $y - by'$, and therefore has a period of 1. Hence we have $y' = b^i$ and $y = b^{i+1}$. Since x is a suffix of $y' = b^i$, $x = b^j$ for some j with $0 < j \le i$. If $j < i$, then $u[j+1:] \in x^{-1}Suffix(u)$, a contradiction. Thus we have $j = i$, i.e., $x = b^i$. On the other hand, $u[1:i] = y' = b^i$ and thus u is of the form $b^\ell s$ such that $\ell \ge i$ and $s \in \Sigma^*$ does not begin with b. We can show that the string s cannot contain an occurrence of $x = b^i$.

Note that we have $x \in Prefix(u)$ in both cases.

2. When $y \notin Prefix(bu)$. We have $y^{-1}Suffix(u) = y^{-1}Suffix(bu) = x^{-1}Suffix(u)$, which implies $x \equiv_u y$. From the choice of x, y must be a suffix of x and $x = \delta y$ with $\delta \in \Sigma^*$. Assume, for a contradiction, that $x^{-1}Suffix(bu) \ne x^{-1}Suffix(u)$. Then there must be a suffix $u[j+1:]$ of u such that $j < i$ and $bu = hxu[j+1:]$ with $h \in \Sigma^*$. Since $x = \delta y$, we have $bu = h\delta yu[j+1:]$, which implies $u[j+1:] \in y^{-1}Suffix(bu)$, a contradiction. Hence we have $x \equiv_{bu} y$. From the choice of y, x must be a suffix of y. Thus we have $x = y$.

$\square$

It should be noted that Case 1 and Case 2 of Lemma   fit to Proposition , whereas Case 3 is irregular in the sense that the two equivalence classes $[x]_u$ and $[y]_{bu}$ have no common member despite $\langle u, [x] \rangle \sim_w \langle bu, [y] \rangle$. See Fig. , which includes instances of Case 1, Case 2, and Case 3.

The *owner* of a node of $MASDAWG(w)$ is defined by $DAWG(w[k:])$ such that k is the smallest integer for which $DAWG(w[k:])$ shares the node. We are now ready to estimate the lower bound of the number of nodes of $MASDAWG(w)$.

Theorem 1. *When $|\Sigma| \ge 2$, the number of nodes of $MASDAWG(w)$ for a string w is $\Theta(|w|^2)$. It is $\Theta(|w|)$ for a unary alphabet.*

Proof. The proof for the case of a unary alphabet $\Sigma = \{a\}$ is not difficult. We can use Lemma . We now prove the lower bound in case $|\Sigma| \ge 2$. Let us consider string $w = (ab)^m(ba)^m$, where a, b are distinct characters from Σ. For each $i = 2, \ldots, m-1$, let $u_i = (ab)^i(ba)^m$. Let $x = (ba)^j$ with $0 < j < i$. It is not difficult to show that $x \ne_{u_i} ax$ and $x \ne_{u_i} b^{-1}x$, and therefore $[x]_{u_i} = \{x\}$. Thus x is the representative of $[x]_{u_i}$, and we can use the above lemma. Since $x \in Prefix(bu_i)$, $x \notin Prefix(u_i)$, and the first character of u_i is not b, none of the three conditions is satisfied, and therefore $DAWG(u_i)$ is the owner of the node corresponding to $[x]_{u_i}$. Thus, the nodes of $MASDAWG(w)$ corresponding to

$$[(ba)^1]_{u_i}, [(ba)^2]_{u_i}, \ldots, [(ba)^{i-1}]_{u_i}$$

are distinct and are owned by $DAWG(u_i)$. For each i with $1 < i < m$, $DAWG(u_i)$ has at least $i-1$ own nodes. Thus, $MASDAWG(w)$ has $\Omega(m^2) = \Omega(|w|^2)$ nodes.

$\square$

3   Construction

Since the construction of the naive $ASDAWG(w)$ takes $O(|w|^2)$ and the minimization can be performed in time linear in the number of edges of the naive $ASDAWG(w)$ (see [ ]), we can build $MASDAWG(w)$ in $O(|w|^2)$ time. On the other hand, we have shown that the number of nodes in $MASDAWG(w)$ is $\Theta(|w|^2)$. We are therefore interested in on-line and direct construction of $MASDAWG(w)$. We obtained the following result.

Theorem 2. *$MASDAWG(w)$ can be constructed directly and on-line in linear time with respect to its size.*

The algorithm for the on-line construction of $MASDAWG(w)$ basically simulates the on-line constructions of the DAWGs for all suffixes of a string w. Fig. illustrates the on-line construction of $MASDAWG(abbab)$.

We present a basic idea of the algorithm together with showing several lemmas which support it.

3.1   Suffix Link

In the construction, the *suffix links* play a key role. One main difference compared with constructing a single DAWG is that a node may have more than one suffix link. This is because $MASDAWG(w)$ may contain two distinct, equivalent nodes $\langle u, [x] \rangle$ and $\langle v, [y] \rangle$ such that the node to which the suffix link of $\langle u, [x] \rangle$ points is not equivalent to the one to which the suffix link of $\langle v, [y] \rangle$ points. We update $MASDAWG(w)$ into $MASDAWG(wa)$ as if the underlying DAWGs for $w[1 :], w[2 :], \ldots$ were updated simultaneously, as follows. Conceptually, we reserve all suffix links of these DAWGs, by associating each suffix link with the corresponding DAWG. Whenever two or more suffix links are duplicated, the corresponding DAWGs are consecutive due to Lemma . Therefore we can handle them at once. This is critical for linear-time performance of our algorithm. We traverse the dag induced by the suffix links rooted from the sink node, in the order of the corresponding DAWGs, and process each encountered node appropriately (creating a new edge to the new sink node, separating the node, or redirecting an edge to the separated node).

3.2   Compact Representation of Node Length Information

In the on-line construction of the DAWG for a single string, there occurs an event so-called *node separation* [ ]. Formally, this event is described as follows. We store in each node $[x]_w$ of $DAWG(w)$ its *length*, namely, the length of the

160   Shunsuke Inenaga et al.

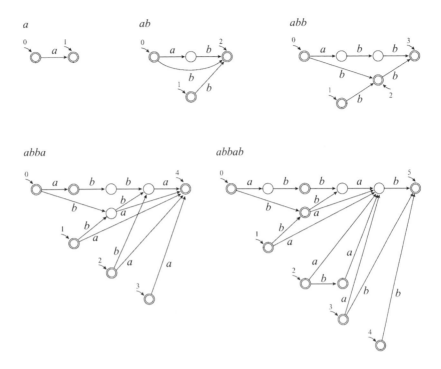

Fig. 3. On-line construction of $MASDAWG(w)$ for $w = abbab$. Each initial node
becomes independent whenever a newly appended character violates the condition of Lemma . Node separation of other type occurs only twice. One happens during the update of $MASDAWG(ab)$ to $MASDAWG(abb)$. The node con-
sisting of $\langle ab, [ab] \rangle$ and $\langle b, [b] \rangle$ is separated into two nodes. This is regarded
as a node separation in $DAWG(abb)$. The other occurs during the update of
$MASDAWG(abba)$ to $MASDAWG(abbab)$. The node consisting of $\langle abba, [abb] \rangle$,
$\langle bba, [bb] \rangle$, and $\langle ba, [b] \rangle$ is separated into two. This is a special case in the sense
that no node separation occurs inside any of $DAWG(abba)$, $DAWG(bba)$, and
$DAWG(ba)$ (See the first case of Lemma .) (Note: Though each accepting node
is double-circled in any step in this figure, we do not maintain it on-line. After
the construction of $MASDAWG(w)$ is completed, we mark every node reachable
by the suffix-links-traversal from the sink node.)

representative of $[x]_w$. Consider updating $DAWG(w)$ to $DAWG(wa)$ where a is
a character. Let z be the longest suffix of wa that also occurs within w. We call
it the *longest repeated suffix* of wa. A node separation happens iff z is not the
representative of $[z]_w$. The node $[z]_w$ can be detected by traversing the suffix link
chain from the sink node of $DAWG(w)$ in order to find its parent node $[z']_w$,
which is the first encountered node on the chain that has an out-going edge
labeled by a. Whenever the length of $[z]_w$ is greater than that of its parent $[z']_w$

plus one, the node $[z]_w$ of $DAWG(w)$ is separated into two nodes $[x]_{wa}$ and $[z]_{wa}$ in $DAWG(wa)$, where x is the representative of $[z]_w$.

Recall that a node of $MASDAWG(w)$ corresponds to an equivalence class under the equivalence relation $\sim_w$, and therefore two or more DAWGs may share a node of $MASDAWG(w)$. We need to know the length of the corresponding node of an arbitrary one of them. Naive solution would be to store into a node of $MASDAWG(w)$ a $(|w|+1)$-tuple of integers, the i-th value of which indicates the length of the corresponding node of the i-th DAWG, where $i = 0, 1, \ldots, |w|$. The overall space requirement is, however, proportional to $|w|^3$. Below we give an idea of compact representation of the tuple.

Lemma 4. *Let* $\langle w[i+1 :], [x_1] \rangle, \ldots, \langle w[i+\ell :], [x_\ell] \rangle$ *be the nodes of the naive* $ASDAWG(w)$ *which are merged into a single node in* $MASDAWG(w)$, *where* $0 \le i$ *and* $i + \ell \le |w| + 1$. *We assume each of the strings* $x_1, \ldots, x_\ell$ *is the representatives of the equivalence class of it. Then, there exists an integer* k *with* $1 \le k \le \ell$ *such that*

$$x_j = \begin{cases} x_k, & \text{if } 1 \le j \le k; \\ x_k[j-k+1 :], & \text{if } k < j \le \ell. \end{cases}$$

(See Fig. .)

Proof. By Lemma .   □

For example, $MASDAWG(abb)$ in Fig.   has a node consisting of $\langle abb, [b] \rangle$ and $\langle bb, [b] \rangle$. Also, $MASDAWG(abba)$ has a node consisting of $\langle abba, [abb] \rangle$, $\langle bba, [bb] \rangle$, and $\langle ba, [b] \rangle$.

It follows from the above lemma that the function, which takes an integer s as an input and returns $|x_s|$ if $1 \le s \le \ell$, can be represented as a quartet $(i, \ell, k, |x_k|)$, which requires only a constant space (or $O(\log |w|)$ space). The update procedure of the quartet for each node is basically apparent, except for the nodes to be separated.

3.3   Node Separation

Recall that two or more DAWGs can share one node in $MASDAWG(w)$, and each of them has a possibility of being separated into two nodes. This seems to complicate the update of $MASDAWG(w)$. However, we can readily show the following lemma.

Lemma 5. *Suppose* $b \in \Sigma$ *and* $u, bu \in Suffix(w)$. *Let* $x \in Sub(u)$ *with* $x \ne \varepsilon$. *Let* $y \in Sub(bu)$ *with* $y \ne \varepsilon$. *Assume* x *and* y *are the representatives of* $[x]_u$ *and* $[y]_{bu}$, *respectively. Suppose* $\langle u, [x] \rangle \sim_w \langle bu, [y] \rangle$. *Let* $a \in \Sigma$, *and let* z *be the longest repeated suffix of* bua. *Suppose* $z \in [y]_{bu}$. *If* $|z| < |y|$, *then* z *is also the longest repeated suffix of* ua, *and* $z \in [x]_u$. *If* $|z| = |y|$, *then* x *is a repeated suffix of* ua *(not necessarily to be the longest).*

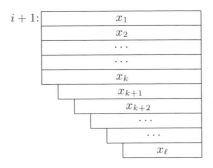

Fig. 4. The representatives x_j of $[x_j]_{w[i+j:]}$ such that the nodes $\langle w[i+j:],[x_j]\rangle$ of the naive $ASDAWG(w)$ are merged into a single node of $MASDAWG(w)$

The next lemma characterizes the node separations that occur during the update of $MASDAWG(w)$ to $MASDAWG(wa)$.

Lemma 6. *Consider the node of $MASDAWG(w)$ stated in Lemma (see Fig.). Let z be the longest repeated suffix of $w[i+j:]a$. Suppose $z \in [x_j]_{w[i+j:]}$.*

1. *When $|z| = |x_k|$: Node separation occurs in none of the DAWGs for the strings $w[i+j:],\ldots,w[i+\ell:]$.*
2. *When $|z| < |x_k|$: Let t be the maximum integer such that z is a proper suffix of x_t. Node separation occurs in each of the DAWGs for the strings $w[i+j:],\ldots,w[i+t:]$. That is, for each $j = 1,\ldots,t$, the node $[x_j]_{w[i+j:]}$ of $DAWG(w[i+j:])$ is separated into $[x_j]_{w[i+j:]a}$ and $[z]_{w[i+j:]a}$ inside $DAWG(w[i+j:]a)$. The nodes $\langle w[i+j:]a,[x_1]\rangle,\ldots,\langle w[i+\ell:]a,[x_\ell]\rangle$ are equivalent under $\sim_{wa}$, and the new nodes $\langle w[i+j:]a,[z]\rangle,\ldots,\langle w[i+t:],[z]\rangle$ are also equivalent under $\sim_{wa}$.*

The node separations of DAWGs characterized in the above lemma lead to a node separation in the update of $MASDAWG(w)$ to $MASDAWG(wa)$. It simultaneously performs the node separations within each DAWG caused by the common z. (For the same z, we can take j as small as possible.)

The remaining problem to be overcome is that there is another kind of node separation in the update of $MASDAWG(w)$.

Lemma 7. *In the update of $MASDAWG(w)$ to $MASDAWG(wa)$, node separation of the following types may occur, where $w \in \Sigma^*$ and $a \in \Sigma$.*

1. *When $w[i+1:]$ is of the form $b^{\ell+1}s$ such that $w[i] \neq b$ or $i = 0$, $\ell \geq 1$, and $s \in \Sigma^*$ does not begin with b nor contain an occurrence of b^ℓ:*
   Assume that d is the largest integer such that s contains an occurrence of b^d. $MASDAWG(w)$ has a node consisting of

$$\langle w[i+j+1:],[b^{d+k}]\rangle, \langle w[i+j+2:],[b^{d+k-1}]\rangle,\ldots,\langle w[i+j+k],[b^{d+1}]\rangle,$$

where $k = \ell - (d + j) + 1$, *for each* $j = 0, 1, \ldots, d$. *If* $|s| > 0$, s *ends with* b^d, *and* $a = b$, *then the node is separated into two nodes, one of which consists of*

$$\langle w[i + j + 1 :]a, [b^{d+k}]\rangle, \langle w[i + j + 2 :]a, [b^{d+k-1}]\rangle, \ldots$$
$$, \langle w[i + j + k - 1]a, [b^{d+2}]\rangle,$$

and the other consists only of $\langle w[i + j + k :]a, [b^{d+1}]\rangle$.

2. *When* $w[i + 1 :]$ *is of the form* b^ℓ *with* $\ell \geq 1$ *such that* $w[i] \neq b$ *or* $i = 0$: $MASDAWG(w)$ *has a node consisting of*

$$\langle b^\ell, [b^j]\rangle, \langle b^{\ell-1}, [b^{j-1}]\rangle, \ldots, \langle b^{\ell-j}, [\varepsilon]\rangle,$$

for each $j = 1, \ldots, \ell$. *Whenever* $b \neq a$, *the node is separated into two nodes, one of which consists of*

$$\langle b^\ell a, [b^j]\rangle, \langle b^{\ell-1}a, [b^{j-1}]\rangle, \ldots, \langle b^{\ell-j+1}a, [b]\rangle,$$

and the other consists only of $\langle b^{\ell-j}a, [\varepsilon]\rangle$.

For an example of the first case of the above lemma, consider the update of $MASDAWG(w)$ to $MASDAWG(wb)$ for $w = bbbbbab$. The naive $ASDAWG(w)$ and the naive $ASDAWG(wb)$ are shown in Fig. , whereas $MASDAWG(w)$ and $MASDAWG(wb)$ are displayed in Fig. .

It should be emphasized that in the node separation mentioned in the above lemma no node separation occurs *inside a DAWG*. This kind of node separation can also be performed during the suffix link traversal started at the sink node.

4   Applications

In this section we show some applications to which the data structure ASDAWG and its variants effectively contribute.

4.1   Finding Beginning-Sensitive Patterns

Definition 3 (Beginning-Sensitive Pattern). *A* beginning-sensitive pattern *(a* BS-pattern *for short) is a pair* $\langle p, i \rangle$ *where* p *is a string in* Σ^* *and* i *is a positive integer.*

Definition 4 (BS-Pattern Matching Problem).
Instance*: a text* w *and a BS-pattern* $\langle p, i \rangle$.
Determine*: whether* p *is a substring of* $w[i :]$.

This is a natural extension of the substring pattern matching problem with $i = 1$. The BS-pattern matching problem is solvable in $O(|p|)$ time for an arbitrary pair $\langle p, i \rangle$, by using $ASDAWG(w)$. For a given text w, we construct $MASDAWG(w)$ with the on-line algorithm proposed in Section . For a BS-pattern $\langle p, i \rangle$, if $i > |w|$, the BS-pattern never matches w. Otherwise, we start with the i-th initial node of $MASDAWG(w)$ and examine whether or not the string p is recognized.

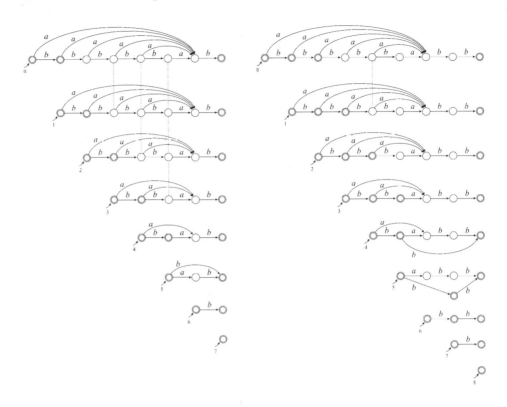

Fig. 5. The naive $ASDAWG(bbbbbab)$ on the left, and the naive $ASDAWG(w)$ on the right. The nodes connected by the dotted lines are equivalent due to Case 3. Recall the value of d mentioned in Lemma   . In string $bbbbbab$ the value of d is 1, whereas in string $bbbbbabb$ $d = 2$ since the new b is added afterward

4.2   Pattern Matching within a Specific Region

Definition 5 (Region-Sensitive Pattern). *A region-sensitive pattern (an RS-pattern for short) is a triple* $\langle p, (i,j) \rangle$ *where* p *is a string in* Σ^* *and* i, j *are positive integers.*

Definition 6 (RS-Pattern Matching Problem).
Instance*: a text* w *and an RS-pattern* $\langle p, (i,j) \rangle$.
Determine*: whether* p *occurs within the region* $w[i : j]$ *in the text* w.

This is a natural extension of the BS-pattern matching problem in which $j = |w|$. For a given text w, we construct $MASDAWG(w)$. We assign each node the integer for the position of the *rightmost occurrence* of the string corresponding to the node. For an RS-pattern $\langle p, (i,j) \rangle$, if $i > |w|$, the RS-pattern never matches w. Otherwise, we start with the i-th initial node of $MASDAWG(w)$ and examine whether or not the string p is recognized. If it is recognized, we compare j with

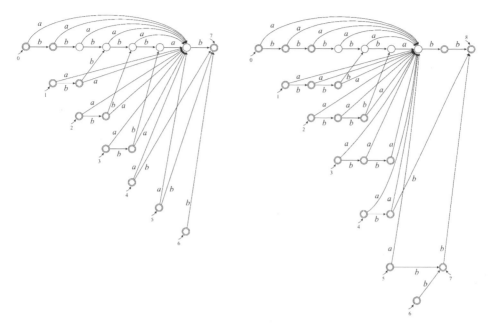

Fig. 6. $MASDAWG(bbbbbab)$ is on the left, and $MASDAWG(bbbbbabb)$ is on the right. Compare the update of $MASDAWG(bbbbbab)$ to $MASDAWG(bbbbbabb)$ with that of the naive $ASDAWG(bbbbbab)$ to the naive $ASDAWG(bbbbbabb)$ shown in Fig.

the integer k stored in the node at which p finally arrived. Then: If $j \leq k$, YES; Otherwise, NO. Obviously, the problem can be solved in $O(|p|)$ time.

4.3   Finding Variable-Length-Don't-Care's Patterns

Definition 7 (Variable-Length-Don't-Care's Pattern). *Let $\Pi = (\Sigma \cup \{\star\})^*$, where $\star$ is a* wildcard *that matches any string. An element $q \in \Pi$ is called a* variable-length-don't-care's pattern *(a VLDC-pattern for short).*

For instance, $\star a \star ab \star ba \star$ is a VLDC-pattern for $a, b \in \Sigma$. We say that a VLDC-pattern q matches a text string $w \in \Sigma^*$ if w can be obtained by replacing $\star$'s in q with some strings. In the running example, the VLDC-pattern $\star a \star ab \star ba \star$ matches text $abababbbaa$ by replacing the $\star$'s with ab, b, b and a, respectively.

Definition 8 (VLDC-Pattern Matching Problem).
Instance: *a text w and a VLDC-pattern q.*
Determine: *whether q matches w.*

The smallest automaton to recognize all possible VLDC-patterns that match a text $w \in \Sigma^*$ is a variant of $MASDAWG(w)$. We call the automaton the *wildcard*

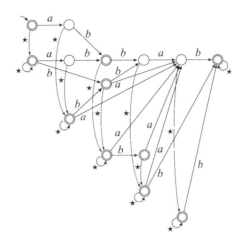

Fig. 7. $WDAWG(w)$ where $w = abbab$

$DAWG$ for w, and write it as $WDAWG(w)$. $WDAWG(abbab)$ is displayed in Fig. . In $WDAWG(w)$, a $\star$-transition is added between each node and the initial node of the "same layer" in $MASDAWG(w)$ (see also $MASDAWG(abbab)$ in Fig.). Note that there exist two additional nodes, one of which is a unique initial node of $WDAWG(abbab)$. They are added in order that VLDC-patterns beginning with a can be recognized. For any $q \in \Pi$, the VLDC-pattern matching problem can be solved in $O(|q|)$ time, by using $WDAWG(w)$.

References

1. D. Angluin. Finding patterns common to a set of strings. *J. Comput. Sys. Sci.*, 21:46–62, 1980.
2. R. A. Baeza-Yates. Searching subsequences (note). *Theoretical Computer Science*, 78(2):363–376, Jan. 1991.
3. A. Blumer, J. Blumer, D. Haussler, A. Ehrenfeucht, M. T. Chen, and J. Seiferas. The smallest automaton recognizing the subwords of a text. *Theoretical Computer Science*, 40:31–55, 1985.   ,
4. A. Blumer, J. Blumer, D. Haussler, R. McConnell, and A. Ehrenfeucht. Complete inverted files for efficient text retrieval and analysis. *J. ACM*, 34(3):578–595, 1987.

5. M. Crochemore. Transducers and repetitions. *Theoretical Computer Science*, 45:63–86, 1986.   ,
6. M. Crochemore and Z. Troníček. Directed acyclic subsequence graph for multiple texts. Technical Report IGM-99-13, Institut Gaspard-Monge, June 1999.
7. M. Crochemore and R. Vérin. On compact directed acyclic word graphs. In J. Mycielski, G. Rozenberg, and A. Salomaa, editors, *Structures in Logic and Computer Science*, volume 1261 of *LNCS*, pages 192–211. Springer-Verlag, 1997.

8. R. Grossi and J. S. Vitter. Compressed suffix arrays and suffix trees with applications to text indexing and string matching. In *Proc. of 32nd ACM Symposium on Theory of Computing (STOC'00)*, pages 397–406, 2000.
9. M. Hirao, H. Hoshino, A. Shinohara, M. Takeda, and S. Arikawa. A practical algorithm to find the best subsequence patterns. In S. Arikawa and S. Morishita, editors, *Proc. The Third International Conference on Discovery Science*, volume 1967 of *LNAI*, pages 141–154. Springer-Verlag, 2000.
10. M. Hirao, S. Inenaga, A. Shinohara, M. Takeda, and S. Arikawa. A practical algorithm to find the best episode patterns. In K. P. Jantke and A. Shinohara, editors, *Proc. The Fourth International Conference on Discovery Science*, volume 2226 of *LNAI*, pages 435–440. Springer-Verlag, 2001.
11. S. Inenaga, H. Hoshino, A. Shinohara, M. Takeda, S. Arikawa, G. Mauri, and G. Pavesi. On-line construction of compact directed acyclic word graphs. In A. Amir and G. M. Landau, editors, *Proc. 12th Annual Symposium on Combinatorial Pattern Matching (CPM'01)*, volume 2089 of *LNCS*, pages 169–180. Springer-Verlag, 2001.
12. J. Kärkkäinen. Suffix cactus: A cross between suffix tree and suffix array. In Z. Galil and E. Ukkonen, editors, *Proc. 6th Annual Symposium on Combinatorial Pattern Matching (CPM'95)*, volume 973 of *LNCS*, pages 191–204. Springer-Verlag, 1995.

13. V. Mäkinen. Compact suffix array. In R. Giancarlo and D. Sankoff, editors, *Proc. 11th Annual Symposium on Combinatorial Pattern Matching (CPM'00)*, volume 1848 of *LNCS*, pages 305–319. Springer-Verlag, 2000.
14. U. Manber and G. Myers. Suffix arrays: A new method for on-line string searches. *SIAM J. Compt.*, 22(5):935–948, 1993.
15. H. Mannila, H. Toivonen, and A. I. Verkamo. Discovering frequent episode in sequences. In U. M. Fayyad and R. Uthurusamy, editors, *Proc. 1st International Conference on Knowledge Discovery and Data Mining*, pages 210–215. AAAI Press, Aug. 1995.
16. E. M. McCreight. A space-economical suffix tree construction algorithm. *J. ACM*, 23(2):262–272, Apr. 1976.
17. D. Revuz. Minimization of acyclic deterministic automata in linear time. *Theoretical Computer Science*, 92(1):181–189, Jan. 1992.
18. K. Sadakane. Compressed text databases with efficient query algorithms based on the compressed suffix array. In *Proc. of 11th International Symposium on Algorithms and Computation (ISAAC'00)*, volume 1969 of *LNCS*, pages 410–421. Springer-Verlag, 2000.
19. Z. Troníček. Episode matching. In A. Amir and G. M. Landau, editors, *Proc. 12th Annual Symposium on Combinatorial Pattern Matching (CPM'01)*, volume 2089 of *LNCS*, pages 143–146. Springer-Verlag, 2001.
20. E. Ukkonen. On-line construction of suffix trees. *Algorithmica*, 14(3):249–260, 1995.
21. P. Weiner. Linear pattern matching algorithms. In *Proc. 14th Annual Symposium on Switching and Automata Theory*, pages 1–11, Oct. 1973.

On the Complexity of Deriving Position Specific Score Matrices from Examples

Tatsuya Akutsu[1,2], Hideo Bannai[3], Satoru Miyano[1,3], and Sascha Ott[3]

[1] Bioinformatics Center, Institute for Chemical Research, Kyoto University
Uji 611-0011, Japan
`takutsu@kuicr.kyoto-u.ac.jp`
[2] Graduate School of Informatics, Kyoto University
Sakyo-ku, Kyoto 606-8501, Japan
[3] Human Genome Center, Institute of Medical Science, University of Tokyo
Minato-ku, Tokyo 108-8639, Japan
`{bannai,miyano,ott}@ims.u-tokyo.ac.jp`

Abstract. PSSMs (Position-Specific Score Matrices) have been applied to various problems in Bioinformatics. We study the following problem: given positive examples (sequences) and negative examples (sequences), find a PSSM which correctly discriminates between positive and negative examples. We prove that this problem is solved in polynomial time if the size of a PSSM is bounded by a constant. On the other hand, we prove that this problem is NP-hard if the size is not bounded. We also prove similar results on deriving a mixture of PSSMs.

1 Introduction

Position-Specific Score Matrices (PSSMs) have been applied to various problems in *Bioinformatics* such as detection of remote homology, identification of DNA regulatory regions and detection of motifs [ ]. Usually, PSSMs are derived from training data. Therefore, how to derive a good PSSM from training data (examples) has been a key issue in these applications. Various methods have been proposed for this purpose. Among them, simple statistical methods based on residue frequencies and local search algorithms (such as Expectation Maximization algorithms) have been widely used [ ]. However, from the algorithmic viewpoint, almost no theoretical studies have been done on the derivation of PSSMs. Therefore, we study the following fundamental version of the problem: *given positive examples (sequences) and negative examples (sequences), find a PSSM which completely discriminates between positive and negative examples.* We prove that this problem is NP-hard in general but can be solved in polynomial time if the size of a PSSM is bounded. It follows from the latter result that derivation of position *non-specific* score matrices can be derived from examples in polynomial time, where hydropathic indices [ ] are well-known examples of position *non-specific* score matrices.

Before reviewing related results, we formally define PSSMs and the derivation problem (see also Fig. 1). Let Σ be an alphabet. Let $POS = \{P^1, P^2, \ldots\}$ and

A. Apostolico and M. Takeda (Eds.): CPM 2002, LNCS 2373, pp. 168–   , 2002.

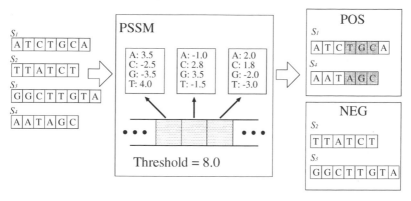

Fig. 1. An example of PSSM. S_1 (resp. S_4) is classified as a positive sequence because $f(\text{TGC}) = 9.3 \geq 8.0$ (resp. $f(\text{AGC}) = 8.8 \geq 8.0$)

$NEG = \{N^1, N^2, \ldots\}$ be sets of strings on Σ, where POS and NEG mean a set of positive examples and a set of negative examples, respectively. For string S, $S[i]$ denotes the i-th letter of S and $S_{i,j}$ denotes the substring $S[i]S[i+1]\ldots S[j]$ of S. For strings S_1 and S_2, $S_1 \cdot S_2$ denotes the concatenation of S_1 and S_2. Let L be a positive integer indicating the length of a motif region to be detected.

Definition 1 (PSSM). *A PSSM is a function $f_k(a)$ from $[1,\ldots,L] \times \Sigma$ to the set of real numbers, where $k \in [1,\ldots,L]$ and $a \in \Sigma$.*

For string S of length L, we define $f(S)$ (the score of S) by $f(S) = \sum_{i=1}^{L} f_i(S[i])$.

Problem 1 (Derivation of a PSSM from examples).
Given Σ, POS, NEG and L, find a PSSM and a threshold Θ which satisfy the following conditions:

- *For all $P^h \in POS$, $f(P^h_{j,j+L-1}) \geq \Theta$ holds for some $j \in [1,\ldots,|P^h|-L+1]$,*
- *For all $N^h \in NEG$ and for all $j = 1,\ldots,|N^h|-L+1$, $f(N^h_{j,j+L-1}) < \Theta$.*

There are many studies on related problems. For example, hardness results and approximation algorithms were obtained for local multiple alignment [,] and the distinguishing string selection problem [ ], and hardness results were obtained for learning string patterns from positive and negative examples [,]. However, techniques used in these papers are not directly applicable to the derivation problem of PSSMs.

We also consider derivation of *a mixture of PSSMs*. Mixtures of PSSMs are also used in Bioinformatics since a single PSSM is not always sufficient for characterizing sequences having common biological properties. As in the above, there are almost no theoretical studies on derivation of a mixture of PSSMs. We prove that this problem can be solved in polynomial time if the size of a PSSM is bounded. We also consider a special case in which the regions to be identified are already known. In this case, derivation of a single PSSM can be done in polynomial time by a naive algorithm based on linear programming. However, we show that derivation of a mixture of two PSSMs is NP-hard even for this restricted case.

Relating to derivation of PSSMs, Akutsu and Yagiura studied the following problem [ ]: given correct alignments and incorrect alignments, find a score function with which the scores of the correct alignments are optimal and the scores of the incorrect alignments are not optimal. They proved that this problem is computationally hard for multiple alignment with SP-scoring, but is polynomial time solvable for pairwise alignment by using a reduction to linear programming. In this paper, we consider the problem of deriving score matrices for pairwise alignment under the condition that each (positive or negative) example consists of a pair of sequences (i.e., alignment results are not given). This definition is reasonable because we can obtain sets of homologous sequences and sets of non-homologous sequences by human knowledge, but it is very difficult to know correct alignments. It should be noted that a score matrix is usually obtained from the results of sequence alignment using another score matrix [, ,]. But, it seems that this approach is not adequate because a circular reasoning method is used. Therefore, we study this problem and prove that it is NP-hard for a general alphabet.

2   Deriving a PSSM from Examples

In this section, we show that Problem 1 is NP-hard in general but can be solved in polynomial time if the size of a PSSM is bounded by a constant.

Theorem 1. *Problem 1 is NP-hard.*

Proof. We use a polynomial time reduction from 3SAT.

Let $C = \{c_1, \ldots, c_m\}$ be a set of clauses over a set of boolean variables $X = \{x_1, \ldots, x_n\}$, where each clause consists of three literals.

From this instance, we construct an instance of the PSSM derivation problem (see Fig. 2). Let $\Sigma = \{0, 1\}$ and $L = 4n$. Let $S(i_1, i_2, \ldots)$ denote the string of length $4n$ such that $S[i] = 1$ for $i = i_1, i_2, \ldots$, otherwise $S[i] = 0$. Then, NEG is defined by

$$NEG = \{ S(), S(4n) \} \cup \{ S(i,j) \mid 1 \le i < j < 4n \} \cup$$
$$\{ S(i) \mid i = 1, \ldots, 4n \} \cup \{ S(2i-1, 2i, 4n) \mid i = 1, \ldots, n \}.$$

It should be noted that the $(2i-1)$-th position and the $(2i)$-th position ($i = 1, \ldots, n$) of each string correspond to literals x_i and $\overline{x_i}$, respectively.

Let $c_i = l_{i_1} \vee l_{i_2} \vee l_{i_3}$, where l_{i_k} is either x_{i_k} or $\overline{x_{i_k}}$. We define $g(i_k)$ by $g(i_k) = 2i_k - 1$ if $l_{i_k} = x_{i_k}$, otherwise $g(i_k) = 2i_k$. Then, P^i is defined by $P^i = S(g(i_1), 4n) \cdot S() \cdot S(g(i_2), 4n) \cdot S() \cdot S(g(i_3), 4n)$. POS consists of $P^1, P^2, \ldots, P^n$.

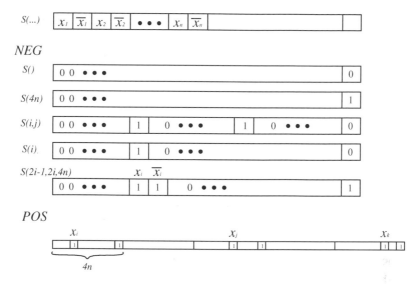

Fig. 2. Construction of POS and NEG in Theorem

First we show that if C is satisfiable then there exists a PSSM f_k satisfying the condition of Problem 1 for $\Theta = 3$. From the truth assignment to X satisfying all the clauses, we construct f_k by:

- for $i = 1, \ldots, n$,
  $f_{2i-1}(0) = 0$, $f_{2i-1}(1) = 1$, $f_{2i}(0) = 0$ and $f_{2i}(1) = -1$ if x_i is true,
  $f_{2i-1}(0) = 0$, $f_{2i-1}(1) = -1$, $f_{2i}(0) = 0$ and $f_{2i}(1) = 1$ otherwise,
- for $i = 2L + 1, \ldots, 4n - 1$, $f_i(0) = f_i(1) = 0$,
- $f_{4n}(0) = 0$ and $f_{4n}(1) = 2$.

Then, it is easy to see that $f(N^h) < 3$ holds for all $N^h \in NEG$, and either $f(P^h_{1,4n}) = 3$, $f(P^h_{8n+1,12n}) = 3$ or $f(P^h_{16n+1,20n}) = 3$ holds for all $P^h \in POS$.

Next we show that if there is a PSSM satisfying the condition of Problem 1, there exists a truth assignment satisfying all the clauses in C. Let $\hat{x}_i = f_{2i-1}(1) - f_{2i-1}(0)$, $\hat{\overline{x}}_i = f_{2i}(1) - f_{2i}(0)$ and $\hat{a} = f_{4n}(1) - f_{4n}(0)$. Let $f(S()) = z$. Then, it is easy to check that for all $P^h \in POS$, all substrings of length L except $P^h_{1,4n}$, $P^h_{8n+1,12n}$ and $P^h_{16n+1,20n}$ appear in NEG. Therefore, the following relations hold if Problem 1 has a solution:

- $z < \Theta$, $z + \hat{a} < \Theta$
- $z + \hat{l}_i + \hat{l}_j < \Theta$ for all $l_i \neq l_j$, where l_i (resp. l_j) is either x_i (resp. x_j) or $\overline{x_i}$ (resp. $\overline{x_j}$),
- $z + \hat{l}_i < \Theta$ for all l_i,
- $z + \hat{x}_i + \hat{\overline{x}}_i + \hat{a} < \Theta$ for all x_i,
- $z + \hat{l}_{i_k} + \hat{a} \geq \Theta$ holds for some $k \in [1, 2, 3]$ for all $c = l_{i_1} \vee l_{i_2} \vee l_{i_3}$.

From f and Θ satisfying the above relations, we construct a truth assignment to X as follows: x_i is true if $z + \hat{x}_i + \hat{a} \geq \Theta$, x_i is false if $z + \overline{\hat{x}_i} + \hat{a} \geq \Theta$, otherwise x_i is arbitrary.

It is sufficient to show that either $z + \hat{x}_i + \hat{a} < \Theta$ or $z + \overline{\hat{x}_i} + \hat{a} < \Theta$ holds. Suppose that $z + \hat{x}_i + \hat{a} \geq \Theta$ held. Then, $\hat{x}_i > 0$ would hold from this inequality and $z + \hat{a} < \Theta$. From $\hat{x}_i > 0$ and $z + \hat{x}_i + \overline{x_i} + a < \Theta$, $z + \overline{x_i} + a < \Theta$ would hold. □

Theorem 2. *Problem 1 can be solved in polynomial time if Σ and L are fixed.*

Proof. We construct an *arrangement* of hyperplanes, where the arrangement is a well-known concept in computational geometry [ ]. We construct the arrangement in the $(|\Sigma|L+1)$-dimensional Euclidean space for the following hyperplanes:

- $f(P^h_{j,j+L-1}) - \Theta = 0$ for $j = 1, \ldots, |P^h| - L + 1$ and for all $P^h \in POS$,
- $f(N^h_{j,j+L-1}) - \Theta = 0$ for $j = 1, \ldots, |N^h| - L + 1$ and for all $N^h \in NEG$.

Then, we pick an arbitrary point (which corresponds to a pair of f and Θ) from each cell and check whether or not the conditions of Problem 1 hold. Since the sign of each function (i.e., $f(P^h_{j,j+L-1}) - \Theta$, $f(N^h_{j,j+L-1}) - \Theta$) does not change within a cell, this algorithm correctly solves Problem 1.

Since the arrangement of hyperplanes in fixed dimensions can be constructed in polynomial time and the combinatorial complexity of the arrangement is also polynomially bounded [ ], the algorithm works in polynomial time. □

It should be noted that if L and Σ are fixed, the number of possible sequences is bounded by a constant and thus Theorem is trivial. However, the proof can be extended for the case where L is not fixed but the size of a PSSM is fixed (i.e., the number of parameters in a PSSM is bounded by a constant). In this case, the number of possible sequences is not necessarily bounded by a constant.

Derivation of hydropathic indices [ ] is such an example. Hydropathic indices have been used for the identification of transmembrane domains of membrane proteins. Usually, hydropathic indices are not position-specific, i.e., $f_i[a] = f_j[a]$ for all $i \neq j$. The algorithm above can also be applied to this case. In this case, the arrangement in the $(|\Sigma| + 1)$-dimensional Euclidean space is constructed. Since $|\Sigma|$ is 4 or 20, we have:

Corollary 1. *Hydropathic indices satisfying the condition of Problem 1 can be derived from examples in polynomial time.*

In most cases of deriving hydropathic indices, positive examples given as training data contain information about the positions of the transmembrane domains. Therefore, various learning algorithms have been applied to derivation of hydropathic indices. But, the above theorem suggests that hydropathic indices can be derived even if the positions of the transmembrane domains are not known. Of course, the time complexity of the algorithm is still too high even for $|\Sigma| = 4$. Thus, an improved algorithm should be developed.

3 Deriving a Mixture of PSSMs from Examples

In this section, we consider the following problem.

Problem 2 (Derivation of a mixture of PSSMs from examples).
Given Σ, POS, NEG, L and N where N denotes the number of PSSMs, find a set of PSSMs with cardinality N and a threshold Θ which satisfy the following conditions:

- *For all $P^i \in POS$, $f^k(P^i_{j,j+L-1}) \geq \Theta$ holds for some $j \in [1, \ldots, |P_i| - L + 1]$ and for some $k \in [1, \ldots, N]$,*
- *For all $N^i \in NEG$, for all $j = 1, \ldots, |N^i| - L + 1$ and for all $k \in [1, \ldots, N]$, $f^k(N^i_{j,j+L-1}) < \Theta$,*

where f^k denotes the score given by the k-th PSSM.

Clearly, Problem 2 is NP-hard from Theorem . As in Theorem , Problem 2 can be solved in polynomial time if the size of a PSSM is fixed (i.e., the number of parameters in a PSSM is bounded by a constant) and N is a constant.

Theorem 3. *Problem 2 can be solved in polynomial time if the size of a PSSM is fixed and N is bounded by a constant.*

Proof. We show the proof for the case of $N = 2$. Extension of the proof to an arbitrary constant N is straight-forward.

As in the proof of Theorem , we construct an arrangement in $(d + 1)$-dimensional Euclidean space, where d is the number of parameters in the PSSM. Each point in the space corresponds to a pair of PSSM and Θ. For each point p in the space, p^θ denotes the value of the coordinate corresponding to Θ. We pick a pair of points (p_1, p_2) from each pair of cells (c_1, c_2) such that $p_1 \in c_1$, $p_2 \in c_2$ and $(p_1)^\theta = (p_2)^\theta$ if such a pair exists. It is easy to check whether or not such a pair exists: we simply compare the maximum and minimum values of the coordinate corresponding to Θ in the cells. Then, we check whether or not the conditions of Problem 2 are satisfied for the pair of PSSMs corresponding to (p_1, p_2). Since the combinatorial complexity of the arrangement is polynomially bounded, this algorithm works in polynomial time. □

Here, we consider a special case of the problem in which the regions to be identified are given. It is a reasonable restriction because the regions are known for training data in several applications. For example, transmembrane domains are usually known for training data of membrane proteins. In such a case, we treat the regions in *POS* as positive examples and we can assume that all of sequences are of the same length L. We denote this special case by **Problem 3**.

Proposition 1. *Problem 3 can be solved in polynomial time if $N = 1$ or $N \geq |POS|$.*

Proof. The case of $N = 1$ is trivial and well-known. We simply construct the linear inequalities: $f^1(P^h) \geq \Theta$ for all $P^h \in POS$, and $f^1(N^h) < \Theta$ for all $N^h \in NEG$. Then, we can obtain a PSSM and Θ by applying any polynomial time algorithm for linear programming.

In the case of $N = |POS|$, we solve the following inequalities: $f^h(P^h) \geq \Theta$ for all $P^h \in POS$, and $f^k(N^h) < \Theta$ for all $N^h \in NEG$ and for all $k \in [1 \ldots N]$.   ⊔

Theorem 4. *Problem 3 is NP-hard even for $N = 2$.*

Proof. We reduce NOT-ALL-EQUAL 3SAT (LO3 in [ ]) to Problem 3.
Let U be a set of variables and C be a set of clauses over U such that each clause $c \in C$ has $|c| = 3$. Given the instance (U, C) for NOT-ALL-EQUAL 3SAT, we define an instance $I(U, C)$ for Problem 3 as follows.

Let L denote the set of literals over U and let $p : L \rightarrow \{1, \ldots, 2|U|\}$ be a bijection such that for each $u \in U$ $p(\overline{u}) = p(u) + 1$ holds. We define POS as the set

$$POS = \{0^i 10^{2|U|-i-1} \mid i = 0, \ldots, 2|U| - 1\}$$

and NEG as

$$NEG = \{ \ 0^i 110^{2|U|-i-2} \mid i = 0, 2, \ldots, 2|U| - 2 \ \} \ \cup \ \{ \ 0^{2|U|} \ \} \ \cup$$
$$\{ \ 0^{p(x)-1} 10^{p(y)-p(x)-1} 10^{p(z)-p(y)-1} 10^{2|U|-p(z)} \mid \{ \ x, y, z \ \} \ \in C,$$
$$p(x) < p(y) < p(z) \ \}.$$

We denote the Problem 3-instance consisting of POS and NEG as $I(U, C)$. We say a string in POS corresponds to a literal in L, iff the 1 appears at the $p(L)$-th position in the string. For a literal $l \in L$, we denote the string corresponding to l with w_l. In the same way, we denote strings $0^i 110^{2|U|-i-2} \in NEG$ as $w_{u,\overline{u}}$ for the variable $u \in U$ with $p(u) = i + 1$.

We have to show that there is a not-all-equal truth assignment for (U, C), iff there is a solution for $I(U, C)$ with 2 matrices.

Let σ be a not-all-equal truth assignment. Call the set of all strings of POS, for which the corresponding literal is satisfied by σ, POS_A, and the set of all other strings of POS POS_B. Let A denote the PSSM, which assigns the value 1 to all occurrences of the character 1 at positions corresponding to satisfied literals, the value -2 to all other occurrences of 1, and the value 0 to all occurrences of 0. In the same way, let B denote the PSSM assigning 1 to the occurrences of 1 at positions corresponding to literals not satisfied by σ and -2 to the other occurrences.

For $\Theta = 1$, A accepts all strings in POS_A and B accepts all strings in POS_B. Furthermore, for every clause $c \in C$, there is a literal in c satisfied by σ and a literal not satisfied by σ. Therefore, for every string $s \in NEG$, $A(s) \leq 0$ and $B(s) \leq 0$, which shows that (A, B) is a solution for $I(U, C)$.

It remains to show, that there is a not-all-equal truth assignment for C, if there is a solution for $I(U, C)$. Let A, B be PSSMs solving $I(U, C)$. Let σ denote the truth assignment satisfying all literals, for which the corresponding string in POS is accepted by A, and unsatisfying all other literals. If there were a variable $u \in U$ with $\sigma(u) = \sigma(\overline{u})$, then $A(w_u) + A(w_{\overline{u}}) \geq 2\Theta$ or $B(w_u) + B(w_{\overline{u}}) \geq 2\Theta$ would hold. Therefore, since $A(w_u) + A(w_{\overline{u}}) = A(0^{2|U|}) + A(w_{u,\overline{u}})$ holds (and analogously for B), $0^{2|U|}$ or $w_{u,\overline{u}}$ would be accepted by either A or B, a contradiction. Thus, σ is well-defined.

To see that σ has the not-all-equal property, assume that there is a clause $\{x, y, z\} \in C$ with $\sigma(x) = \sigma(y) = \sigma(z)$. If $\sigma(x) = 1$, then we have

$$A(w_x) + A(w_y) + A(w_z) \geq 3\Theta.$$

Moreover,

$$A(w_x) + A(w_y) + A(w_z) = 2A(0^{2|U|}) + A(w_{x,y,z})$$

holds, where $w_{x,y,z}$ denotes the string of NEG corresponding to clause $\{x, y, z\}$. Therefore, since $A(w_{x,y,z}) < \Theta$ holds, we have $A(0^{2|U|}) > \Theta$, contradicting $0^{2|U|} \in NEG$. If $\sigma(x) = 0$, the same contradiction follows for B.   □

4   Deriving a Score Matrix for Pairwise Alignment

Let S_1 and S_2 be sequences over Σ. An *alignment* of S_1 and S_2 is obtained by inserting *gap symbols* (denoted by '$-$') into or at either end of S_1 and S_2 such that the two resulting sequences S_1' and S_2' are of the same length l [ ]. Let $f(x, y)$ be a function from $\Sigma' \times \Sigma'$ to $\mathcal{R}$ that satisfies $f(x, y) = f(y, x)$ and $f(x, -) = f(-, y) = g$ for all $x, y \in \Sigma$ and $f(-, -) = -\infty$, where $\Sigma' = \Sigma \cup \{-\}$. Note that we consider a linear gap penalty [ ] and g denotes the penalty per gap. The score of an alignment is defined by $\sum_{i=1}^{l} f(S_1'[i], S_2'[i])$. The *optimal alignment* between S_1 and S_2 is the alignment with the *maximum score*. Let $s(S_1', S_2')$ denote the score of alignment (S_1', S_2') and let $opt(S_1, S_2)$ denote the score of the optimal alignment between S_1 and S_2.

In this case, we assume that each example is a pair of sequences (i.e., POS and NEG are sets of sequence pairs over Σ).

Problem 4.
Given POS and NEG over Σ, find $f(x, y)$ and Θ which satisfy the following conditions:

- *For all $(S_i, S_j) \in POS$, $opt(S_i, S_j) \geq \Theta$,*
- *For all $(S_i, S_j) \in NEG$, $opt(S_i, S_j) < \Theta$.*

Theorem 5. *Problem 4 is NP-hard for a general alphabet.*

Proof. We use a reduction from 3SAT.
As in the proof of Theorem , let C be a set of clauses over a set of variables X. We let $\Sigma = \{x_i, \overline{x_i} | x_i \in X\} \cup \{\alpha, \beta\}$. Then, POS and NEG are defined by

$$POS = \{(\alpha, \alpha), (\alpha\alpha, \alpha\alpha), (\beta, \beta), (\beta\beta, \beta\beta)\} \ \cup$$

$$\{(\alpha, x_i\overline{x_i}) \mid x_i \in X\} \cup \{(\beta\alpha, \beta l_{i_1}l_{i_2}l_{i_3}) \mid l_{i_1} \vee l_{i_2} \vee l_{i_3} \in C\},$$
$$NEG = \{(\alpha, \alpha\alpha), (\beta, \beta\beta)\} \cup \{(\alpha\alpha, x_i\overline{x_i}) \mid x_i \in X\} \cup$$
$$\{(\beta\alpha, \beta\alpha\alpha\alpha)\} \cup \{(\beta, x_i), (\beta, \overline{x_i}) \mid x_i \in X\}$$

First we show that if C is satisfiable then there exists a score matrix $f(x, y)$ satisfying the conditions of Problem 4 for $\Theta = 1$. We define $f(x, y)$ by

- $f(\alpha, \alpha) = f(\beta, \beta) = 1$ and $g = -1$,
- $f(\alpha, x_i) = 2$ and $f(\alpha, \overline{x_i}) = -2$ if x_i is true,
- $f(\alpha, x_i) = -2$ and $f(\alpha, \overline{x_i}) = 2$ if x_i is false,
- $f(x, y) = -1$ for the other pairs (x, y).

Then, it is easy to see that the conditions are satisfied.

Next we show that if there is a score matrix satisfying the conditions of Problem 4, there exists a truth assignment satisfying all the clauses in C.

From $(\alpha, \alpha), (\alpha\alpha, \alpha\alpha) \in POS$ and $(\alpha, \alpha\alpha) \in NEG$, we have $g < 0$ and $g < f(\alpha, \alpha)$. Similarly, we have $g < f(\beta, \beta)$. From $opt(\alpha, x_i\overline{x_i}) \geq \Theta > opt(\alpha, \alpha\alpha)$ and $opt(\alpha\alpha, x_i\overline{x_i}) < \Theta \leq opt(\alpha\alpha, \alpha\alpha)$, either $f(\alpha, x_i) > f(\alpha, \alpha) > f(\alpha, \overline{x_i})$ or $f(\alpha, x_i) < f(\alpha, \alpha) < f(\alpha, \overline{x_i})$ must hold for each x_i. Then, we construct an assignment to X by:

$$x_i \text{ is true iff. } f(\alpha, x_i) > f(\alpha, \alpha).$$

Since $opt(\beta\alpha, \beta l_{i_1}l_{i_2}l_{i_3}) > opt(\beta\alpha, \beta\alpha\alpha\alpha)$, $f(\beta, \beta) > f(\beta, l_j)$ and $f(\alpha, \alpha) > f(\beta, l_j)$ for all j, $f(\alpha, l_k) > f(\alpha, \alpha)$ must hold for some $k \in \{i_1, i_2, i_3\}$. Therefore, we can satisfy all the clauses. $\qquad\square$

This result is interesting because this general case can be solved in polynomial time if alignment results are given [ ]. It is not yet known whether Problem 4 is NP-hard for a fixed alphabet.

5 Concluding Remarks

In this paper, we have shown that derivation of a PSSM is NP-hard in general but is polynomial time solvable if the size of the PSSM is bounded by a constant. We also showed that derivation of a mixture of two PSSMs is NP-hard even if the regions to be identified are known. Development of approximation algorithms for the NP-hard problems and development of faster algorithms for derivation of bounded-size PSSMs are important future work.

Acknowledgement

This work was supported in part by a Grant-in-Aid for Scientific Research on Priority Areas (C) for "Genome Information Science" from the Ministry of Education, Culture, Sports, Science and Technology (MEXT) of Japan. Tatsuya Akutsu was also partially supported by HITOCC (Hyper Information Technology Oriented Corporation Club) and Grant-in-Aid #13680394 from MEXT, Japan.

References

1. Akutsu, T., Yagiura, M.: On the complexity of deriving score functions from examples for problems in molecular biology. Proc. ICALP'98. Lecture Notes in Computer Science, Vol. 1443. Springer-Verlag, Berlin Heidelberg New York (1998) 832–843
   ,
2. Akutsu, T., Arimura, H., Shimozono, S.: On approximation algorithms for local multiple alignment. Proc. 4th ACM Int. Conf. Computational Molecular Biology (2000) 1–7
3. Durbin, R., Eddy, S., Krogh, A., Mitchison, G.: Biological Sequence Analysis. Probabilistic Models of Proteins and Nucleic Acids. Cambridge University Press (1998)
   ,   ,
4. Edelsbrunner, H.: Algorithms in Combinatorial Geometry. Springer-Verlag, Berlin Heidelberg New York (1987)
5. Garey, M. R., Johnson, D. S.: Computers and Intractability. Freeman (1979)
6. Henikoff, S., Henikoff, J. G.: Amino acid substitution matrices from protein blocks. Proc. National Academy of Sciences of the USA **89** (1992) 10915–10919
7. Jiang, T., Li, M.: On the complexity of learning strings and sequences. Proc. 4th ACM Workshop on Computational Learning Theory (1991) 367–274
8. Kann, M., Qian, B., Goldstein, R. A.: Optimization of a new score function for detection of remote homologs. Proteins **41** (2000) 498–503
9. Kyte, J., Doolittle, R. F.: A simple method for displaying the hydropathic character of a protein. J. Molecular Biology **157** (1982) 105–132   ,
10. Lanctot, K., Li, M., Ma, B., Wang, S., Zhang, L.: Distinguishing string selection problems. Proc. 10th ACM-SIAM Symp. Discrete Algorithms (1999) 633–642
11. Li, M., Ma, B., Wang, L.: Finding similar regions in many strings. Proc. 31st ACM Symp. Theory of Computing (1999) 473–482
12. Miyano, S., Shinohara, A., Shinohara, T.: Which classes of elementary formal systems are polynomial-time learnable. Proc. 2nd Workshop on Algorithmic Learning Theory (1991) 139–150

Three Heuristics for δ-Matching: δ-BM Algorithms

Maxime Crochemore[1] [*], Costas S. Iliopoulos[2,3], Thierry Lecroq[4],
Wojciech Plandowski[5], and Wojciech Rytter[6,7]

[1] Institut Gaspard Monge, Université Marne-la-Vallée, Cité Descartes
5 Bd Descartes, Champs-Sur-Marne, 77454 Marne-la-Vallée CEDEX 2, France
mac@univ-mlv.fr
http://www-igm.univ-mlv.fr/~mac
[2] Dept. of Computer Science, King's College London
London WC2R 2LS, U.K.
[3] School of Computing, Curtin University of Technology
GPO Box 1987 U, WA. Australia csi@dcs.kcl.ac.uk
http://www.dcs.kcl.ac.uk/staff/csi
[4] LIFAR–ABISS, Faculté des Sciences et Techniques, Université de Rouen
76821 Mont-Saint-Aignan CEDEX, France
Thierry.Lecroq@univ-rouen.fr
http://www-igm.univ-mlv.fr/~lecroq
[5] Instytut Informatyki, Uniwersytet Warszawski
ul. Banacha 2, 02-097, Warszawa, Poland
W.Plandowski@mimuw.edu.pl
URL: http://www.mimuw.edu.pl/~wojtekpl
[6] Instytut Informatyki, Uniwersytet Warszawski
ul. Banacha 2, 02-097, Warszawa, Poland
[7] Department of Computer Science, Liverpool University
Peach Street, Liverpool L69 7ZF, U.K.
W.Rytter@mimuw.edu.pl
http://www.mimuw.edu.pl/~rytter

Abstract. We consider a version of pattern matching useful in processing large musical data: δ-matching, which consists in finding matches which are δ-approximate in the sense of the distance measured as maximum difference between symbols. The alphabet is an interval of integers, and the distance between two symbols a, b is measured as $|a - b|$. We present δ-matching algorithms fast on the average providing that the pattern is "non-flat" and the alphabet interval is large. The pattern is "flat" if its structure does not vary substantially. We also consider (δ, γ)-matching, where γ is a bound on the total number of errors. The algorithms, named δ-BM1, δ-BM2 and δ-BM3 can be thought as members of the generalized Boyer-Moore family of algorithms. The algorithms are fast on average. This is the first paper on the subject, previously only "occurrence heuristics" have been considered. Our heuristics are much stronger and refer to larger parts of texts (not only to single positions).

[*] The work of the three first authors was partially supported by NATO grant
PST.CLG.977017.

A. Apostolico and M. Takeda (Eds.): CPM 2002, LNCS 2373, pp. 178–   , 2002.
© Springer-Verlag Berlin Heidelberg 2002

We use δ-versions of suffix tries and subword graphs. Surprisingly, in the context of δ-matching subword graphs appear to be superior compared with compacted suffix trees.

1   Introduction

A musical score can be viewed as a string: at a very rudimentary level, the alphabet could simply be the set of notes in the chromatic or diatonic notation, or the set of intervals that appear between notes (e.g. pitch may be represented as MIDI numbers and pitch intervals as number of semitones). Approximate matching in one or more musical works play a crucial role in discovering similarities between different musical entities and may be used for establishing "characteristic signatures" [ ]. Such algorithms can be particularly useful for melody identification and musical retrieval. The approximate matching problem has been used for a variety of musical applications (see overviews in [, , ,]). It is known that exact matching cannot be used to find occurrences of a particular melody. Approximate matching should be used in order to allow the presence of errors. The amount of error allowed will be referred to as δ. This paper focuses in one special type of approximation that arise especially in musical information retrieval, i.e. δ-approximation. Most computer-aided musical applications adopt an absolute numeric pitch representation (most commonly MIDI pitch and pitch intervals in semitones; duration is also encoded in a numeric form). The absolute pitch encoding, however, may be insufficient for applications in tonal music as it disregards tonal qualities of pitches and pitch-intervals (e.g. a tonal transposition from a major to a minor key results in a different encoding of the musical passage and thus exact matching cannot detect the similarity between the two passages). One way to account for similarity between closely related but non-identical musical strings is to use δ-approximate matching. In δ-approximate matching, equal-length patterns match if each corresponding integer differs by not more than δ. For example, a C-major $\{60, 64, 65, 67\}$ and a C-minor $\{60, 63, 65, 67\}$ sequence can be matched if a tolerance of $\delta = 1$ is allowed in the matching process. For a given sequence of length m, the total amount of error is bounded by $O(\delta \cdot m)$. This increases dramatically for large values of δ, and therefore, we can restrict it to a maximum of γ. This further restriction will be referred as (δ, γ)-approximate matching. In [ ], a number of efficient algorithms for δ-approximate and (δ, γ)-approximate matching were presented (i.e. the SHIFT-AND algorithm and the SHIFT-PLUS algorithm, respectively). These algorithms use the bit-wise technique. In [ ] exact string matching algorithms are adapted to δ-approximation using heuristics on the alphabet symbols. Here, we present three new algorithms: δ-BM1, δ-BM2 and δ-BM3. They can be thought as members of the Boyer-Moore family of algorithms. The two first algorithms implement a heuristic based on a suitable generalization of the suffix trees data structure. The third algorithm uses a heuristic that considers fingerprints for selected substrings of the pattern and compares them with corresponding fingerprints of substrings of the text to be processed. The algorithms are fast on average. We provide experimental results and observations on the suitability of the heuristics. Our algorithms are particularly efficient for "non-flat" patterns over large alphabet intervals, and many patterns are of this kind.

2 δ-Approximate Dictionaries

Let Σ be an alphabet of integers and δ an integer. Two symbols a, b of Σ are said to be δ-approximate, denoted $a \stackrel{\delta}{=} b$, iff $|a - b| \le \delta$. We say that two strings x, y are δ-approximate, denoted $x \stackrel{\delta}{=} y$ if and only iff $|x| = |y|$, and $x[i] \stackrel{\delta}{=} y[i]$, for each $i \in \{1..|x|\}$. For a given integer γ we say that two strings x, y are γ-approximate, denoted $x \stackrel{\gamma}{=} y$ if and only if

$$|x| = |y|, \quad \text{and} \quad \textstyle\sum_{i=1}^{|x|} |x[i] - y[i]| \le \gamma \ .$$

Two strings x, y are (δ, γ)-approximate, denoted $x \stackrel{\delta,\gamma}{=} y$, if and only if x and y satisfy both conditions above. The Boyer-Moore type algorithms are very efficient on average since scanning a small segment of size k allows, on average, to make large shifts of the pattern. Eventually this gives sublinear average time complexity. This general idea has many different implementations, see [ ]. Our approach to δ-matching is similar, we scan a segment of size k in the text. If this segment is not δ-approximate with any subword of the pattern we know that no occurrence of the pattern starts at $m - k$ positions to the left of the scanned segment. This allows to make a large shift of size $m - k$. The choice of k affects the complexity. In practice small k would suffice. Hence the first issue, with this approach, is to have a data structure which allows to check *fast* if a word of size k is δ-approximate to a subword of *pat*. We are especially interested in the answer "no" which allows to make a large shift, so an important parameter is the *rejection ratio*, denote by *Exact-RR*. It is the probability that a randomly chosen k-subword is not δ-approximate with a subword of *pat*. If this ratio is high then our algorithms would work much faster on average. However another parameter is the time to check if the answer is "no". It should be proportional to k. We do a compromise: build a data structure with smaller rejection ratio but with faster queries about subwords of size k. Smaller rejection ratio means that sometimes we have answer "yes" though it should be "no", however if the real answer is "no" then our data structure outputs "no" also. This is the negative answer which speeds up Boyer-Moore type algorithms. The positive answer has small effect. The data structure is an approximate one, its rejection ratio is denoted by *RR*, and it is hard to analyze it exactly. Hence we rather deal with heuristics. The performance depends on particular structure, the parameter k and class of patterns. Another important factors are rejection ratios: *Exact-RR* and *RR*. If *Exact-RR* is too small we cannot expect the algorithms to be very fast. On the other hand we need to have *RR* as close to *Exact-RR* as possible. The applicability is verified in practice. The starting structure is the suffix trie, it is effective in searching but it could be too large theoretically, though in practice k is small and k-truncated suffix trie is also small. Surprisingly we do not have linear size equivalent of (compact) suffix trees, but we have a linear size equivalent of subword graphs: δ-subword graphs, this shows that suffix trees and subword graphs are very different in the context of δ-matching. Below we give formal definition of our data structures and rejection ratios. Denote by $SUB(x, k)$ the set of all substrings of x of size k. Denote also:

$$\delta\text{-}SUB(x, k) = \{z \mid z \overset{\delta}{=} y \text{ for some } y \in SUB(x, k)\}$$

An *approximate dictionary* for a given string x is the data structure $\mathcal{D}_x$ which answers the queries:

$$\mathcal{D}_x(z): \text{ "} z \in \delta\text{-}SUB(x, k) \text{ ?"}$$

Let $\mathcal{D}_x(z)$ be the result (*true* or *false*) of such query for a string z given by the data structure $\mathcal{D}_x$. By $\mathcal{D}_x^1$ we denote the corresponding data structure for the queries involving the equality $z \overset{\delta,\gamma}{=} y$. In order for our data structure to work fast we allow that the answers could be incorrect. By an efficiency of $\mathcal{D}_x$ we understand the *rejection-ratio* proportion:

$$RR_k(\mathcal{D}_x) = \frac{|\{z \in \Sigma^k \mid \mathcal{D}_x(z) = false\}|}{|\Sigma|^k} .$$

Optimal efficiency is the exact *rejection-ratio* for x:

$$Exact\text{-}RR_k(x) = 1 - \frac{|\delta\text{-}SUB(x, k)|}{|\Sigma|^k} .$$

In other words the efficiency is the probability that a random substring z of length k is not accepted by $\mathcal{D}_x$ or (in the case of *Exact-RR*) it is not an element of $\delta\text{-}SUB(x, k)$. Our data structures $\mathcal{D}$ are *partially correct*:

$$\delta\text{-}SUB(x, k) \subseteq \{z \mid \mathcal{D}_x(z) = true\}$$

Denote $i \ominus \delta = \max\{i - \delta, \min\{\Sigma\}\}$ and $i \oplus \delta = \min\{i + \delta, \max\{\Sigma\}\}$. We define δ-*suffix* tries and δ-*subword* graphs algorithmically. The δ-*suffix* trie of a pattern x is built as follows:

- build the trie $T = (V, E)$ recognizing all the suffixes of x where V is the set of nodes and $E \subseteq V \times \Sigma \times V$ is the set of edges of T;
- replace each edge $(p, a, q) \in E$ by $(p, [\max\{0, a - \delta\}, \min\{\max\{\Sigma\}, a + \delta\}], q)$;
- for all the nodes $v \in V$, if there are two edges $(v, [a, b], p), (v, [c, d], q) \in E$ such that $[a, b] \cap [c, d] \neq \emptyset$ then merge p and q into a single new node s and replace $(v, [a, b], p)$ and $(v, [c, d], q)$ by one edge $(v, [\min\{a, c\}, \max\{b, d\}], s)$.

We have an equivalence relation on the set of vertices: two vertices are equivalent iff they are roots of isomorphic subtrees. In the δ-suffix trie construction we process nodes by taking at each *large* step all vertices which are in a same equivalence class C. Then in this step we process all edges outgoing from vertices from C. All these vertices are independent and we can think that it is done in parallel. The construction terminates when the trie stabilizes. The δ-subword graph of a sequence x is obtained by minimizing its δ-*suffix* trie. This means that each equivalence class of vertices is merged into a single vertex. Figure shows an example of δ-*suffix* trie and δ-subword graph.

Theorem 1.   *The numbers of nodes and of edges of δ-subword graph for the string x are $O(|x|)$.*

Proof. The number of equivalence classes of the initial suffix trie is at most $2n$. In the process of merging edges the nodes which are equivalent initially will remain equivalent until the end. Hence the number of equivalence classes of intermediate δ-suffix trie (after processing all edges outgoing from nodes in a same equivalence class) is at most $2n$, which gives the upper bound on the number of nodes of the δ-subword graph. The bound on the number of edges can be shown similarly as for standard subword graphs.   □

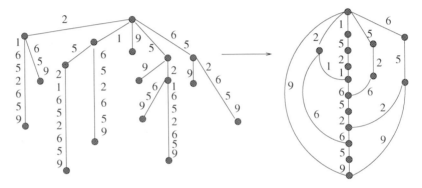

Fig. 1. The suffix trie and subword graph for the word $w = 1521652659$

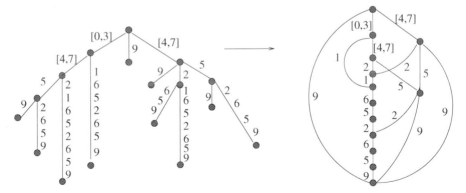

Fig. 2. The *δ-suffix trie* and the δ-subword graph for the sequence $w = 1521652659$ with $\delta = 1$ and $\Sigma = [0..9]$. A single integer means i means the interval $[i \ominus \delta, i \oplus \delta]$

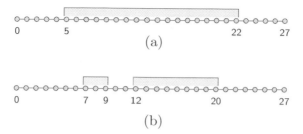

Fig. 3. (a) The family of intervals $\mathcal{F}_\delta(x,k)$ and (b) the family $\mathcal{M}_\gamma(x,k)$, for the string 1 5 2 9 2 8 3 with $\delta = \gamma = 1$, $k = 3$ and $\Sigma = [0..9]$

For each subword $y \in SUB(x,k)$ of x, denote by $hash(y)$ the sum of the symbols of y. For each $k < |x|$ we introduce the following families of intervals (overlapping and adjoined intervals are "glued together") of the interval $[\min\{\Sigma\}, k\cdot\max\{\Sigma\}]$ which represents respectively the sets:

$$\mathcal{F}_\delta(x,k) = \bigcup_{y \in SUB(x,k)} [hash(y) \ominus k\delta,\; hash(y) \oplus k\delta]$$

$$\mathcal{M}_\gamma(x,k) = \bigcup_{y \in SUB(x,k)} [hash(y) \ominus \min\{k\delta, \gamma\},\; hash(y) \oplus \min\{k\delta, \gamma\}] .$$

Clearly $\mathcal{M}_{k\delta}(x,k) = \mathcal{F}_\delta(x,k)$. Figure   presents an example.

Lemma 1.
(a) If $z \overset{\delta}{=} y$ for some $y \in SUB(x,k)$ then $hash(z) \in \mathcal{F}_\delta(x,k)$.
(b) If $z \overset{\delta,\gamma}{=} y$ for some $y \in SUB(x,k)$ then $hash(z) \in \mathcal{M}_\gamma(x,k)$.

The efficiency of the family $\mathcal{I}$ of intervals can be measured as $RR(\mathcal{I}) = 1 - Prob(hash(x) \in \mathcal{I})$ where x is a random string of length k. In other words it is the probability that an integer is not in any interval of the family. Observe that $\mathcal{I}$ in our case is always represented as a family of disjoint intervals, overlapping and adjoined ones have been glued together.

3   Three δ-BM Algorithms

We show in this section how the data structures of Section 2 are used in δ-matching. We now want to find all the δ-occurrences of a pattern *pat* of length m in a text *text* of length n. We apply a very simple greedy strategy: place the pattern over the text such that the right end of the pattern is over position i in the text. Then check if the suffix *suf* of length k (k may depend on *pat*) of text ending at i is "sensible". If not the pattern is shifted by a large amount and many position of text are never inspected at all. If *suf* is sensible then a naive search in a "window" of text is performed.

> **Algorithm δ-BM1;**
> $i \leftarrow m$;
> **while** $i \leq n$
> $\quad$ **if** $text[i - k + 1...i] \in \delta\text{-}Suffix\text{-}Trie(pat)$
> $\quad\quad$ **then** $NAIVE(i, i + m - k - 1)$;
> $\quad$ $i \leftarrow i + m - k$;

We denote here by $NAIVE(p, q)$ a procedure checking directly if pat ends at positions in the interval $[p..q]$, for $p < q$.

We design an improved version of δ-BM1 using δ-subword graphs instead tries. Let (Σ, V, v_0, F, E) be the δ-subword graph of the reverse pattern, where Σ is the alphabet, V is the set of states, $v_0 \in V$ is the initial state, $F \subseteq V$ is the set of final states and $E \subseteq V \times \Sigma \times V$ is the set of transitions. Let $\delta\text{-}per(x)$ be the δ-period of the word x defined by

$$\delta\text{-}per(x) = \min\{p \mid \forall 1 \leq i \leq m - p \ \ x[i] \overset{\delta}{=} x[i + p]\}.$$

Then it is possible to adopt the same strategy as the Reverse Factor algorithm [ ] for exact string matching to δ-approximate string matching. When the pattern pat is compared with $text[i - m + 1 \ldots i]$ the symbols of $text[i - m + 1 \ldots i]$ are parsed through the δ-subword graph of the reverse pattern from right to left starting with the initial state. If transitions are defined for every symbol of $text[i - m + 1 \ldots i]$, it means that a δ-occurrence of the pattern has been found and the pattern can be shifted by $\delta\text{-}per(pat)$ positions to the right. Otherwise the pattern can be shifted by m minus the length of the path, in the δ-subword graph, from the initial state and the last final state encountered while scanning $text[i - m + 1 \ldots i]$ from right to left. Indeed the δ-subword graph of the reverse pattern recognizes at least all the δ-suffixes of the reverse pattern from right to left and thus at least all the δ-prefixes of the pattern from left to right.

> **Algorithm δ-BM2;**
> $i \leftarrow m$;
> **while** $i \leq n$
> $\quad$ $q \leftarrow v_0;\ j \leftarrow i;\ b \leftarrow 0$;
> $\quad$ **while** $(q, text[j], p) \in E$
> $\quad\quad$ $q \leftarrow p;\ j \leftarrow j - 1$;
> $\quad\quad\quad$ **if** $q \in F$ **then** $b \leftarrow i - j$;
> $\quad\quad$ **if** $i - j > m$ **then** check and report
> $\quad\quad\quad\quad$ δ-occurrence at position $i - m + 1$;
> $\quad\quad\quad\quad\quad$ $i \leftarrow i + \delta\text{-}per(pat)$;
> $\quad\quad$ **else** $i \leftarrow i + m - b$;

The value $\delta\text{-}per(pat)$ can be approximated using the δ-subword graph of the reverse pattern.

Our last algorithm can be used also for (δ, γ)-approximate string matching. We apply the data structure of interval families.

```
Algorithm δ-BM3;
         i ← m;
    while i ≤ n
                if hash(text[i − k + 1...i]) ∈ M_{δ,γ}(pat, k)
                    then NAIVE(i, i + m − k − 1);
                i ← i + m − k;
```

4   Average Time Analysis of Algorithms δ-BM1 and δ-BM3

Denote $p = Prob(x \overset{\delta}{=} y)$ where x and y are random symbols of Σ and $q_{k,pat} = RR_k(\mathcal{D}_{pat})$.

Lemma 2. *The overall average time complexity of δ-BM1 and δ-BM3 algorithms is at most*

$$\frac{n}{m-k}\left(\left(q_{k,pat} + (1 - q_{k,pat})\frac{p}{1-p}\right)k + (1 - q_{k,pat})m + (1 - q_{k,pat})\frac{1}{(1-p)^2}\right)$$

Proof. Each iteration of the algorithms tries to find occurrences of the pattern ending at positions $i..i + m − k − 1$. We call those positions a window. The window for next iteration starts just behind the window for the previous iteration. The probability that the pattern is moved to the next window after at most k comparisons is exactly $q_{k,pat} = RR_k(\mathcal{D}_{pat})$. Now it is enough to prove that the average number of comparisons made by the naive algorithm in one iteration is bounded by

$$m + \frac{p}{1-p}k + \frac{1}{(1-p)^2}.$$

Let p_j, for $j = 1..m − k − 1$ be the average number of comparisons made by the naive algorithm at position $i + j − 1$ of the window. Since the we cannot assume that text symbols at positions $i − k + 1..i$ are random (because $\mathcal{D}_{pat}(text[i − k + 1..i]) = true$), we assume that the probability of matching those symbols with the symbols of the pattern during the naive algorithm is 1. Hence

$$p_1 \leq (1 − p)(k + 1) + p(1 − p)(k + 2) + p^2(1 − p)(k + 3) + \cdots +$$
$$+p^{m-k-1}(1 − p)(m − 1) + p^{m-k}m.$$

Similarly the average number of comparisons made by the naive algorithm at position $i + 1$ is at most

$$p_2 \leq (1-p) + p(1-p)(k+1) + p^2(1-p)(k+2) + \cdots + p^{m-k-1}(1-p)(m-1) + p^{m-k}m.$$

Similarly the average number of comparisons made by the naive algorithm at position $m + i − 1$ is at most

$$p_i \leq (1 − p) + p(1 − p) + \cdots + p^{i-2}(1 − p)+$$
$$+p^{i-1}(1 − p)(k + 1) + p^i(1 − p)(k + 2) + \ldots p^{m-k-1}(1 − p)(m − 1) + p^{m-k}m$$

It can be proved that $p_1 \leq k + \frac{1}{1-p}$ and $p_2 \leq (1-p) + p(k + \frac{1}{1-p})$ and generally $p_i \leq (1-p) + p(1-p) + \cdots + p^{i-2}(1-p) + p^{i-1}(k + \frac{1}{1-p})$. Hence, after some calculations we get

$$\sum p_i \leq m + \frac{p}{1-p}k + \frac{1}{(1-p)^2}.$$

This completes the proof. $\square$

As a corollary of previous lemma we have.

Theorem 2. *Let $k \leq 0.99m$ and $p \leq 0.99$. The average time complexity of the algorithms δ-BM1 and δ-BM3 is*

$$O(\frac{n}{m}(k + (1 - q_{k,pat})m)) .$$

Observe here that our analysis does not depend on the data structure $\mathcal{D}$. The only thing it assumes is that the scheme of the algorithm matches the structure of the algorithms δ-BM1 and δ-BM3. Clearly, the efficiency of such algorithms depends heavily on the choice of k and the efficiency of $\mathcal{D}$. For instance, for $\delta = 0$, (ie. we consider string matching without errors) we may choose $k = 2 \log_{|\Sigma|} m$. Then, for δ-BM1, $1 - q_{k,pat}$ is the probability that a random string of length k is not a subword of pat. The number of subwords of length k of pat is at most m and the number of all words of length k is m^2 so $1 - q_{k,pat} \leq \frac{1}{m}$ thus the average time complexity is $O(\frac{n}{m} \log m)$ the best possible. Moreover k may depend also on the pattern pat itself. If pat is "good" then k may be chosen small and when it is "bad" k may be chosen bigger. In particular we may increase k up to the moment when $1 - q_{k,pat}$ decreases below an acceptable level.

5   Experimental Results

We computed experimentally the values RR and $Exact$-RR for our approximate dictionaries for various values of k and different sizes of the alphabet. These efficiencies correspond to average case complexity of our δ-BM algorithms. We compared the values of RR and $Exact$-RR with average running time for sufficiently large sample of random inputs. We counted the average number of text character inspections for the following algorithms: δ-Tuned-Boyer-Moore, δ-Skip-Search, δ-Maximal-Shift [ ] and δ-BM1, δ-BM2 and δ-BM3. All the algorithms have been implemented in C in a homogeneous way such as to keep their comparison significant. The text used is composed of $500,000$ symbols and were randomly built. The size of the alphabet is 100. The target machine is a PC, with a AMD-K6 II processor at 330MHz running Linux kernel 2.2. The compiler is gcc. For each pattern length m, we searched per one hundred patterns randomly built. We counted the number c of text character inspections for one text character. The results are presented in Figures   to   . For $\delta = 1$ the best results for δ-BM1 algorithm have been obtained with $k = \log_2 m$. The best results for the δ-BM3 algorithm have always been obtained with $k = 2$. For small

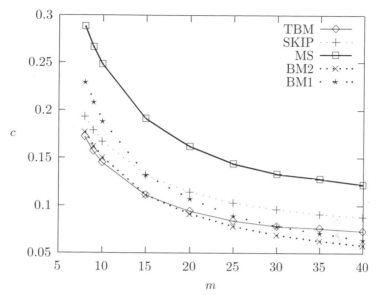

Fig. 4. Results for δ = 1

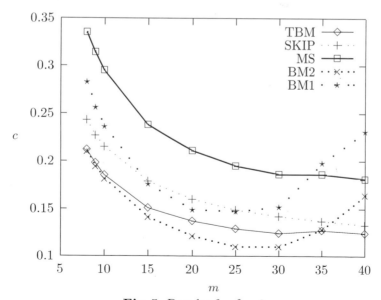

Fig. 5. Results for δ = 2

values of δ, δ-BM1 and δ-BM2 algorithms and better than δ-Tuned-Boyer-Moore algorithm (which is the best among the known algorithms) for large values of m ($m \geq 20$). For larger values of δ (up to 5) δ-BM1 and δ-BM2 algorithms are better than δ-Tuned-Boyer-Moore algorithm for small values of m ($m \leq 12$). For larger values of δ, the δ-Tuned-Boyer-Moore algorithm is performing better than the other algorithms. In conclusion the algorithms introduced in this article are of particular practical interest for large alphabets, short patterns and small values of δ. Alphabets used for music representations are typically very large. A "bare" absolute pitch representation can be base-7 (7 symbols), base-12, base-40 or 120 symbols for midi. But meaningful alphabets that will allow us to do in-depth music analysis use symbols that in reality is set of parameters. A typical symbol could be $(a_1, a_2, a_3, ...a_k)$, where a_1 represents the pitch, a_3 represents the duration, a_4 the accent etc. A typical pattern ("motif") in musical sequence is 15-20 notes but an alphabet can have thousands of symbols. Thus the need of algorithms that perform well for small patterns and large alphabets.

References

1. E. Cambouropoulos, T. Crawford and C. S. Iliopoulos, Pattern Processing in Melodic Sequences: Challenges, Caveats and Prospects, In G. Wiggins, editor, *Proceedings of the Artificial Intelligence and Simulation of Behaviour Symposium*, The Society for the Study of Artificial Intelligence and Simulation of Behaviour, Edinburgh, UK, pp 42-47, 1999.
2. E. Cambouropoulos, M. Crochemore, C. S. Iliopoulos, L. Mouchard and Y. J. Pinzon, Algorithms for computing approximate repetitions in musical sequences. In R. Raman and J. Simpson, editors, *Proceedings of the 10th Australasian Workshop On Combinatorial Algorithms*, Perth, WA, Australia, pp 129–144, 1999.
3. T. Crawford, C. S. Iliopoulos and R. Raman, String Matching Techniques for Musical Similarity and Melodic Recognition, *Computing in Musicology* **11** (1998) 73–100.
4. M. Crochemore, A. Czumaj, L. Gąsieniec, S. Jarominek, T. Lecroq, W. Plandowski and W. Rytter, Speeding-up two string matching algorithms, *Algorithmica* **12** (4/5) (1994) 247–267.
5. M. Crochemore, C. S. Iliopoulos, T. Lecroq and Y. J. Pinzon, Approximate string matching in musical sequences, In M. Balík and M. Simánek, editors, *Proceedings of the Prague Stringology Conference'01*, Prague, Tcheque Republic, 2001, Annual Report DC–2001–06, 26–36.
6. V. Fischetti, G. Landau, J.Schmidt and P. Sellers, Identifying periodic occurrences of a template with applications to protein structure, *Proceedings of the 3rd Combinatorial Pattern Matching*, Lecture Notes in Computer Science 644, pp. 111–120, 1992.
7. S. Karlin, M. Morris, G. Ghandour and M.-Y. Leung, Efficient algorithms for molecular sequences analysis, *Proc. Natl. Acad. Sci. USA* **85** (1988) 841–845.
8. P. McGettrick, MIDIMatch: Musical Pattern Matching in Real Time, MSc Dissertation, York University, UK, 1997.
9. A. Milosavljevic and J. Jurka, Discovering simple DNA sequences by the algorithmic significance method, *Comput. Appl. Biosci.* **9** (1993) 407–411.

10. P. A. Pevzner and W. Feldman, Gray Code Masks for DNA Sequencing by Hybridization, *Genomics* **23** (1993) 233-235.
11. P. Y. Rolland and J. G. Ganascia, Musical Pattern Extraction and Similarity Assessment, In E. Miranda, editor, *Readings in Music and Artificial Intelligence*, Harwood Academic Publishers, 1999.

A Better Method for Length Distribution Modeling in HMMs and Its Application to Gene Finding

Broňa Brejová and Tomáš Vinař

School of Computer Science, University of Waterloo
ON N2L 3G1, Canada
{bbrejova,tvinar}@math.uwaterloo.ca

Abstract. Hidden Markov models (HMMs) have proved to be a useful abstraction in modeling biological sequences. In some situations it is necessary to use generalized HMMs in order to model the length distributions of some sequence elements because basic HMMs force geometric-like distributions. In this paper we suggest the use of an arbitrary length distributions with geometric tails to model lengths of elements in biological sequences. We give an algorithm for annotation of a biological sequence in $O(ndm^2\Delta)$ time using such length distributions coupled with a suitable generalization of the HMM; here n is the length of the sequence, m is the number of states in the model, d is a parameter of the length distribution, and Δ is a small constant dependent on model topology (compared to previously proposed algorithms with $O(n^3m^2)$ time [ ]). Our techniques can be incorporated into current software tools based on HMMs.

To validate our approach, we demonstrate that many length distributions in gene finding can be accurately modeled with geometric-tail length distribution, keeping parameter d small.

Keywords: computational biology, hidden Markov models, gene finding, length distribution

1 Introduction

Hidden Markov models (HMMs) have been used in bioinformatics for such tasks as gene finding, prediction of protein secondary structure, and sequence alignment. HMMs have many features suitable for modeling biological sequences, such as a sound probabilistic basis, and ease in expressing signals and content preferences. Efficient algorithms exist for making predictions and for training HMMs on unannotated or only partially annotated sequences.

One disadvantage of Hidden Markov models is their deficiency in modeling regions with different types of length distribution. If an HMM can stay in one state for more than one step in a row, the probability that it stays in this state for l steps is $(1 - p)p^{l-1}$ for some constant p. This gives rise to a geometric probability distribution on the length of the region generated by that state.

A. Apostolico and M. Takeda (Eds.): CPM 2002, LNCS 2373, pp. 190–   , 2002.

However, many elements of biological sequences have length probability distributions that are inadequately approximated by a geometric distribution. Still, such elements may need to be represented in a model by a state or a small group of states. Several approaches have been used to overcome this problem, but none of them is completely satisfactory.

1.1   Current Approaches

Durbin et al. [ ] discuss several ways to model non-geometric length distributions by replacing a single state by a group of states. Transitions are added inside this group so that the probability of staying within the group for l steps is close to the probability that the modeled feature has length l. This technique does not require any changes in HMM modeling language or in algorithms for training and decoding. However, the running time and needed memory will increase with the number of states, and the conceptual complexity of the model can become unwieldy.

Another disadvantage of this approach is that the additional transitions introduce many paths through the model corresponding to the same sequence annotation. As pointed out in [ ], this affects the accuracy of the standard Viterbi algorithm for decoding, which considers only the most probable path through the model. While a natural response would be to look for the most probable annotation instead of the most probable path, this problem was recently proved NP-hard [ ].

Rabiner [ ] presents generalized HMMs to address the problem of length distributions for speech recognition. A state in a generalized HMM does not generate one character at a time but a region of arbitrary length. The number of characters produced in each state is determined according to a given arbitrary length distribution. Applications of this approach in bioinformatics include the gene prediction tools Genie [ ] and GENSCAN [,].

Generalized Hidden Markov models allow for non-geometric length distributions, but give slower decoding algorithms. For a pure HMM, the running time of decoding (prediction) grows linearly with the length of the sequence. For each state and each position, the probability of generating the symbol at that position is evaluated once. In a generalized HMM, a single state can produce an arbitrarily long region. There are $\Theta(n^2)$ regions in a sequence of length n. If the time needed to evaluate a probability of generating one region in a given state grows linearly with the region's length, this gives an $O(n^3m^2)$ running time. In some cases it is possible to organize the computation so that the probability for one region is obtained in constant time, but the running time is still quadratic, which is infeasible for computation on genome-size sequences.

Two approaches have been taken to reduce the running time. First, the running time can be restricted by imposing an upper bound of d on the number of characters produced by each state [ ], which gives an algorithm with running time $O(ndm^2)$ or $O(nd^2m^2)$ depending on the type of model used. In speech recognition applications, it is usually possible to keep the bound d relatively small, again yielding a practical running time. However, such assumption cannot be made in the case of elements of biological sequences.

A different approach is taken by the gene finder GENSCAN [ ]. Here, practical running time is achieved by using biological knowledge to limit the number of regions that need to be considered for each generalized state. A disadvantage of this approach is that only some elements of the sequence (exons) can be characterized in this way. For the other elements (introns, intergenic regions), geometric distribution is used to achieve a practical running time.

1.2   Our Approach

In this paper, we introduce a technique for modeling length distribution in HMMs that addresses some of the problems of the existing techniques mentioned above. We propose a geometric-tail (GT) distribution which can model length distributions of many elements of biological sequences more accurately than the geometric distribution, while still being handled efficiently in decoding algorithms. In the GT distribution long regions have geometric distribution and the probability of shorter lengths can be arbitrary. We describe a general class of models, which we call boxed HMMs, which can be used in conjunction with the GT length distribution. Then, we give an $O(ndm^2\Delta)$ time decoding algorithm, where n is the length of the sequence, m is the number of states of the HMM, d is a parameter of the GT distributions used, and Δ is a small constant dependent on model topology .

The rest of the article is organized as follows. In Section 2 we define the GT distribution and provide an algorithm to estimate its parameters from a given sample. In Section 3 we introduce boxed HMM models. In Section 4 we give $O(ndm^2\Delta)$ modification of Viterbi algorithm to decode a boxed HMM with the GT distribution. Finally, in Section 5 we give a brief overview of experimental results that validate our approach and discuss the application of our approach to gene finding.

2   The GT Length Distribution and Its Estimation

The *geometric-tail length distribution* (GT distribution) joins two distributions: the first part is an arbitrary length distribution, and the second part is a geometric tail (see example in Figure a).

Definition 1 *A GT distribution δ of order t is defined by a t-tuple $(d_1, d_2, \ldots, d_t)$ and a parameter q, where $0 \leq q < 1$ and $\left(\sum_{i=1}^{t-1} d_i\right) + d_t \cdot \frac{1}{1-q} = 1$. In this distribution,*

$$\delta(x) = Pr[X = x] = \begin{cases} d_x, & \text{if } x \in \{1, \ldots, t\}, \\ d_t \cdot q^{x-t}, & \text{if } x > t. \end{cases}$$

---

[1] In our gene finding experiments $d \approx 187$, and $\Delta = 3$.

This variety of distribution is suitable for modeling the lengths of many biological sequence elements that have most of the probability mass concentrated in relatively short lengths, but where much longer sequences do sporadically occur. The geometric tail assigns a non-zero probability to a region of any length, so we do not exclude any region by setting the upper bound too small. Also, the geometric tail has only one parameter q, so we do not need dense data to estimate it. On the other hand, densely populated medium length region of the distribution can be estimated simply by the smoothed empirical distribution. If appropriate, we can also stipulate a lower bound on the length of the region by assigning zero probability to all smaller lengths.

Assume we have a training sample of lengths, in which length i occurs $\gamma(i)$ times. We want to use this sample to estimate the parameters $(d_1, \ldots, d_t)$ and q of the GT distribution δ of order t. We want to optimize the parameters subject to the following criteria: First, the relative frequency of observing a length at least t is the same in the distribution δ and the observed sample γ. Second, the likelihood of independently generating lengths in the sample γ from distribution δ is maximized.

In particular, let $x = [\sum_{i=t}^{\infty} \gamma(i)]/[\sum_{i=1}^{\infty} \gamma(i)]$ be the relative frequency of observing length at least t. Then we want to maximize $\prod_{i=t}^{\infty} \delta(i)^{\gamma(i)}$ subject to $x = \sum_{i=t}^{\infty} \delta(i)$. We can easily show that the maximum is achieved when

$$q = \left[\sum_{i=0}^{\infty} i \cdot \gamma(i+t) \right] \bigg/ \left[\sum_{i=0}^{\infty} (i+1) \cdot \gamma(i+t) \right] \qquad (1)$$

$$d_t = x \cdot (1 - q) \qquad (2)$$

Parameters $d_1, \ldots, d_{t-1}$ can be set to the relative frequency of observing corresponding lengths in training data. However, it is better to smooth these values to avoid overfitting to training data. We used a smoothing procedure introduced in GENSCAN [ ]. In this method, every point from the training data set is replaced by a normal distribution centered in this point. We obtain the resulting smooth distribution as a sum of these normal distributions.

As we will show in Section , the running time for decoding of an HMM, where lengths of the sequence elements are modeled by GT distribution, grows linearly with t. Therefore, for small values of t, the algorithm runs fast, but the distribution is close to geometric, which may not fit the data very well. For large values of t, the decoding will be slower, but the distribution might fit the data better. On the other hand, increasing the value of t increases the number of parameters to be estimated (parameters $d_1, \ldots, d_{t-1}$), but the amount of data available remains the same. Therefore the length distribution obtained for large values of t may not generalize well to examples not present in the training set.

In our tests, to compute a good value of t for a given training data, we first divided data randomly into k equal size parts. In each step, we set aside one part as a testing data set, and we used the $k - 1$ remaining parts to estimate parameters of a GT distribution for all values of t. The parameters were then tested on the testing data set and the likelihood score of the testing data was

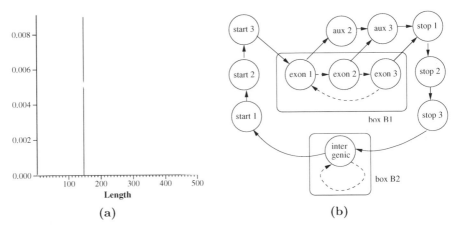

(a)   (b)

Fig. 1. (a) **An example of a GT distribution** consisting of an arbitrary length distribution for lengths up to 146 and a geometric tail starting at length 147. (b) **A simple boxed HMM model** for DNA sequences consisting of single-exon genes on the forward strand only. Internal transitions are marked by dashed lines. All transitions have probability 1. Box B_1 models the region of exon between the start and stop codons, box B_2 models intergenic regions. Both boxes can use non-geometric length distributions. Here a boxed HMM is a more accurate model than the corresponding pure HMM

computed. We repeated this process k times, where in each step, one of the parts served as testing data. Finally, we chose the value of t with the best overall performance.

In practice, such process can yield large values of t, so of course one can specify an upper bound T based on constraints on running time of the decoding algorithm.

3   HMMs and Boxed HMMs

A hidden Markov model is a generative probabilistic model, consisting of *states* and *transitions*. Each state u is a Markov chain of order o_u. Let $e_u(x_1, \ldots, x_{o_u}, x)$ be the emission probability of generating character x, provided that the last o_u symbols generated are $x_1, \ldots, x_{o_u}$. After a symbol is generated, one of the transitions is followed to another state. We denote the probability of a transition from state u to state v as $a_{u,v}$.

Typically each sequence element is modeled by a state or a group of states. For example, in gene finding, groups of states correspond to introns, exons and intergenic regions. Given a biological sequence, we can compute the path through the model which has the highest probability of generating the sequence. This can be done using the well-known Viterbi algorithm (see [ ]) in $O(nm^2)$ time, where n is the length of the sequence, and m is the number of states in the HMM. This most probable path predicts the positions of the sequence elements.

HMMs impose a geometric distribution on the lengths of the sequences generated by individual states. This is inappropriate in many applications, and therefore generalized Hidden Markov models are often used. In a generalized HMM each state produces several characters instead of one. The number of characters produced in each state is determined according to a given arbitrary length distribution.

We will consider the following generalization of HMMs, which we call *boxed HMMs*. States of a boxed HMM are organized in boxes. Each state corresponds to a Markov chain of some order. A box is a subset of states; each state belongs to at most one box (some states may be unboxed). Denote $B(u)$ the box containing state u. Each box of a boxed HMM is assigned a length distribution for how long the HMM stays in the box. Denote δ_B the length distribution associated with box B.

Each state located in a box has two sets of outgoing transitions: *internal* transitions which lead only into states grouped in the same box, and *external* transitions which can lead to any state. The probability of an internal transition from state u to state v is denoted by $a'_{u,v}$, and the probability of an external transition is denoted $a_{u,v}$. For every state u, $\sum_v a_{u,v} = 1$ and for every boxed state u, $\sum_v a'_{u,v} = 1$. Unboxed states, of course, have only external transitions.

The boxed HMM as a generative model works as follows. Unboxed states are treated as in ordinary HMMs. Once a box B is entered, the length m is generated according to the length distribution δ_B. For the next $m - 1$ steps the internal set of transitions is used in each state (i.e. the HMM will remain in box B for m steps). In the m-th step the external set of transitions is used to take the model to a new box or unboxed state (or to the same box, if that is chosen). A simple example of a boxed HMM is given in Figure b.

It might seem that the actual length distribution is a combination of the explicit distribution δ_B associated with a box B and the implicit geometric-like distribution induced by the internal transitions inside the box. However, for a given length m and start state in box B, the sum of probabilities over all strings of length m generated by box B is 1, and the internal transitions do not constitute any implicit distribution of lengths.

Boxes of a boxed HMM correspond to states in other classes of generalized HMMs. However, each state inside a box can have a different set of external transitions and transitions entering a box specify which state of the box is to be used as the first. This feature is very useful for example for modeling frame dependencies between exons in genes.

---

[2] This implicitly assumes that each state inside the box has at least one external transition. In some cases this might be undesirable. However, there are several ways to relax this restriction to satisfy modeling needs, and the algorithms we present can easily be modified to handle such extensions.

4   Algorithms for Boxed HMMs

In this section we introduce an efficient algorithm which can be used to decode boxed HMMs with GT length distributions. The algorithm is a modification of the well-known Viterbi algorithm. Our modifications can be applied in similar way to other generalized HMM models (and thus incorporated in other software tools based on HMMs), as well as to other algorithms related to HMMs (e.g. computing forward and backward probabilities).

Let the input sequence be $x = x_1 x_2 \ldots x_n$. The Viterbi algorithm uses dynamic programming to compute, for each position in the sequence i and each state u, the probability of the most probable path generating the sequence $x_1 x_2 \ldots x_i$, and ending in state u. In our algorithm, let $P(u, i)$ be the probability of such a path with additional constraint that if u is a part of a box B, then the segment generated by B ends at position u.

In addition to $P(u, i)$, we compute two auxiliary quantities Q and p to make the algorithm more efficient. Let δ_B be the GT distribution associated with box B, and t_B be the order of this distribution. The value $Q(u, i)$ is defined similarly to $P(u, i)$, except that we consider only paths, in which the segment generated by B has length at least t_B ($Q(u, i)$ is undefined if state u is not part of any box). The value of $Q(u, i)$, representing all lengths greater than t_B, can be computed in a single step, rather than checking all such lengths one-by-one.

If states u and v are in the same box, let $p(u, v, i, j)$ be the probability of the most probable path generating sequence $x_i x_{i+1} \ldots x_j$ starting in state u and ending in state v, in which only internal transitions are used (this value is undefined if u and v are not part of the same box). The values of p allow us to quickly evaluate probabilities of regions generated inside a single box.

Figure   shows recurrence formulas for obtaining the values of P, Q, and p. Using these formulas, we can order the dynamic programming computation so that no recursive evaluation is needed, as follows:

initialize P and Q **Time**

for $i = 1 \ldots n$, **for** all states u
| **if** u is a boxed state
| | **for** $l = 0 \ldots t_{B(u)} - 1$, **for** all states v, where $B(v) = B(u)$
| | | compute $p(v, u, i - l, i)$ (*) $O(|B(u)|)$
| | compute $Q(u, i)$ $O(m \cdot |B(u)|)$
| compute $P(u, i)$ $O(m \cdot |B(u)| \cdot t_{B(u)})$
| values computed in (*) can be discarded now

Note, that if a model does not contain any boxes, our algorithm is identical to the Viterbi algorithm.

The algorithm has time complexity $n \sum_u (m \cdot |B(u)| \cdot t_{B(u)} + |B(u)|^2 \cdot t_{B(u)}) = O(ndm^2\Delta)$, where $d = \sum_B \frac{|B|}{m} t_B$ (weighted average of start of geometric tail in GT distribution), and Δ is the maximum number of states located in a single box. For example, in our gene finding experiments, we have used a model with $d \approx 187$ and $\Delta = 3$. The algorithm requires $O(nm)$ memory, which is the same as Viterbi algorithm for ordinary HMMs. By contrast, the algorithm presented in [ ] runs in $O(nD^2m^2)$ time, where D is an upper bound on the length of duration of one state, or $O(n^3m^2)$, if such upper bound cannot be guaranteed.

The algorithm can be made more efficient if the topology of states in each box satisfies further restrictions. For example, if the states in each box are organized as in Figure b Box 1 (a k-periodic Markov chain), we can reduce the time complexity to $O(ndm^2)$, because for a fixed i, j, and u, only one of the values $p(*, u, i, j)$ is non-zero.

Notation. $e_u(i) = e_u(x_{i-o_u}, \ldots, x_i)$.
All values not defined otherwise are considered to be 0.

Computation of $p(u, v, i, j)$:

If $1 \leq i < j$, $u, v \in B$: $p(u, v, i, j) = \max_{w \in B} e_u(i) \cdot a'_{u,w} \cdot p(w, v, i+1, j)$

If $u = v \in B$, $1 \leq i = j$: $p(u, v, i, j) = e_u(i)$

Computation of $Q(u, i)$:

If $u \in B$, $i \geq 1$: $Q(u, i) = \max \begin{cases} \max\limits_{v \in B, w} P(w, i - t_B) \cdot a_{w,v} \cdot \delta_B(t_B) \cdot p(v, u, i - t_B + 1, i) \\ \max\limits_{v \in B} Q(v, i-1) \cdot a'_{v,u} \cdot q_B \cdot e_u(i) \end{cases}$

Computation of $P(u, i)$:

If $u \in B$, $i \geq 1$: $P(u, i) = \max \begin{cases} \max\limits_{\substack{j=1 \ldots t_B - 1 \\ v \in B, w}} P(w, i - j) \cdot a_{w,v} \cdot \delta_B(j) \cdot p(v, u, i - j + 1, i) \\ Q(u, i) \end{cases}$

If u is not in a box, $i \geq 1$: $P(u, i) = \max\limits_{w} P(w, i-1) \cdot a_{w,u} \cdot e_u(i)$

$P(\text{start}, 0) = 1$

Fig. 2. Recursive formulas used in the modified Viterbi algorithm to compute values of P, Q, and p

5   Application to Gene Finding

Gene finding is the process of locating genes in DNA sequences. A gene is a region of DNA used in a cell as a template for producing a protein. A gene consists of several coding regions called exons separated by non-coding regions called introns. The DNA between genes is called "intergenic". The goal of gene finding is to label parts of the sequence as being exons, introns or intergenic regions. HMMs and generalized HMMs have been used for this task [, , ,].

We have conducted experiments to verify that the GT distributions are suitable for modeling length distributions occurring in the context of gene finding. For this purpose, we have extracted the lengths of introns, intergenic regions and exons from annotated DNA sequences. We distinguish among the first exon in a gene, internal exons and the last exon in a gene, as the first and the last exon of a gene are processed differently in a cell and their length distribution is different.

At the end of the section we also discuss the effects of incorporation of GT distributions into an existing gene finder.

5.1   Fitting GT Distribution to Gene Structure Elements

We have performed a simple analysis to determine how well the geometric-tail distribution can fit the different elements of a gene structure and what are suitable values of t (start of the geometric tail). The analysis was done using the Sanger Centre annotation of human chromosome 22 [ ]. We have extracted the observed length distribution for each of the gene-structure elements from the annotated sequences.

The results are presented in Figure . The length distribution for internal, first, and last exons can be modeled quite accurately by a GT distribution with small values of t (we have used values of t up to 150). Note, that in the case of length distribution for last exons, there is a significant local maximum centered

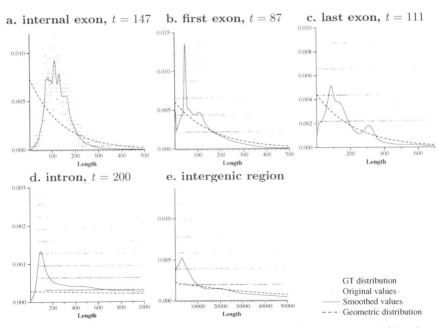

a. internal exon, $t = 147$ **b. first exon,** $t = 87$ **c. last exon,** $t = 111$

d. intron, $t = 200$ **e. intergenic region**

GT distribution
· · · · · Original values
——— Smoothed values
- - - Geometric distribution

Fig. 3. Analysis of length distributions for human chromosome 22 (the Sanger Centre annotation Release 2)

at position approximately 300 which was not captured by the GT distribution. In order to capture this local maximum well, we would need to choose t at least 370. We have observed similar behavior of last exon length distribution in other species as well (arabidopsis, rice), however this phenomena is less pronounced there (results not shown). This may point to a presence of several types of last exons in a genome.

The observed length distribution for introns appears to contain a very long tail that decreases slower than exponentially. Thus the intron tail starts below the observed values to compensate for the rest of the tail (using a very high geometric coefficient $q \approx 0.9997$). This behavior of the intron length distribution can be explained by presence of two types of introns – short and long – which are spliced by different splicing mechanisms (the distribution of short introns is centered around 100bp, while the distribution of long introns is centered around 1000bp [ ]).

To capture the mode of the distribution for intergenic region it would be necessary to use impractically large values of t. Thus in this case we have retreated to a geometric distribution.

We have conducted similar experiments on other species (arabidopsis, rice) with similar results (data not shown). In general, the geometric-tail distribution works well with practical values of t for elements where large mass probability is assigned to smaller lengths (exons and introns). However, to model the length of intergenic regions accurately, we would need to use impractically high values of t, so the GT distributions are not suitable for modeling the length of intergenic regions.

5.2   Generalization to New Data

The parameters of a GT distribution are estimated by optimizing the fit to the training data. However, if the resulting distribution fits training data well, it does not mean that this would generalize to testing data.

In the second set of experiments we performed a 5-fold evaluation on the data set used in the previous section. First, the data set was split into 5 parts. In each step, four parts served as a training data and one part as a testing data. We have used our algorithm to fit the GT distribution to the training data for different values of t. Likelihood score was then evaluated for testing data set. We repeated this process 5 times, each time with a different part serving as a testing data set.

For comparison, we have also computed the likelihood score for the distribution obtained by the smoothing function from GENSCAN [ ], and for the "ideal fit" distribution that is obtained by using the same data set as both training and testing data. The results are presented in Figure .

In all cases, the likelihood of testing data grows rapidly for small values of t. Since a geometric distribution corresponds to value $t = 1$, even small values of t give much better results than a geometric distribution. In the case of first and last exons, the GT distribution outperformed the GENSCAN smoothing

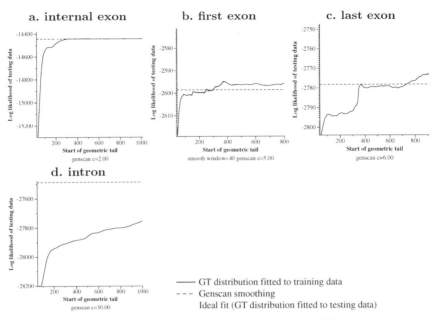

Fig. 4. Generalization capacity of GT distribution for different lengths of tail. Compare with GENSCAN smoothing algorithm and geometric distribution (GT with $t_B = 1$)

for large values of t because the GENSCAN smoothing overfit the data in the sparsely populated tail sections of the length distributions.

In the case of internal and first exons, we achieve performance close to GEN-SCAN smoothing function very fast (for values of t less than 150). For the last exon the likelihood grows in two phases. In the first phase it achieves a local maximum at $t = 111$; however, the likelihood is still significantly smaller than that of GENSCAN smoothing. The likelihood of GENSCAN smoothing is achieved at a second local maximum at $t = 370$. This behavior is explained by a local maximum centered around length of 300 in the length distribution of last exons as observed in Figure c. This local maximum cannot be captured by a GT distribution for $t < 300$.

For intron length distribution, the likelihood of testing data grows steadily even for large values of t (several thousands – data not shown). In this case we need to balance running time versus computation precision.

In all the cases, we see that even small values of t help significantly compared to a geometric distribution ($t = 1$). Moderate values of t allow fast running time, while yielding good results.

5.3   Incorporating GT Distribution into a Gene Finder

We have implemented our algorithm for boxed HMMs and created a simple gene finder. Our preliminary experiments show that using GT distributions can improve the prediction accuracy compared to a geometric distribution. Sensitivity/specificity in terms of exact exons was improved by 5%/3%. The benchmark dataset by Burset and Guigó [ ] was used in this experiment. The rate of improvement is consistent with results presented in [ ].

However, this rate of improvement cannot be expected after incorporating GT distribution to gene finders based on generalized HMMs, such as GENSCAN [ ], which is one of the most successful gene finders. Still, our technique can improve both speed and modeling accuracy of such a gene finder.

GENSCAN uses generalized states to model length distribution of exons. Generalized states generate exon-like sequences with the smoothed length distribution observed in the training data. This, in general, leads to a quadratic running time in the length of the sequence (which is impractical in case of long DNA sequences). However, in case of exons it is possible to use boundaries observed in DNA sequences (so called open reading frames) to limit the search in Viterbi algorithm as exons cannot extend beyond these regions. This is still slow in the case of long open reading frames.

By using GT distributions, it is possible to limit the search also by parameter t of the distribution. As we have shown before, the parameter t can be quite small (up to 200), while still modeling exon length distribution accurately. This means that the decoding can be sped up by our technique without sacrificing prediction accuracy.

GENSCAN's technique cannot be extended reasonably to other elements of DNA sequence (introns and intergenic regions). This is because there are no sequence features such as open reading frames which could be used to limit the search. Therefore in GENSCAN the authors used geometric distributions for introns and intergenic regions.

However, we have shown that a GT distribution can model intron lengths much better than a geometric distribution with reasonably small values of t (such as 200). Modeling intron lengths with GT distributions could thus improve the accuracy of GENSCAN. Still, GT distributions are impractical for intergenic regions, as we would need large values of t (more than 10000).

6   Conclusion

We have studied the problem of modeling length distributions in HMMs for annotation of biological sequences. We have described geometric-tail distributions and boxed HMMs—a class of generalized HMMs that work well with GT distributions. We have demonstrated how to estimate the parameters of a GT distribution from training data and how to modify the classical algorithms for HMMs to handle boxed HMMs with GT distributions efficiently. Our framework is flexible, and several used types of generalized HMMs can be expressed as a boxed HMM. Previously known methods are either too slow to be applied to biological data, or the algorithms have been tailored to one particular application, and hard to generalize for other cases or models.

We have also studied an application of our methods to gene finding. We provide evidence that the GT distribution is indeed a reasonable model for length distribution of introns and exons in genes. We showed that the GT distribution that fits the training data generalizes well to unseen testing data sets. We have discussed the possibility of incorporating GT distributions to current HMM based gene finders.

Acknowledgements

We would like to thank Dan Brown and Ming Li for many useful suggestions. Broňa Brejová is supported by Ontario Graduate Scholarship. Both authors are supported by NSERC research grant.

References

1. C. Burge and S. Karlin. Prediction of complete gene structures in human genomic DNA. *Journal of Molecular Biology*, 268(1):78–94, 1997.   ,
2. C. B. Burge. *Identification of Genes in Human Genomic DNA*. PhD thesis, Department of Mathematics, Stanford University, March 1997.   ,   ,   ,
      ,   ,
3. M. Burset and R. Guigó. Evaluation of gene structure prediction programs. *Genomics*, 34(3):353–357, 1996.
4. I. Dunham et al. The DNA sequence of human chromosome 22. *Nature*, 402(6761):489–495, 1999.
5. R. Durbin, S. Eddy, A. Krogh, and G. Mitchison. *Biological sequence analysis: Probabilistic models of proteins and nucleic acids*. Cambridge University Press, 1998.   ,
6. A. Krogh. Two methods for improving performance of an HMM and their application for gene finding. In *Proceedings of the 5th International Conference on Intelligent Systems for Molecular Biology (ISMB)*, pages 179–186, 1997.   ,
7. D. Kulp, D. Haussler, M. G. Reese, and F. H. Eeckman. A generalized hidden Markov model for the recognition of human genes in DNA. In *Proceedings of the 4th International Conference on Intelligent Systems for Molecular Biology (ISMB)*, pages 134–142, 1996.   ,
8. L. P. Lim and C. B. Burge. A computational analysis of sequence features involved in recognition of short introns. *Proceedings of the National Academy of Sciences USA*, 98(20):11193–11198, 2001.   ,
9. R. B. Lyngsø and C. N. S. Pedersen. Complexity of comparing hidden Markov models. Technical Report ALCOMFT-TR-01-144, Algorithms and Complexity – Future Technologies Project (ALCOM-FT), June 2001.
10. L. R. Rabiner. A tutorial on Hidden Markov models and selected applications in speech recognition. *Proceedings of the IEEE*, 77(2):257–285, 1989.   ,   ,

Faster Bit-Parallel Approximate String Matching

Heikki Hyyrö[1]* and Gonzalo Navarro[2]**

[1] Dept. of Computer and Information Sciences, University of Tampere, Finland
[2] Dept. of Computer Science, University of Chile

Abstract. We present a new bit-parallel technique for approximate string matching. We build on two previous techniques. The first one [Myers, J. of the ACM, 1999], searches for a pattern of length m in a text of length n permitting k differences in $O(mn/w)$ time, where w is the width of the computer word. The second one [Navarro and Raffinot, ACM JEA, 2000], extends a sublinear-time exact algorithm to approximate searching. The latter technique makes use of an $O(kmn/w)$ time algorithm [Wu and Manber, Comm. ACM, 1992] for its internal workings. This algorithm is slow but flexible enough to support all the required operations. In this paper we show that the faster algorithm of Myers can be adapted to support all those operations. This involves extending it to compute edit distance, to search for any pattern suffix, and to detect in advance the impossibility of a later match. The result is an algorithm that performs better than the original version of Navarro and Raffinot and that is the fastest for several combinations of m, k and alphabet sizes that are useful, for example, in natural language searching and computational biology.

1 Introduction

Approximate string matching is one of the main problems in classical string algorithms, with applications to text searching, computational biology, pattern recognition, etc. Given a text of length n, a pattern of length m, and a maximal number of differences permitted, k, we want to find all the text positions where the pattern matches the text up to k differences. The differences can be substituting, deleting or inserting a character. We call $\alpha = k/m$ the *difference ratio*, and σ the size of the alphabet Σ. All the average case figures in this paper assume random text and uniformly distributed alphabet.

In this paper we consider online searching, that is, the pattern can be preprocessed but the text cannot. The classical solution to the problem is based on filling a dynamic programming matrix and needs $O(mn)$ time [ ]. Since then, many improvements have been proposed (see [ ] for a complete survey). These can be divided into four types.

---

* Supported by the Academy of Finland and Tampere Graduate School in Information Science and Engineering.
** Partially supported by Fondecyt Project 1-020831.

A. Apostolico and M. Takeda (Eds.): CPM 2002, LNCS 2373, pp. 203– , 2002.
© Springer-Verlag Berlin Heidelberg 2002

The first type is based on dynamic programming and has achieved $O(kn)$ worst case time [,]. These algorithms are not really practical, but there exist also practical solutions that achieve, on the average, $O(kn)$ [ ] and even $O(kn/\sqrt{\sigma})$ time [ ].

The second type reduces the problem to an automaton search, since approximate searching can be expressed in that way. A deterministic finite automaton (DFA) is used in [ ] so as to obtain $O(n)$ search time, which is worst-case optimal. The problem is that the preprocessing time and the space is $O(3^m)$, which makes the approach practical only for very small patterns. In [ ] they trade time for space using a Four Russians approach, achieving $O(kn/\log s)$ time on average and $O(mn/\log s)$ in the worst case, assuming that $O(s)$ space is available for the DFAs.

The third approach filters the text to quickly discard large text areas, using a necessary condition for an approximate occurrence that is easier to check than the full condition. The areas that cannot be discarded are verified with a classical algorithm [, , , ,]. These algorithms achieve "sublinear" expected time in many cases for low difference ratios, that is, not all text characters are inspected. However, the filtration is not effective for higher ratios. The typical average complexity is $O(kn \log_\sigma m/m)$ for $\alpha = O(1/\log_\sigma m)$. The optimal average complexity is $O((k+\log_\sigma m)n/m)$ for $\alpha < 1 - O(1/\sqrt{\sigma})$ [ ], which is achieved in the same paper. The algorithm, however, is not practical.

Finally, the fourth approach is bit-parallelism [,], which consists in packing several values in the bits of the same computer word and managing to update all them in a single operation. The idea is to simulate another algorithm using bit-parallelism. The first bit-parallel algorithm for approximate searching [ ] parallelized an automaton-based algorithm: a nondeterministic finite automaton (NFA) was simulated in $O(k\lceil m/w \rceil n)$ time, where w is the number of bits in the computer word. We call this algorithm BPA (for Bit-Parallel Automaton) in this paper. BPA was improved to $O(\lceil km/w \rceil n)$ [ ] and finally to $O(\lceil m/w \rceil n)$ time [ ]. The latter simulates the classical dynamic programming algorithm using bit-parallelism, and we call it BPM (for Bit-Parallel Matrix) in this paper.

Currently the most successful approaches are filtering and bit-parallelism. A promising approach combining both [ ] will be called ABNDM in this paper (for Approximate BNDM). The original ABNDM was built on BPA because the latter is the most flexible for the particular operations needed. The faster BPM was not used at that time yet because of the difficulty in modifying it to be suitable for ABNDM.

In this paper we extend BPM in several ways so as to permit it to be used in the framework of ABNDM. The result is a competitive approximate string matching algorithm. In particular, the algorithm turns out to be the fastest for a range of m and k that includes interesting cases of natural language searching and computational biology applications.

2   Basic Concepts

2.1   Notation

We will use the following notation on strings: $|x|$ will be the length of string x; ε will be the only string of length zero; string positions will start at 1; substrings will be denoted as $x_{i...j}$, meaning taking from the i-th to the j-th character of x, both inclusive; x_i will denote the single character at position i in x. We say that x is a prefix of xy, a suffix of yx, and a substring or factor of yxz.

Bit-parallel algorithms will be described using C-like notation for the operations: bitwise "and" (&), bitwise "or" (|), bitwise "xor" ($\wedge$), bit complementation ($\sim$), and shifts to the left ($<<$) and to the right ($>>$), which are assumed to enter zero bits both ways. We also perform normal arithmetic operations ($+$, $-$, etc.) on the bit masks, which are treated as numbers in this case. Constant bit masks are expressed as sequences of bits, the first to the right, using exponentiation to denote bit repetition, for example $10^3 = 1000$ has a 1 at the 4-th position.

2.2   Problem Description

The problem of approximate string matching can be stated as follows: given a (long) text T of length n, and a (short) pattern P of length m, both being sequences of characters from an alphabet Σ of size σ, and a maximum number of differences permitted, k, find all the segments of T whose *edit distance* to P is at most k. Those segments are called "occurrences", and it is common to report only their start or end points.

The *edit distance* between two strings x and y is the minimum number of *differences* that would transform x into y or vice versa. The allowed differences are deletion, insertion and substitution of characters. The problem is non-trivial for $0 < k < m$. The *difference ratio* is defined as $\alpha = k/m$.

Formally, if $ed()$ denotes the edit distance, we may want to report start points (i.e. $\{|x|, \ T = xP'y, \ ed(P,P') \le k\}$) or end points (i.e. $\{|xP'|, \ T = xP'y, \ ed(P,P') \le k\}$) of occurrences.

2.3   Dynamic Programming

The oldest and still most flexible (albeit slowest) algorithm to solve the problem is based on dynamic programming [ ]. We first show how to compute the edit distance between two strings x and y. To compute $ed(x,y)$, a matrix $M_{0..|x|,0..|y|}$ is filled, where $M_{i,j} = ed(x_{1..i}, y_{1..j})$, so at the end $M_{|x|,|y|} = ed(x,y)$. This matrix is computed as follows

$$M_{i,0} \leftarrow i, \qquad M_{0,j} \leftarrow j,$$
$$M_{i,j} \leftarrow \text{if } (x_i = y_j) \text{ then } M_{i-1,j-1} \text{ else } 1 + \min(M_{i-1,j}, M_{i,j-1}, M_{i-1,j-1})$$

where the formula accounts for the three allowed operations. This matrix is usually filled columnwise left to right, and each column top to bottom. The time to compute $ed(x, y)$ is then $O(|x||y|)$.

This is easily extended to approximate searching, where $x = P$ and $y = T$, by letting an occurrence start anywhere in T. The only change is the initial condition $M_{0,j} \leftarrow 0$. The time is still $O(|x||y|) = O(mn)$. The space can be reduced to $O(m)$ by storing only one column of the matrix at a time, namely, the one corresponding to the current text position (going left to right means examining the text sequentially).

In this case it is more appropriate to think of a column vector $C_{0...m}$, which is initialized at $C_i \leftarrow i$ and updated to C' after reading text character T_j using

$$C'_i \leftarrow \text{ if } (P_i = T_j) \text{ then } C_{i-1} \text{ else } 1 + \min(C'_{i-1}, C_i, C_{i-1})$$

for all $i > 0$, and hence we report every end position j where $C_i \leq k$.

Several properties of the matrix M are discussed in [ ]. The most important for us is that adjacent cells in M differ at most by 1, that is, both $M_{i,j} - M_{i\pm1,j}$ and $M_{i,j} - M_{i,j\pm1}$ are in the range $\{-1, 0, +1\}$. Also, $M_{i+1,j+1} - M_{i,j}$ is in the range $\{0, 1\}$.

Fig. shows examples of edit distance computation and approximate string matching.

		s	u	r	g	e	r	y
	0	1	2	3	4	5	6	7
s	1	0	1	2	3	4	5	6
u	2	1	0	1	2	3	4	5
r	3	2	1	0	1	2	3	4
v	4	3	2	1	1	2	3	4
e	5	4	3	2	2	1	2	3
y	6	5	4	3	3	2	2	**2**

		s	u	r	g	e	r	y
	0	0	0	0	0	0	0	0
s	1	0	1	1	1	1	1	1
u	2	1	0	1	2	2	2	2
r	3	2	1	0	1	2	2	3
v	4	3	2	1	1	2	3	3
e	5	4	3	2	2	1	2	3
y	6	5	4	3	3	**2**	**2**	**2**

Fig. 1. The dynamic programming algorithm. On the left, to compute the edit distance between "survey" and "surgery". On the right, to search for "survey" in the text "surgery". The bold entries show the cell with the edit distance (left) and the end positions of occurrences for $k = 2$ (right)

2.4 The Cutoff Improvement

In [ ] they consider the dynamic programming algorithm and observe that column values larger than k can be assumed to be $k + 1$ without affecting the output of the computation. Cells of C with value not exceeding k are called *active*. In the algorithm, the index ℓ of the last active cell (i.e., largest i such that $C_i \leq k$) is maintained. All the values $C_{\ell+1...m}$ are assumed to be $k + 1$, so C needs to be updated only in the range $C_{1...\ell}$. Later [ ] it was shown that, on average, $\ell = O(k)$ and therefore the algorithm is $O(kn)$.

The value ℓ has to be updated throughout the computation. Initially, $\ell = k$ because $C_i = i$. It is shown that, at each new column, the last active cell can be incremented at most by one, so we check whether $C_{\ell+1} \leq k$ and in such a case we increment ℓ. However, it is also possible that which was the last active cell becomes inactive now, that is, $C_\ell > k$. In this case we have to search upwards for the new last active cell. Despite that this search can take $O(m)$ time at a given column, we cannot work more than $O(n)$ overall, because there are at most n increments of ℓ in the whole process, and hence there are no more than $n + k$ decrements. Hence, the last active cell is maintained at $O(1)$ amortized cost per column.

2.5   An Automaton View

An alternative approach is to model the search with a non-deterministic automaton (NFA) [ ]. Consider the NFA for $k = 2$ differences shown in Fig. . Each of the $k+1$ rows denotes the number of differences seen (the first row zero, the second row one, etc.). Every column represents matching a pattern prefix. Horizontal arrows represent matching a character. All the others increment the number of differences (i.e., move to the next row): vertical arrows insert a character in the pattern, solid diagonal arrows substitute a character, and dashed diagonal arrows delete a character of the pattern. The initial self-loop allows an occurrence to start anywhere in the text. The automaton signals (the end of) a match whenever a rightmost state is active.

It is not hard to see that once a state in the automaton is active, all the states of the same column and higher-numbered rows are active too. Moreover, at a given text position, *if we collect the smallest active rows at each column, we obtain the vector C of the dynamic programming* (in this case $[0, 1, 2, 3, 3, 3, 2]$, compare to Fig.).

Note that the NFA can be used to compute edit distance by simply removing the self-loop, although it cannot distinguish among different values larger than k.

2.6   A Bit-Parallel Automaton Simulation (BPA)

The idea of BPA [ ] is to simulate the NFA of Fig. using bit-parallelism, so that each row i of the automaton fits in a computer word R_i (each state is represented by a bit). For each new text character, all the transitions of the automaton are simulated using bit operations among the $k+1$ computer words.

The update formula to obtain the new R'_i values at text position j from the current R_i values is as follows:

$$R'_0 \leftarrow ((R_0 << 1) \mid 0^{m-1}1) \ \& \ B[T_j]$$
$$R'_{i+1} \leftarrow ((R_{i+1} << 1) \ \& \ B[T_j]) \mid R_i \mid (R_i << 1) \mid (R'_i << 1)$$

where $B[c]$ is a precomputed table of σ entries such that the r-th bit of $B[c]$ is set whenever $P_r = c$. We start the search with $R_i = 0^{m-i}1^i$. In the formula

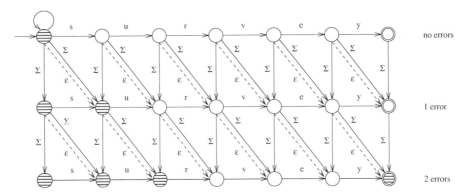

Fig. 2. An NFA for approximate string matching of the pattern "survey" with two differences. The shaded states are those active after reading the text "surgery"

for R'_{i+1} are expressed, in that order, horizontal, vertical, diagonal and dashed diagonal arrows.

If $m > w$ we need $\lceil m/w \rceil$ computer words to simulate every R_i mask, and have to update them one by one. The cost of this simulation is thus $O(k \lceil m/w \rceil n)$. The algorithm is flexible, for example in order to remove the initial self-loop one has to change the update formula for R_0 to $R'_0 \leftarrow (R_0 << 1) \;\&\; B[T_j]$.

2.7 Myers' Bit-Parallel Matrix Simulation (BPM)

A better way to parallelize the computation [ ] is to represent the differences between consecutive rows or columns of the dynamic programming matrix instead of the NFA states. Let us call

$$\Delta h_{i,j} = M_{i,j} - M_{i,j-1} \in \{-1, 0, +1\}$$
$$\Delta v_{i,j} = M_{i,j} - M_{i-1,j} \in \{-1, 0, +1\}$$
$$\Delta d_{i,j} = M_{i,j} - M_{i-1,j-1} \in \{0, 1\}$$

the horizontal, vertical, and diagonal differences among consecutive cells. Their range of values come from the properties of the dynamic programming matrix [ ].

We present a version [ ] that differs slightly from that of [ ]: Although both perform the same number of operations per text character, the one we present is easier to understand and more convenient for our purposes.

Let us introduce the following boolean variables. The first four refer to horizontal/vertical positive/negative differences and the last to the diagonal difference being zero:

$$VP_{i,j} \equiv \Delta v_{i,j} = +1 \qquad VN_{i,j} \equiv \Delta v_{i,j} = -1$$
$$HP_{i,j} \equiv \Delta h_{i,j} = +1 \qquad HN_{i,j} \equiv \Delta h_{i,j} = -1$$

$$D0_{i,j} \equiv \Delta d_{i,j} = 0$$

Note that $\Delta v_{i,j} = VP_{i,j} - VN_{i,j}$, $\Delta h_{i,j} = HP_{i,j} - HN_{i,j}$, and $\Delta d_{i,j} = 1 - D0_{i,j}$. It is clear that these values completely define $M_{i,j} = \sum_{r=1\ldots i} \Delta v_{r,j}$.

The boolean matrices HN, VN, HP, VP, and $D0$ can be seen as vectors indexed by i, which change their value for each new text position j, as we traverse the text. These vectors are kept in bit masks with the same name. Hence, for example, the i-th bit of the bit mask HN will correspond to the value $HN_{i,j}$. The index $j-1$ refers to the previous value of the bit mask (before processing T_j), whereas j refers to the new value, after processing T_j. By noticing some dependencies among the five variables [,], one can arrive to identities that permit computing their new values (at j) from their old values (at $j-1$) fast.

Fig.   gives the pseudo-code. The value $diff$ stores $C_m = M_{m,j}$ explicitly and is updated using $HP_{m,j}$ and $HN_{m,j}$.

```
BPM (P_{1...m}, T_{1...n}, k)
1.     Preprocessing
2.         For c ∈ Σ Do B[c] ← 0^m
3.         For i ∈ 1...m Do B[P_i] ← B[P_i] | 0^{m-i}10^{i-1}
4.         VP ← 1^m, VN ← 0^m
5.         diff ← m
6.     Searching
7.         For j ∈ 1...n Do
8.             X ← B[T_j] | VN
9.             D0 ← ((VP + (X & VP)) ∧ VP) | X
10.            HN ← VP & D0
11.            HP ← VN | ∼(VP | D0)
12.            X ← HP << 1
13.            VN ← X & D0
14.            VP ← (HN << 1) | ∼(X | D0)
15.            If HP & 10^{m-1} ≠ 0^m Then diff ← diff + 1
16.            If HN & 10^{m-1} ≠ 0^m Then diff ← diff - 1
17.            If diff ≤ k Then report an occurrence at j
```

Fig. 3. BPM bit-parallel simulation of the dynamic programming matrix

This algorithm uses the bits of the computer word better than previous bit-parallel algorithms, with a worst case of $O(\lceil m/w \rceil n)$ time. However, the algorithm is more difficult to adapt to other related problems, and this has prevented it from being used as an internal tool of other algorithms.

2.8   The ABNDM Algorithm

Given a pattern P, a *suffix automaton* is an automaton that recognizes every suffix of P. This is used in [ ] to design a simple exact pattern matching algorithm called BDM, which is optimal on average ($O(n \log_\sigma m/m)$ time). To search for a pattern P in a text T, the suffix automaton of $P^r = P_m P_{m-1} \ldots P_1$ (i.e. the pattern read backwards) is built. A window of length m is slid along the text, from left to right. The algorithm scans the window backwards, using the suffix automaton to recognize a factor of P. During this scan, if a final state is reached that does not correspond to the entire pattern P, the window position is recorded in a variable *last*. This corresponds to finding a *prefix* of the pattern starting at position *last* inside the window and ending at the end of the window, because the suffixes of P^r are the reverse prefixes of P. This backward search ends in two possible forms:

1. We fail to recognize a factor, that is, we reach a letter a that does not correspond to a transition in the suffix automaton (Fig.). In this case we shift the window to the right so as to align its starting position to the position *last*.

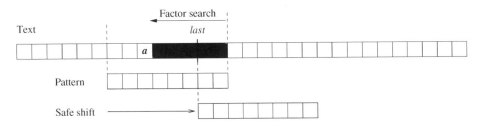

Fig. 4. BDM search scheme

2. We reach the beginning of the window, and hence recognize P and report the occurrence. Then, we shift the window exactly as in case 1 (to the previous *last* value).

In BNDM [ ] this scheme is combined with bit-parallelism so as to replace the construction of the deterministic suffix automaton by the bit-parallel simulation of a nondeterministic one. The scheme turns out to be flexible and powerful, and permits other types of search, in particular approximate search. The resulting algorithm is ABNDM.

We modify the NFA of Fig. so that it recognizes not only the whole pattern but also any suffix thereof, allowing up to k differences. Fig. illustrates the modified NFA. Note that we have removed the initial self-loop, so it does not search for the pattern but recognizes strings at edit distance k or less from the pattern. Moreover, we have built it on the reverse pattern. We have also added an initial state "I", with ϵ-transitions leaving it. These allow the automaton to recognize, with up to k differences, any suffix of the pattern.

In the case of approximate searching, the length of a pattern occurrence ranges from $m - k$ to $m + k$. To avoid missing any occurrence, we move a

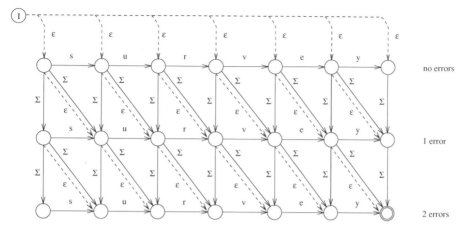

Fig. 5. An NFA to recognize suffixes of the pattern `"survey"` reversed

window of length $m - k$ on the text, and scan backwards the window using the NFA described above.

Each time we move the window to a new position we start the automaton with all its states active, which represents setting the initial state to active and letting the ϵ-transitions propagate this activation to all the automaton (the states in the lower-left triangle are also activated to allow initial insertions). Then we start reading the window characters backward.

We recognize a prefix and update *last* whenever the final NFA state is activated. We stop the backward scan when the NFA is out of active states.

If the automaton recognizes a pattern prefix at the initial window position, then it is possible (but not necessary) that the window starts an occurrence. The reason is that strings of different length match the pattern with k differences, and all we know is that we have matched a prefix of the pattern of length $m - k$.

Therefore, in this case we need to *verify* whether there is a pattern occurrence starting exactly at the beginning of the window. For this sake, we run the traditional automaton that computes edit distance (i.e., that of Fig.   without initial self-loop) from the initial window position in the text. After reading at most $m + k$ characters we have either found a match starting at the window position (that is, the final state becomes active) or determined that no match starts at the window beginning (that is, the automaton runs out of active states).

So we need two different automata in this algorithm. The first one makes the *backward scanning*, recognizing suffixes of P^r. The second one makes the *forward scanning*, recognizing P.

The automata can be simulated in a number of ways. In [ ] they choose BPA [ ] because it is easy to adapt to the new scenario. To recognize all the suffixes we just need to initialize $R_i \leftarrow 1^m$. To make it compute edit distance, we remove the self-loop as explained in Sec.   . The final state is active when $R_k \And 10^{m-1} \neq 0^m$. The NFA is out of active states whenever $R_k = 0^m$.

Other approaches were discarded: an alternative NFA simulation [ ] is not practical to compute edit distance, and BPM [   ] cannot easily tell when the corresponding automaton is out of active states, or which is the same, when all the cells of the current dynamic programming column are larger than k.

Fig.   shows the algorithm.

ABNDM $(P_{1...m}, T_{1...n}, k)$
1.   Preprocessing
2.   Build forward and backward NFA simulations ($fNFA$ and $bNFA$)
3.   Searching
4. $pos \leftarrow 0$
5.   **While** $pos \leq n - (m - k)$ **Do**
6. $j \leftarrow m - k, last \leftarrow m - k$
7.   Initialize $bNFA$
8.   **While** $j \neq 0$ AND $bNFA$ has active states **Do**
9.   Feed $bNFA$ with T_{pos+j}
10. $j \leftarrow j - 1$
11.   **If** $bNFA$'s final state is active **Then** /* prefix recognized */
12.   **If** $j > 0$ **Then** $last \leftarrow j$
13.   **Else** check with $fNFA$ a possible
14.   occurrence starting at $pos + 1$
15. $pos \leftarrow pos + last$

Fig. 6. The generic **ABNDM** algorithm

The algorithm is shown to be good for moderate m, low k and small σ, which is an interesting case, for example, in DNA searching. However, the use of BPA for the NFA simulation limits its usefulness to very small k values. Our purpose in this paper is to show that BPM can be extended for this task, so as to obtain a faster version of ABNDM that works with larger k.

Average Case Analysis of ABNDM. We show that ABNDM inspects on average $O(kn \log_\sigma(m)/m)$ text positions. This is better than what was previously obtained [ ]. Using previous results [ ], we have that the total number of strings that match a suffix of a pattern of length m with k errors is at least $\binom{m}{k}(\sigma - 1)^k$ (assuming only replacements and matching only the whole pattern) and at most $m\binom{m}{k}^2 \sigma^k$ (counting every suffix as if it had length m and assuming all the strings different). If we inserted those strings in a trie, the resulting height would be logarithmic (base σ) in the number of strings inserted. This means a height of $\Theta(k + m \log_\sigma m - k \log_\sigma k - (m - k) \log_\sigma(m - k))$, which can be factorized as $\Theta\left(k + m \log_\sigma \frac{m}{m-k} + k \log_\sigma \frac{m-k}{m}\right)$. We have $m \log_\sigma \frac{m}{m-k} = m \log_\sigma\left(1 + \frac{k}{m-k}\right) \leq \frac{m}{m-k}k \leq 2k$. The latter is because we are interested in the case $k < m/2$, as otherwise the algorithm cannot be sublinear time: the window is of length $m - k$ and we read at least $k+1$ characters before the NFA can run out of active states.

Hence we have that the height is $\Theta(k + k\log_\sigma(m/k))$. This is $\Theta(k\log_\sigma m)$, for example consider the case $k = m^\lambda$.

Traversing the window backwards until the NFA runs out of active states is equivalent to entering the above trie with the reverse window. On average, we reach the end of the trie in $\Theta(k\log_\sigma m)$ steps. Then we shift the window forward in $m - \Theta(k\log_\sigma m)$ positions. Overall, we inspect $O(kn\log_\sigma(m)/m)$ text positions, for $\alpha < 1/2$. If we use BPA, the complexity is $O(k^2 n\log_\sigma(m)/w)$. If we manage to use BPM, this goes down to $O(kn\log_\sigma(m)/w)$.

3   Forward Scanning with the BPM Simulation

We first focus on how to adapt the BPM algorithm to perform the forward scanning required by the ABNDM algorithm. Two modifications are necessary. The first is to make the algorithm compute edit distance instead of performing text searching. The second is making it able to determine when it is not possible to obtain edit distance $\leq k$ by reading more characters.

3.1   Computing Edit Distance

We recall that BPM implements the dynamic programming algorithm of Sec.   in such a way that differential values, rather than absolute ones, are stored. Therefore, we must consider which is the change required in the dynamic programming matrix in order to compute edit distance. As explained in Sec.   , the only change is that $M_{0,j} = j$. In differential terms (Sec.   ), this means $\Delta h_{0,j} = 1$ instead of zero.

When $\Delta h_{0,j} = 0$, its value does not need to be explicitly present in the BPM algorithm. The value makes a difference only when HP or HN is shifted left, which happens on lines 12 and 14 of the algorithm (Fig.   ). On these occasions the assumed bit zero enters automatically from the right, thereby implicitly using a value $\Delta h_{0,j} = 0$. To use a value $\Delta h_{0,j} = 1$ instead, we change line 12 of the algorithm to $X \leftarrow (HP << 1) \mid 0^{m-1}1$.

Since we will use this technique several times from now on, we give in Fig.   the code for a single step of edit distance computation.

3.2   Preempting the Computation

Albeit in the forward scan we could always run the automaton through $m + k$ text characters, stopping only if $diff \leq k$ to signal a match, it is also possible to determine that $diff$ will always be larger than k in the characters to come. This happens when all the cells of the vector C_i are larger than k, because there is no way in the recurrence to introduce a value smaller than the current ones. In the automaton view, this is the same as the NFA running out of active states (since an active state at column i and row r would mean $C_i = r \leq k$).

This is more difficult in the dynamic programming matrix simulation of BPM. The only column value that is explicitly stored is $diff = C_m$. The others are

```
BPMStep (Bc)
1.        X ← Bc | VN
2.        D0 ← ((VP + (X & VP)) ∧ VP) | X
3.        HN ← VP & D0
4.        HP ← VN | ∼ (VP | D0)
5.        X ← (HP << 1) | 0^{m-1}1
6.        VN ← X & D0
7.        VP ← (HN << 1) | ∼ (X | D0)
```

Fig. 7. Single step of the adaptation of BPM to compute edit distance. It receives the bit mask B of the current text character and shares all the other variables with the calling process

implicitly represented as $C_i = \sum_{r=1...i}(VP_r - VN_r)$. It is not easy to check whether $\forall i$, $C_i > k$ using this incremental representation.

Our solution is inspired by the cutoff algorithm of Sec.   . This algorithm permits knowing all the time the largest ℓ such that $C_\ell \leq k$, at constant amortized time per text position. Although designed for text searching, the technique can be applied without any change to the edit distance computation algorithm. Clearly $\exists i$, $C_i \leq k \iff \ell \geq 0$.

So we have to figure out how to compute ℓ using BPM. Initially, since $C_i = i$, we set $\ell \leftarrow k$. Later, we have to update ℓ for each new text character read. Recall that, given that neighboring cells in M differ by at most one, and that by definition $M_{\ell+1,j-1} > k$, we have that $M_{\ell,j-1} = k$.

Since ℓ can increase at most by one at the new text position, we start by effectively increasing it. This increment is correct when $M_{\ell+1,j} \leq k$ *before* doing the increment. Since $M_{\ell+1,j} - M_{\ell,j-1} = \Delta d_{\ell+1,j} \in \{0,1\}$, we have that it was correct to increase ℓ if and only if the bit $D0_{\ell,j}$ is set *after* the increment. If it was not correct to increase ℓ, we decrease it as much as necessary to obtain $M_{\ell,j} \leq k$. In this case we know that $M_{\ell,j} = k+1$, and so we obtain $M_{\ell-1,j} = M_{\ell,j} - VP_{\ell,j} + VN_{\ell,j}$, and so on with $\ell-2$, $\ell-3$, etc. If we reach $\ell = 0$ and still $M_{\ell,j} > k$, then all the rows are larger than k and we stop the scanning process.

The above arguments assume $\ell < m$. Note that, as soon as $\ell = m$, we have that $C_m \leq k$ and therefore the forward scan will then terminate because we have found an occurrence.

Fig.   shows the forward scanning algorithm. It scans from text position j and determines whether there is an occurrence starting at j. Instead of P, the routine receives the mask B already computed (see Fig.   ). Note that for efficiency ℓ is maintained in unary.

```
BPMFwd (B, T_{j...n}, k)
  1.        VP ← 1^m, VN ← 0^m
  2.        ℓ ← 0^{m-k}10^{k-1}
  3.        While j ≤ n Do
  4.            BPMStep (B[T_j])
  5.            ℓ ← ℓ << 1
  6.            If D0 & ℓ = 0^m Then
  7.                val ← k + 1
  8.                While val > k Do
  9.                    If ℓ = 0^{m-1}1 Then Return FALSE
 10.                    If VP & ℓ ≠ 0^m Then val ← val − 1
 11.                    If VN & ℓ ≠ 0^m Then val ← val + 1
 12.                    ℓ ← ℓ >> 1
 13.            Else If ℓ = 10^{m-1} Then Return TRUE
 14.            j ← j + 1
 15.        Return FALSE
```

Fig. 8. Adaptation of BPM to perform a forward scan from text position j and return whether there is an occurrence starting at j

4   Backward Scanning with the BPM Simulation

The backward scan has the particularity that all the NFA states start active. This is equivalent to initializing C as $C_i = 0$ for all i. The place where this initialization is expressed in BPM is on line 4 of Fig. . Since $VP = 1^m$, we have $C_i = i$. We change it to $VP \leftarrow 0^m$ and obtain the desired effect. Also, like in forward scanning, $M_{0,j} = j$, so we apply to line 12 the same change as with forward scanning in order to use the value $\Delta h_{0,j} = 1$.

With these tools at hand, we could simply apply the forward scan algorithm with B built on P^r and reading the window backwards. We could use ℓ to determine when the NFA is out of active states. Every time $\ell = m$ we know that we have recognized a prefix and hence update $last$. There are a few changes, though: (i) we start with $\ell = m$ because $M_{i,0} = 0$; and (ii) we have to deal with the case $\ell = m$ when updating ℓ, because now we do not stop the backward scanning in that case but just update $last$.

The latter problem is solved as follows. As soon as $\ell = m$ we stop tracking ℓ and initialize $diff \leftarrow k$ as the known value for C_m. We keep updating $diff$ using HP and HN just as in Fig. , until $diff > k$. At this moment we switch to updating ℓ again, moving it upwards as necessary.

The above scheme works correctly but it is terribly slow. The reason is that ℓ starts at m and it has to reach zero before we can leave the window. This requires m shifting operations $\ell \leftarrow \ell >> 1$, which is a lot considering that on average one traverses $O(k \log_\sigma m)$ characters in the window. The $O(k + n)$ complexity given in Sec.   becomes here $O(m + k \log_\sigma m)$. So, the problem is that all the column values reach a value larger than k quite soon, but we take too much time traversing all them to determine that this has happened.

We present two solutions to determine fast that all the C_i values have surpassed k.

4.1   Bit-Parallel Counters

In the original BPM algorithm the integer value $diff = C_m$ is explicitly maintained in order to determine which text positions match. The way is to use the m-th bit of HP and HN to keep track of C_m. This part of the algorithm is not bit-parallel, so in principle one cannot do the same with all the C_i values and still hope to update all of them in a single operation.

However, it is possible to store several such counters in the same computer word MC and use them as upper bound to the others. Since the C_i values start at zero and the window is of length $m - k$, we need in principle $\lceil \log_2(m - k) \rceil$ bits to store any C_i value (their value after reading the last window character is not important). Hence we have space for $O(m/\log m)$ counters in MC.

To determine the minimum number Q of bits needed for each counter we must look a bit ahead in our algorithm. We will need to determine that all the counters have exceeded $k' = k + \lfloor Q/2 \rfloor$. For this sake, we initialize the counters at a value b that makes sure that their last bit will be activated when they surpass this threshold. So we need that (i) $b + k' + 1 = 2^{Q-1}$. On the other hand, we have to ensure that the Q-th bit is always set for any counter value up to the $(m - k)$-th step in the backward scan, and that Q bits are still enough to represent the counter. So we need that (ii) $b + m - k < 2^Q$. Finally, we need (iii) $b \geq 0$. By replacing (i) in (ii) we get (i') $b = 2^{Q-1} - k' - 1$ and (ii') $m - k - k' \leq 2^{Q-1}$. By (iii) and (i') we get (iii') $k' + 1 \leq 2^{Q-1}$. Hence the solution to the new system of inequalities is $Q = 1 + \lceil \log_2(\max(m - k - k', k' + 1)) \rceil$, and $b = 2^{Q-1} - k' - 1$.

The problem with the above solution is that $k' = k + \lfloor Q/2 \rfloor$, so the solution is indeed a recurrence for Q. Fortunately, it is easy to solve. Since $(X + Y)/2 \leq \max(X, Y) \leq X + Y$ for any nonnegative X and Y, if we call $X = m - k - k'$ and $Y = k' + 1$, we have that $X + Y = m - k + 1$. So $Q \leq 1 + \lceil \log_2(m - k + 1) \rceil$, and $Q \geq 1 + \lceil \log_2((m - k + 1)/2) \rceil = \lceil \log_2(m - k + 1) \rceil$. This gives a 2-integer range for the actual Q value. If $Q = \lceil \log_2(m - k + 1) \rceil$ does not satisfy (ii') and (iii'), we use $Q + 1$. This scheme works correctly as long as $X, Y \geq 0$, that is, $\lfloor Q/2 \rfloor \leq m - 2k$, or $m - k \geq k'$. As it becomes clear later, we cannot use this method if this does not hold, because the algorithm will have to verify every text window.

We decide to store $t = \lceil m/Q \rceil$ counters, for $C_m, C_{m-Q}, C_{m-2Q}, \ldots,$ $C_{m-(t-1)Q}$. The counter for C_{m-rQ} uses the bits of the region $m - rQ \ldots m - rQ + Q - 1$, both inclusive. This means that we need $m + Q - 1$ bits for MC .

The counters can be used as follows. Since every cell is at distance at most $\lfloor Q/2 \rfloor$ to some represented counter and the difference between consecutive cells

---

[1] If sticking to m bits is necessary we can store C_m separately in the $diff$ variable, at the same complexity but more cost in practice.

is at most 1, it is enough that all the counters are $\geq k' + 1 = k + \lfloor Q/2 \rfloor + 1$, to be sure that all the cells of C exceed k.

So the idea is to traverse the window until all the counters exceed k' and then shift the window. We will examine a few more cells than if we had controlled exactly all the C values: The backward scan will behave as if we permitted $k' = k + \lfloor Q/2 \rfloor$ differences, so the number of characters inspected is $\Theta(n(k + \log m) \log_\sigma m/m)$. Note that we have only m/Q suffixes to test but this does not affect the complexity. Note also that the amount of shifting is not affected because we have C_m correctly represented.

We have to face two problems. The first one is how to update all the counters in a single operation. This is not hard because counter C_{m-rQ} can be updated from its old to its new value by considering the $(m - rQ)$-th bits of HP and HN. That is, we define a mask $sMask = (0^{Q-1}1)^t 0^{m+Q-1-tQ}$ and update MC using $MC \leftarrow MC + (HP \ \& \ sMask) - (HN \ \& \ sMask)$.

The second problem is how to determine that all the counters have exceeded k'. For this sake we have defined b and Q so that the Q-th bits of the counters get activated when they exceed k'. If we define $eMask = (10^{Q-1})^t 0^{m+Q-1-tQ}$, then we can stop the scanning whenever $MC \ \& \ eMask = eMask$.

Finally, note that our assumption that every cell in C is at distance at most $\lfloor Q/2 \rfloor$ to a represented cell may not be true for the first $\lfloor Q/2 \rfloor$ cells. However, we know that, at the j-th iteration, $C_0 = j$, so we may assume there is an implicit counter at row zero. Moreover, since this counter is always incremented, it is larger than any other counter, so it will surely surpass k' when other counters do. The initial $\lfloor Q/2 \rfloor$ cells are close enough to this implicit counter.

Fig.   shows the pseudocode of the algorithm. All the bit masks are of length m, except $sMask$, $eMask$ and MC, which are of length $m + Q - 1$.

In case our upper bound turns out to be too loose, we can use several interleaved sets of counters, each set in its own bit-parallel counter. For example we could use two interleaved MC counters and hence the limit would be $Q/4$. In general we could use c counters and have a limit of the form $Q/2^c$. The cost would be $O(nc(k + \log(m)/2^c) \log_\sigma m/m)$, which is optimized for $c = \log_2(\log_\sigma(m)/k)$, where the complexity is $O(nk \log_\sigma m \log \log_\sigma m/m)$ for $k = o(\log_\sigma m)$ and the normal $O(nk \log_\sigma m/m)$ otherwise.

4.2   Bit-Parallel Cutoff

The previous technique, although simple, has the problem that it changes the complexity of the search and inspects more cells than necessary. We can instead produce, using a similar approach, an algorithm with the same complexity as the basic version. This time the idea is to mix the bit-parallel counters with a bit-parallel version of the cutoff algorithm (Sec.   ).

Consider regions $m - rQ - Q + 1 \ldots m - rQ$ of length Q. Instead of having the counters fixed at the end of each region (as in the previous section), we let the counters "float" inside their region. The distance between consecutive counters is still Q, so they all float together and all are at the same distance δ to the end

```
ABNDMCounters (P₁...ₘ, T₁...ₙ, k)
1.     Preprocessing
2.         For c ∈ Σ Do Bf[c] ← 0ᵐ , Bb[c] ← 0
3.         For i ∈ 1...m Do
4.             Bf[Pᵢ] ← Bf[Pᵢ] | 0ᵐ⁻ⁱ10ⁱ⁻¹
5.             Bb[Pᵢ] ← Bb[Pᵢ] | 0ⁱ⁻¹10ᵐ⁻ⁱ
6.         Q ← ⌈log₂(m − k + 1)⌉
7.         If 2^{Q−1} < max(m − 2k − ⌊Q/2⌋, k + 1 + ⌊Q/2⌋) Then Q ← Q + 1
8.         b ← 2^{Q−1} − k − ⌊Q/2⌋ − 1
9.         t ← ⌈m/Q⌉
10.        sMask ← (0^{Q−1}1)ᵗ0^{m+Q−1−tQ}
11.        eMask ← (10^{Q−1})ᵗ0^{m+Q−1−tQ}
12.    Searching
13.        pos ← 0
14.        While pos ≤ n − (m − k) Do
15.            j ← m − k, last ← m − k
16.            VP ← 0ᵐ, VN ← 0ᵐ
17.            MC ← [b]ᵗ_Q0^{m+Q−1−tQ}
18.            While j ≠ 0 AND MC & eMask ≠ eMask Do
19.                BPMStep (Bb[T_{pos+j}])
20.                MC ← MC + (HP & sMask) − (HN & sMask)
21.                j ← j − 1
22.                If MC & 10^{m+Q−2} ≠ 0^{m+Q−1} Then /* prefix recognized */
23.                    If j > 0 Then last ← j
24.                    Else If BPMFwd (Bf, T_{pos+1...n}) Then
25.                        Report an occurrence at pos + 1
26.            pos ← pos + last
```

Fig. 9. The **ABNDM** algorithm using bit-parallel counters. The expression $[b]_Q$ denotes the number b seen as a bit mask of length Q. Note that **BPMFwd** can share its variables with the calling code because these are not needed any more at that point

of their regions. We use $sMask$ and $eMask$ with the same meanings as before, but they are displaced so as to be all the time aligned to the counters.

The invariant is that the counters will be as close as possible to the end of their regions, as long as all the cells past the counters exceed k. That is,

$$\delta = \min\{d \in 0...Q, \forall r \in \{0...t − 1\}, \gamma \in \{0...d − 1\}, C_{m−rQ−\gamma} > k\}$$

where we assume that C yields values larger than k when accessed at negative indexes. When δ reaches Q, this means that all the cell values are larger than k and we can suspend the scanning. Prefix reporting is easy since no prefix can match unless $\delta = 0$, as otherwise $C_m = C_{m−0\cdot Q} > k$, and if $\delta = 0$ then the last floating counter has exactly the value C_m.

The floating counters are a bit-parallel version of the cutoff technique, where each counter cares of its region. Consequently the way of moving the counters up and down resembles the cutoff technique. We first move down and use $D0$ to determine if we should have moved down. If not, we move up as necessary using VP and VN. To determine if we should have moved down, we need to know whether there is a counter that exceeds k. We compute Q as in Section   , except that $k = k'$ and hence no recurrence arises. We use $eMask$ and, in order to increment and decrement the counters, $sMask$. We have to deal with the case where the counters are at the end of their region and hence cannot move down further. In this case we update them using HP and HN.

It is possible that the upmost counter goes out of bounds while shifting the counters, which in effect results in that counter being removed. For this to happen, however, all the area in C covered by the upmost counter must have values larger than k, and it is not possible that a cell in this area gets a value $\leq k$ later. So this counter can be safely removed from the set, and hence we remove it from $eMask$ as soon as it gets out of bounds for the first time. Note that ignoring this fact leads to inspecting slightly more characters (an almost negligible amount) but one instruction is saved, which in practice is convenient.

As for the case of a single counter, we work $O(1)$ amortized time per text position. More specifically, if we read u window characters then we work $O(u+Q)$ because we have to move from $\delta = 0$ to $\delta = Q$. But $O(u + Q) = O(k \log_\sigma m)$ on average because $Q = O(\log m)$, and therefore the classical complexity is not altered.

We also tried a practical version of using cutoff, in which the counters are not shifted. Instead they are updated in a similar fashion to the algorithm of Fig.   , and when all counters have a value $> k$, we try to shift a *copy* of them up until either a cell with value $\leq k$ is found or $Q - 1$ consecutive shifts are made. In the latter case we can stop the search, since then we have covered checking the whole column C. This version has a worse complexity, $O(Qk \log m) = O(k \log^2 m)$, as at each processed character it is possible to make $O(Q)$ shifts. But in practice it turned out to be very similar to the present cutoff algorithm.

Fig.   shows the algorithm. The counters are not physically shifted, we use δ instead.

5   Experimental Results

We compared our BPM-based ABNDM against the original BPA-based AB-NDM, as well as those other algorithms that, according to a recent survey [   ], are the best for moderate pattern lengths. We tested with random patterns and text over uniformly distributed alphabets. Each individual test run consisted of searching for 100 patterns a text of size 10 Mb. We measured total elapsed times.

The computer used in the tests was a 64-bit Alphaserver ES45 with four 1 Ghz Alpha EV68 processors, 4 GB of RAM and Tru64 UNIX 5.1A operating system. All test programs were compiled with the DEC CC C-compiler and full optimization. There were no other active significant processes running on the computer during the tests. All algorithms were set to use a 64 KB text buffer. The tested algorithms were:

ABNDMCutoff $(P_{1...m},\ T_{1...n},\ k)$

1.   Preprocessing
2.   **For** $c \in \Sigma$ **Do** $Bf[c] \leftarrow 0^m$, $Bb[c] \leftarrow 0$
3.   **For** $i \in 1 \ldots m$ **Do**
4. $Bf[P_i] \leftarrow Bf[P_i] \mid 0^{m-i}10^{i-1}$
5. $Bb[P_i] \leftarrow Bb[P_i] \mid 0^{i-1}10^{m-i}$
6. $Q \leftarrow 1 + \lceil \log_2(\max(m-2k, k+1)) \rceil$
7. $b \leftarrow 2^{Q-1} - k - 1$
8. $t \leftarrow \lceil m/Q \rceil$
9. $sMask \leftarrow (0^{Q-1}1)^t 0^{m+Q-1-tQ}$
10. $eMask \leftarrow (10^{Q-1})^t 0^{m+Q-1-tQ}$
11.   Searching
12. $pos \leftarrow 0$
13.   **While** $pos \leq n - (m-k)$ **Do**
14. $j \leftarrow m-k,\ last \leftarrow m-k$
15. $VP \leftarrow 0^m,\ VN \leftarrow 0^m$
16. $MC \leftarrow [b]_Q^t 0^{m+Q-1-tQ}$
17. $\delta \leftarrow 0$
18.   **While** $j \neq 0$ AND $\delta < Q$ **Do**
19.   **BPMStep** $(Bb[T_{pos+j}])$
20.   **If** $\delta = 0$ **Then** $MC \leftarrow MC + ((HP\ \&\ sMask) -$
21. $(HN\ \&\ sMask)$
22.   **Else**
23. $\delta \leftarrow \delta - 1$
24. $MC \leftarrow MC + (\sim (D0 << \delta)\ \&\ sMask)$
25.   **While** $\delta < Q$ AND $MC\ \&\ eMask = eMask$ **Do**
26. $MC \leftarrow MC - ((VP << \delta)\ \&\ sMask) + ((VN << \delta)\ \&$
27. $sMask)$
28. $\delta \leftarrow \delta + 1$
29.   **If** $\delta = m - (t-1)Q$ **Then** $eMask \leftarrow eMask\ \&$
30. $1^{(t-1)Q} 0^{m+2Q-1-tQ}$
31. $j \leftarrow j - 1$
32.   **If** $\delta = 0$ AND $MC\ \&\ 10^{m+Q-2} \neq 0^{m+Q-1}$ **Then**
33.   /* prefix recognized */
34.   **If** $j > 0$ **Then** $last \leftarrow j$
35.   **Else If BPMFwd** $(Bf, T_{pos+1...n})$ **Then**
36.   Report an occurrence at $pos + 1$
37. $pos \leftarrow pos + last$

Fig. 10. The **ABNDM** algorithm using bit-parallel cutoff. The same comments of Fig.   apply

ABNDM/BPA(regular): ABNDM implemented on BPA [   ], using a generic implementation for any k.

ABNDM/BPA(special code): Same as before but especially coded for each value of k to avoid using an array of bit masks.

ABNDM/BPM(counters): ABDNM implemented using BPM and counters (Sec.   ). The implementation differed slightly from Fig.   due to optimizations.

ABNDM/BPM(cutoff): ABDNM implemented using BPM and cutoff (Sec   ). The implementation differed slightly from Fig.   due to optimizations.

BPM: The sequential BPM algorithm [ ]. The implementation was from us and used the slightly different (but practically equivalent in terms of performance) formulation from [ ].

BPP: A combined heuristic using pattern partitioning, superimposition and hierarchical verification, together with a diagonally bit-parallelized NFA [,]. The implementation was from the original authors.

EXP: Partitioning the pattern into $k + 1$ pieces and using hierachical verification with a diagonally bit-parallelized NFA in the checking phase [ ]. The implementation was from the original authors.

Fig.   shows the test results for $\sigma = 4$, 13 and 52 and $m = 30$ and 55. This is only a small part of our complete tests, which included $\sigma = 4, 13, 20, 26$ and 52, and $m = 10, 15, 20, \ldots, 55$. We chose $\sigma = 4$ because it behaves like DNA, $\sigma = 13$ because it behaves like English, and $\sigma = 52$ to show that our algorithms are useful even on large alphabets.

First of all it can be seen that ABNDM/BPM(cutoff) is always faster than ABNDM/BPM(counters) by a nonnegligible margin.

It can be seen that our ABNDM/BPM versions are often faster than ABNDM/BPA(special code) when $k = 4$, and always when $k > 4$. Compared to ABNDM/BPA(regular), our version is always faster for $k > 1$. We note that writing down a different procedure for every possible k value, as done for ABNDM/BPA(special code), is hardly a real alternative in practice.

With moderate pattern length $m = 30$, our ABNDM/BPM versions are competitive for low error levels. However, BPP is better for small alphabets and EXP is better for large alphabets. In the intermediate area $\sigma = 13$, we are the best for $k = 4 \ldots 6$. This area is rather interesting when searching natural language text.

When $m = 55$, our ABNDM/BPM versions become much more competitive, being the fastest in many cases: For $k = 5 \ldots 9$ with $\sigma = 4$, and for $k = 4 \ldots 11$ both with $\sigma = 13$ and $\sigma = 52$, with the single exception of the case $\sigma = 52$ and $k = 9$, where EXP is faster (this seems to be a variance problem, however).

6   Conclusions

The most successful approaches to approximate string matching are bit-parallelism and filtering. A promising algorithm combining both is ABNDM [ ]. However, ABNDM uses a slow $O(kmn/w)$ time bit-parallel algorithm (BPA [ ]) for its internal working because no other alternative exists with the necessary flexibility. In this paper we have shown how to extend BPM [ ] to replace BPA. Since BPM is $O(mn/w)$ time, we obtain a much faster version of ABNDM.

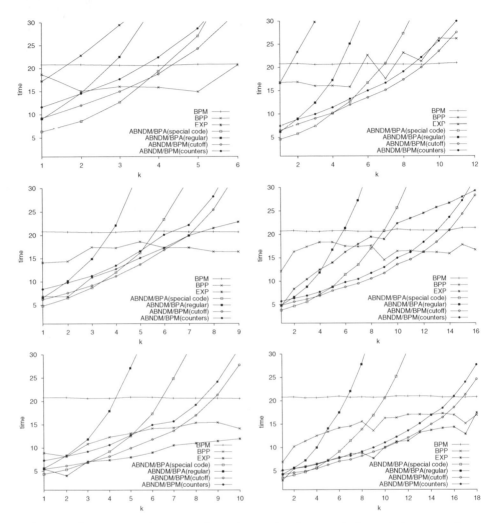

Fig. 11. Comparison between algorithms, showing total elapsed time as a function of the number of differences permitted, k. From top to bottom row we show $\sigma = 4$, 13 and 52. On the left we show $m = 30$ and on the right $m = 55$

For this sake, BPM was extended to permit backward scanning of the window and forward verification. The extensions involved making it compute edit distance, making it able to recognize any suffix of the pattern with k differences, and, the most complicated, being able to tell in advance that a match cannot occur ahead, both for backward and forward scanning. We presented two alternatives for the backward scanning: a simple one that may read more characters than necessary, and a more complicated (and more costly per processed character) that reads exactly the required characters.

The experimental results show that our new algorithm beats the old AB-NDM, even when BPA is especially coded with a different procedure for every possible k value, often for $k = 4$ and always for $k > 4$, and that it beats a general BPA implementation for $k \geq 2$. Moreover it was seen that our version of ABNDM becomes the fastest algorithm for many cases with moderately long pattern and fairly low error level, provided the counters fit in a single computer word. This includes several interesting cases in searching DNA, natural language text, protein sequences, etc.

References

1. R. Baeza-Yates. Text retrieval: Theory and practice. In *12th IFIP World Computer Congress*, volume I, pages 465–476. Elsevier Science, 1992.
2. R. Baeza-Yates. A unified view of string matching algorithms. In *Proc. Theory and Practice of Informatics (SOFSEM'96)*, LNCS 1175, pages 1–15, 1996.
3. R. Baeza-Yates and G. Navarro. Faster approximate string matching. *Algorithmica*, 23(2):127–158, 1999.
4. W. Chang and J. Lampe. Theoretical and empirical comparisons of approximate string matching algorithms. In *Proc. 3rd Combinatorial Pattern Matching (CPM'92)*, LNCS 644, pages 172–181, 1992.
5. W. Chang and T. Marr. Approximate string matching and local similarity. In *Proc. 5th Combinatorial Pattern Matching (CPM'94)*, LNCS 807, pages 259–273, 1994.
6. M. Crochemore and W. Rytter. *Text Algorithms*. Oxford University Press, Oxford, UK, 1994.
7. Z. Galil and K. Park. An improved algorithm for approximate string matching. *SIAM Journal on Computing*, 19(6):989–999, 1990.
8. H. Hyyrö. Explaining and extending the bit-parallel algorithm of Myers. Technical Report A-2001-10, University of Tampere, Finland, 2001.
9. G. Landau and U. Vishkin. Fast parallel and serial approximate string matching. *Journal of Algorithms*, 10:157–169, 1989.
10. G. Myers. A fast bit-vector algorithm for approximate string matching based on dynamic progamming. *Journal of the ACM*, 46(3):395–415, 1999.

11. G. Navarro. A guided tour to approximate string matching. *ACM Computing Surveys*, 33(1):31–88, 2001.
12. G. Navarro and R. Baeza-Yates. Very fast and simple approximate string matching. *Information Processing Letters*, 72:65–70, 1999.
13. G. Navarro and R. Baeza-Yates. Improving an algorithm for approximate string matching. *Algorithmica*, 30(4):473–502, 2001.
14. G. Navarro and M. Raffinot. Fast and flexible string matching by combining bit-parallelism and suffix automata. *ACM Journal of Experimental Algorithmics (JEA)*, 5(4), 2000.
15. G. Navarro and M. Raffinot. *Flexible Pattern Matching in Strings – Practical on-line search algorithms for texts and biological sequences*. Cambridge University Press, 2002. To appear.
16. P. Sellers. The theory and computation of evolutionary distances: pattern recognition. *Journal of Algorithms*, 1:359–373, 1980.

17. E. Sutinen and J. Tarhio. On using q-gram locations in approximate string matching. In *Proc. European Symposium on Algorithms (ESA '95)*, LNCS 979, pages 327–340, 1995.
18. J. Tarhio and E. Ukkonen. Approximate Boyer-Moore string matching. *SIAM Journal on Computing*, 22(2):243–260, 1993.
19. E. Ukkonen. Algorithms for approximate string matching. *Information and Control*, 64:100–118, 1985.   ,
20. E. Ukkonen. Finding approximate patterns in strings. *Journal of Algorithms*, 6:132–137, 1985.   ,
21. S. Wu and U. Manber. Fast text searching allowing errors. *Comm. of the ACM*, 35(10):83–91, 1992.   ,   ,   ,   ,
22. S. Wu, U. Manber, and G. Myers. A sub-quadratic algorithm for approximate limited expression matching. *Algorithmica*, 15(1):50–67, 1996.

One-Gapped q-Gram Filters
for Levenshtein Distance

Stefan Burkhardt[1]* and Juha Kärkkäinen[2]**

[1] Center for Bioinformatics, Saarland University
Postfach 151150, 66041 Saarbrücken, Germany
stburk@mpi-sb.mpg.de
[2] Max-Planck-Institut für Informatik
Stuhlsatzenhausweg 85, 66123 Saarbrücken, Germany
juha@mpi-sb.mpg.de

Abstract. We have recently shown that q-gram filters based on gapped
q-grams instead of the usual contiguous q-grams can provide orders of
magnitude faster and/or more efficient filtering for the Hamming distance. In this paper, we extend the results for the Levenshtein distance,
which is more problematic for gapped q-grams because an insertion or
deletion in a gap affects a q-gram while a replacement does not. To keep
this effect under control, we concentrate on gapped q-grams with just
one gap. We demostrate with experiments that the resulting filters provide a significant improvement over the contiguous q-gram filters. We
also develop new techniques for dealing with complex q-gram filters.

1 Introduction

Given a *pattern* string P and a *text* string T, the *approximate string matching
problem* is to find all substrings of the text (*matches*) that are within a *distance k*
of the pattern P. The most commonly used distance measure is the *Levenshtein
distance*, the minimum number of single character insertions, deletions and replacements needed to change one string into the other. A simpler variant is the
Hamming distance, that does not allow insertions and deletions, i.e., it is the
number of nonmatching characters for strings of the same length. The *indexed*
version of the problem allows preprocessing the text to build an index while the
online version does not. Good surveys are given in [,].

Filtering is a way to speed up approximate string matching, particularly
in the indexed case but also in the online case. A *filter* is an algorithm that
quickly discards large parts of the text based on some *filter criterium*, leaving
the remaining part to be checked with a proper (online) approximate string
matching algorithm. A filter is *lossless* if it never discards an actual match; we
consider only lossless filters. The ability of a filter to reduce the text area is
called its (*filtration*) *efficiency*.

---

* Supported by the DFG 'Initiative Bioinformatik' grant BIZ 4/1-1.
** Partially supported by the Future and Emerging Technologies programme of the
EU under contract number IST-1999-14186 (ALCOM-FT).

A. Apostolico and M. Takeda (Eds.): CPM 2002, LNCS 2373, pp. 225–   , 2002.
© Springer-Verlag Berlin Heidelberg 2002

Many filters are based on *q-grams*, substrings of length q. The q-gram similarity (defined as a distance in [ ]) of two strings is the number of q-grams shared by the strings. The q-gram filter is based on the *q-gram lemma*:

Lemma 1 ([ ]). *Let P and S be strings with (Levenshtein or Hamming) distance k. Then the q-gram similarity of P and S is at least $t = |P| - q + 1 - kq$.*

The value t in the lemma is called the *threshold* and gives the minimum number of q-grams that an approximate match must share with the pattern, which is used as the filter criterium. There are actually many possible ways to count the number of shared q-grams offering different tradeoffs between speed and filtration efficiency (see, e.g., [, , ,]). However, in all cases the value of the threshold is the one given by the lemma.

A generalization of the q-gram filter uses *gapped* q-grams, subsets of q characters of a fixed non-contiguous *shape*. For example, the 3-grams of shape ##-# in the string ACAGCT are ACG, CAC and AGT. In [ ], we showed that the use of gapped q-grams can significantly improve the filtration efficiency and/or speed of the q-gram filter *for the Hamming distance*. The result cannot be trivially extended to the Levenshtein distance due to the effect of insertions and deletions on the gaps. In the above example, a replacement of G would leave the 3-gram CAC unaffected but the deletion of G or an insertion of a character before G would change all 3-grams.

In this paper, we study gapped q-gram filters for the Levenshtein distance based on an idea already suggested in [ ]. The idea is to use multiple shapes that differ only in the length of the gap(s), for example, the shapes ##--#, ##-# and ###. We restrict our consideration to q-grams with only one gap, which minimizes the number of different shapes needed. Our experimental results show that with the right choice of the shape a significant improvement over conventional q-gram filters can be achieved for the Levenshtein distance, too.

Even with the restriction to one-gapped q-grams, a major obstacle in implementing the filters is determining the value of the threshold. As already noted in [ ], for gapped q-gram filters, there is no simple threshold formula like the one given by Lemma . Indeed, even defining the threshold precisely is far from trivial. The definition we give here not only solves the problem for the one-gapped q-gram filters but provides a framework for solving it for other even more complicated filters.

Gapped q-grams have also been used in [, ,]. In [,], the motivation is to increase the filtration efficiency by considering multiple shapes. Pevzner and Waterman [ ] use q-grams containing every $(k + 1)$st character together with contiguous q-grams for the Hamming distance. Califano and Rigoutsos [ ] describe a *lossy* filter for the Levenshtein distance that uses as many as 40 different random shapes. Their approach is effective for high k but they need a huge index (18GB for a 100 million nucleotide DNA database). The Grampse system of Lehtinen et al. [ ] uses a shape containing every hth character for some h (similar to [ ]) for exact matching. Their motivation of using gapped q-grams is to reduce dependencies between the characters of a q-gram.

2   Filter Algorithm

The basic idea behind the gapped q-gram filters for the Levenshtein distance was already suggested in [ ]. The filters use a basic shape with only one gap and two other shapes formed from the basic shape by increasing and decreasing the length of the gap by one. For example, with the basic shape ##-# we would also use the shapes ##--# and ###. The filter compares the q-grams of all three shapes in the pattern to the q-grams of the basic shape in the text. Then matching q-grams are found even if there is an insertion or a deletion in the gap.

Otherwise the filter algorithm is similar to the one described in [ ] for the contiguous q-grams. Let us define a *hit* as a pair (i, j) such that a q-gram (of any of the three shapes) starting at position i in P matches a q-gram (of the basic shape) starting at position j in T. The *diagonal* of a hit (i, j) is $j - i$. The diagonal represents the approximate starting position of the corresponding substring of T, or more accurately, the diagonal of the dynamic programming matrix corresponding to the beginning of the q-gram. The following result was shown for contiguous q-grams in [ ] and it trivially extends to gapped q-grams.

Lemma 2. *If a substring S of T is within distance k of P, then S and P share at least t q-grams such that the diagonals of the corresponding hits differ by at most k.*

For contiguous q-grams, the threshold t in the lemma is the one in Lemma . For gapped q-grams, the threshold is defined in the next section.

Based on the lemma, it is enough to find all sets of $k + 1$ adjacent diagonals that contain at least t hits. We call such a set of diagonals a *match*. The matches can be computed by finding all the hits with a q-gram index of the text, sorting them by diagonal, and scanning the sorted list. If radix sort is used, the time requirement is $O(q|P| + h)$, where h is the number of hits.

3   Threshold

For contiguous q-grams, there is a simple formula (Lemma) for computing the threshold, but this is not the case for gapped q-grams. In fact, even defining the threshold precisely is nontrivial. In this section, we give a formal definition of the threshold. The algorithm we used for computing the threshold is described in [ ].

Following [ ], we define an *edit transcript* as a string over the alphabet M(atch), R(eplace), I(nsert) and D(elete), describing a sequential character-by-character transformation of one string to another. For two strings P and S,

---

[1] An alternative would be to use two shapes, for example ##-# and ##--#, for both strings. The main advantage of the asymmetric approach is that it requires a text index for only one shape.

[2] Matches that overlap are merged into one larger match by the algorithm and are counted as one match in the experiments.

let $\mathcal{T}(P,S)$ denote the set of all transcripts transforming P to S. For example, $\mathcal{T}(\texttt{actg},\texttt{acct})$ contains MMRR, MMIMD, MIMMD, IRMMD, IDIMDID, etc.. For a transcript $\tau \in \mathcal{T}(P,S)$, the source length $slen(\tau)$ of τ is the length of P, i.e., the number of non-insertions in τ. The Levenshtein cost $c_L(\tau)$ is the number of non-matches. The Hamming cost $c_H(\tau)$ is infinite if τ contains insertions or deletions and the same as Levenshtein cost otherwise. The Levenshtein distance and Hamming distance of P and S are $d_L(P,S) = \min_{\tau \in \mathcal{T}(P,S)} c_L(\tau)$ and $d_H(P,S) = \min_{\tau \in \mathcal{T}(P,S)} c_H(\tau)$, respectively.

Here we defined distance measures for strings using cost functions for edit transcripts. Similarly, we define the q-gram similarity measures for strings using *profit functions* for edit transcripts. Then we can define the threshold using edit transcripts as follows.

Definition 1. *The* threshold *for a cost function c and a profit function p of edit transcripts is*

$$t_p^c(m,k) = \min_{\tau}\{p(\tau) \mid slen(\tau) = m, c(\tau) \leq k\}.$$

The following lemma gives the filter criterium.

Lemma 3. *Let c be a cost function and p a profit function for edit transcripts. Define a distance d of two strings P and S as $d(P,S) = \min_{\tau \in \mathcal{T}(P,S)} c(\tau)$ and a similarity s as $s(P,S) = \max_{\tau \in \mathcal{T}(P,S)} p(\tau)$. Now, if $d(P,S) \leq k$, then $s(P,S) \geq t_p^c(|P|,k)$.*

The lemma holds for any choice of cost c and profit p. The cost functions of interest to us were defined above. The profit functions that define the q-gram based similarity measures are described next.

Let I be a set of integers. The *span* of I is $span(I) = \max I - \min I + 1$, i.e., the size of the minimum contiguous interval containing I. The *position* of I is $\min I$, and the *shape* of I is the set $\{i - \min I \mid i \in I\}$. An integer set Q with position zero is called a *shape*. For any shape Q and integer i, let Q_i denote the set with shape Q and position i, i.e., $Q_i = \{i + j \mid j \in Q\}$. Let $Q_i = \{i_1, i_2, \ldots, i_q\}$, where $i = i_1 < i_2 < \cdots < i_q$, and let $S = s_1 s_2 \ldots s_m$ be a string. For $1 \leq i \leq m - span(Q) + 1$, the Q-*gram* at position i in S, denoted by $S[Q_i]$, is the string $s_{i_1} s_{i_2} \ldots s_{i_q}$. For example, if $S = \texttt{acagagtct}$ and $Q = \{0,2,3,6\}$, then $S[Q_1] = S[Q_3] = \texttt{aagt}$ and $S[Q_2] = \texttt{cgac}$.

A *match alignment* M_τ of a transcript τ is the set of pairs of positions that are matched to each other. For example, $M_{\texttt{MIMRDMR}} = \{(1,1),(2,3),(5,5)\}$. For a set I of integers, let $M_\tau(I)$ be the set to which M_τ maps I, i.e., $M_\tau(I) = \{j \mid i \in I \text{ and } (i,j) \in M_\tau\}$. A Q-*hit* in a transcript τ is a pair (i,j) of integers such that $M_\tau(Q_i) = Q_j$. The Q-*profit* $p_Q(\tau)$ of a transcript τ is the number of its Q-hits, i.e., $p_Q(\tau) = |\{(i,j) \mid M_\tau(Q_i) = Q_j\}|$. Using p_Q as the profit function defines the Q-*similarity* of two strings P and S as $s_Q(P,S) = \max_{\tau \in \mathcal{T}(P,S)} p_Q(\tau)$.

For any $b_1, g, b_2 > 0$, let (b_1, g, b_2) denote the *one-gap shape* $\{0, \ldots, b_1 - 1, b_1 + g, \ldots, b_1 + g + b_2 - 1\}$. For a one-gap shape $Q = (b_1, g, b_2)$, let $Q^{+1} = (b_1, g+1, b_2)$ and $Q^{-1} = (b_1, g-1, b_2)$ (or $Q^{-1} = \{0, \ldots, b_1 + b_2 - 1\}$ if $g = 1$).

Then, a $Q \pm 1$-*hit* in a transcript τ is a pair (i,j) of integers such that $Q_j \in \{M_\tau(Q_i^{-1}), M_\tau(Q_i), M_\tau(Q_i^{+1})\}$. The $Q \pm 1$-*profit* of τ, $p_{Q\pm 1}(\tau)$, is the number of $Q\pm 1$-hits in τ, i.e., $p_{Q\pm 1}(\tau) = |\{(i,j) \mid Q_j \in \{M_\tau(Q_i^{-1}), M_\tau(Q_i), M_\tau(Q_i^{+1})\}\}|$. The $Q\pm 1$-*similarity* of two strings P and S is $s_{Q\pm 1}(P,S) = \max_{\tau \in \mathcal{T}(P,S)} p_{Q\pm 1}(\tau)$.

For $c = c_H$ and $p = p_Q$, Definition   gives the thresholds that were used by the Hamming distance filters in [ ]. If Q is the contiguous shape $\{0, 1, \ldots, q-1\}$, the threshold is the same as given by Lemma   and also used in the Levenshtein distance filter. For $c = c_L$ and $p = p_{Q\pm 1}$, Definition   gives the thresholds used by the one-gapped q-gram filters for the Levenshtein distance.

The filters compute the number of matching q-grams which is an upper bound of the corresponding similarity defined here. For example, if $\tau \in \mathcal{T}(P,S)$, then a Q-hit (i,j) in τ implies $P[Q_i] = S[Q_j]$. Therefore, the number of matching Q-grams between P and S is at least $s_Q(P,S)$, but may be higher. For example, aca and cac have two matching $\{0,1\}$-grams (ac and ca) even though $s_{\{0,1\}}(\text{aca}, \text{cac}) = 1$. The exact value of $s_Q(P,S)$ could be computed by a more careful analysis of the matching q-grams, but it may not be worth the extra effort in practice. Since the filter computes an upper bound of the similarity, its value is at least the threshold if $d(P,S) \leq k$. However, a higher threshold cannot be used without possibly making the filter lossy, at least as long as the threshold is a function of only the pattern length and the distance k.

4   Minimum Coverage

The two main properties of a filter are filter speed and filtration efficiency. When using an index the dominating factor for the filter speed is the total number of hits that have to be accessed and processed by the filter. Predicting the number of hits is straighforward using the number of shapes in P and T, and the value q of matching characters per shape with the following equation:

$$hits \approx \frac{1}{\Sigma^q} 3(|P| - span(Q) + 1)(|T| - span(Q) + 1)$$

The factor of 3 comes in for one-gap shapes since we use 3 shapes per starting position in P. In practice, all one-gapped shapes with the same value of q are essentially equivalent with respect to the filter speed.

The filtration efficiency of a filter, i.e., the probability of having a random match at a fixed position depends in a complex way on the basic shape, defining the *quality* of the shape. In [ ], we used the notion of minimum coverage to predict the quality of shapes. This was vital due to the large number of available shapes. When only using shapes with one gap, the number of possible choices is dramatically reduced, which allows experimental evaluation of all candidate shapes. However, we also wanted to analyse the general concept of quality prediction for this case, especially since it was easily verifiable with the experimental results.

For the case of the Hamming distance, the *minimum coverage* was defined as the size of the smallest union of t shapes with different positions. This corresponds to the minimum number of 'fixed' characters required to match between a pattern and a text substring for there to be t matching Q-grams. The probability of matching a certain arrangement of t Q-grams depends exponentially on the number of fixed characters in that arrangement. To exactly compute the probability of a random match one would have to take into account all possible arrangements of t or more shapes. However, since those arrangements with the lowest or close to the lowest number of fixed characters are much more likely, one can compute a good approximation using only the most probable arrangements, i.e. those with the minimum coverage. The expected number of matches given a minimum coverage mc would therefore be roughly proportional to $1/\Sigma^{mc}$.

Since we still use only one shape for matching in the text, the minimum coverage and the number of fixed characters remain the same here. The additional shapes used in the pattern can however increase the number of possible pattern substrings that match the fixed characters in the text. For cases where t shapes are arranged with a large overlap, this increase is negligible. In general, this is the case for the minimum coverage. We evaluated the correlation between minimum coverage and the experimental results for all shapes and display them in Figure . When compared to the Hamming distance, there is a small loss in correlation but the overall predictive capability of the minimum coverage remains intact.

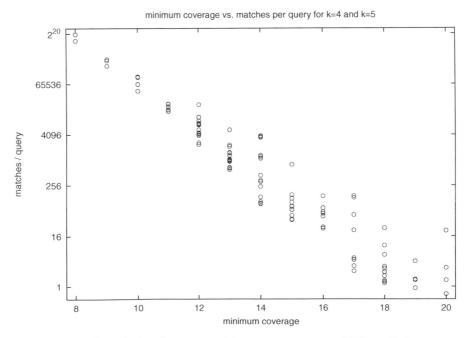

Fig. 1. Correlation between minimum coverage and filter efficiency

5   Experiments

To test the one-gap q-grams in practice, we performed some experiments on a randomly generated DNA database with 50 million basepairs (even and independent distribution of characters). The queries we used were 1000 random strings of length 500. The query length was chosen since it is a typical query length in many computational biology applications.

It should be noted that the threshold used in filtering was computed for $m = 50$ which is a typical value for finding local matches in DNA. This difference between query and window length has the effect that, while the filter is still guaranteed to report all positions where there is an approximate occurrence of a substring of length 50 of the query, it will also lead to a moderate increase in the number of potential matches reported due to the increase in the number of shapes for which one searches.

For the edit distance k we used values of 4 and 5, making the experimental setting correspond to typical high similarity local alignment problems in shotgun sequencing [ ] and EST clustering [,]. The database contained no actual matches of this quality, i.e., all potential matches reported by the filter were false positives.

Like in our earlier paper we compared the gapped shapes with the contiguous shapes used in the classic q-gram lemma. For $k = 4$ and $k = 5$ we tested all shapes with $q \geq 8$. For $k = 4$ there are 87 such shapes and for $k = 5$ there are 35. From these shapes we picked, for each value of q, those with the best experimental filtration efficiency and compared them with the filter based on the classic q-gram lemma. The best gapped shapes are shown in Table . Figure compares them to contiguous q-grams both in theory (expected number of hits vs. minimum coverage) and in practice (hits vs. matches).

The top plot shows the values used to predict filtration efficiency (the minimum coverage) and filter speed (the expected number of hits) for both contiguous q-grams and the best one-gap shapes. The bottom graph contains the actual experimental results showing the average number of hits as well as the average number of matches per query (averaged across all 1000 queries). The expected number of hits for one-gap shapes in the first plot was computed taking into account the fact that we use 3 different shapes for each possible starting position in the query.

Looking at Figure   one can observe that for the Levenshtein distance the shapes with one gap show a better performance than ungapped shapes. In general, they allow a substantial increase in the possible values of q and/or the filtration efficiency. The higher values of q make them prime candidates for index–based implementations. The comparison between predicted and actual

---

[3] A better filter efficiency can be achieved by counting the hits separately for each substring of length 50. This should not have a significant effect on the relative performance of the different shapes and we can therefore use this approximation for comparing them. Looking at the results one should keep in mind that the absolute values for filtration efficiency are not the best possible but slightly worse.

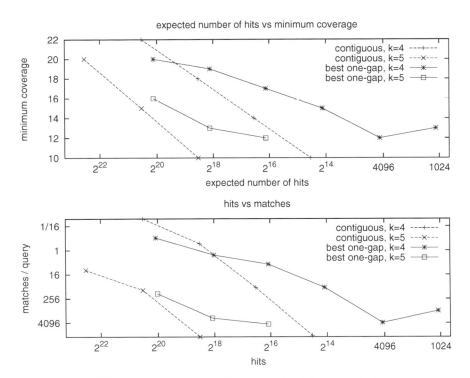

Fig. 2. Comparison of ungapped with gapped shapes

q	best for k=4
8	####---###
9	#######----##
10	#######---###
11	#######--####
12	#########--###
13	#########-####
q	best for k=5
8	######---##
9	######---###
10	#######-###

Table 1. The gapped shapes used in Figure

performance reinforces the correlation described in Section   and with it the predictive capability of the minimum coverage. It is obvious that the value of q affects the filtration speed and efficiency and provides a tradeoff between the two. Which choice is the best for a certain application depends on the actual speed of the filtration and verification algorithms.

Another point we want to make is that the proper choice of the shapes used for filtering is very important. To illustrate this we want to mention that for a fixed set of parameters the best shapes had filtration efficiencies that differed by as much as a factor of 10^6 from the worst. The difference in filtration efficiency between the median and the best shape was still up to a factor of 50.

6   Concluding Remarks

We have shown that suitably chosen one-gap q-grams combined with a simple technique to compensate for insertions and deletions in the gap can significantly improve the performance of the basic q-gram filter for the Levenshtein distance. This demonstrates that they are worth studying in further research on the problem of approximate string matching using the Levenshtein distance.

Aside from looking at shapes with only one gap it might be interesting to look at shapes with more than one gap. It remains to be seen whether the added number of shapes is worth the potential increase in filter quality. Also, techniques like generating word neighborhoods for q-grams, which are for example used in BLAST [ ], could perhaps be adapted to gapped q-grams. Other possibilities include combining two or more different shapes into one filter. The framework for computing thresholds for more complex filters has been provided with the definitions in this paper.

References

1. S. F. Altschul, W. Gish, W. Miller, E. W. Myers, and D. J. Lipman. Basic local alignment search tool. *Journal of Molecular Biology*, 215:403–410, 1990.
2. S. Burkhardt, A. Crauser, P. Ferragina, H.-P. Lenhof, E. Rivals, and M. Vingron. *q*-gram based database searching using a suffix array (QUASAR). In *Proc. 3rd Annual International Conference on Computational Molecular Biology (RECOMB)*, pages 77–83. ACM Press, 1999.   ,
3. S. Burkhardt and J. Kärkkäinen. Better filtering with gapped q-grams. In *Proc. 12th Annual Symposium on Combinatorial Pattern Matching*, volume 2089 of *LNCS*, pages 73–85. Springer, 2001.   ,   ,
4. A. Califano and I. Rigoutsos. FLASH: A fast look-up algorithm for string homology. In *Proc. 1st International Conference on Intelligent Systems for Molecular Biology*, pages 56–64. AAAI Press, 1993.
5. D. Gusfield. *Algorithms on Strings, Trees and Sequences: Computer Science and Computational Biology*. Cambridge University Press, 1997.
6. N. Holsti and E. Sutinen. Approximate string matching using *q*-gram places. In *Proc. 7th Finnish Symposium on Computer Science*, pages 23–32, 1994.   ,

7. P. Jokinen and E. Ukkonen. Two algorithms for approximate string matching in static texts. In *Proc. 16th Symposium on Mathematical Foundations of Computer Science*, volume 520 of *LNCS*, pages 240–248. Springer, 1991.
8. J. Kärkkäinen. Computing the threshold for q-gram filters. In *Proc. 8th Scandinavian Workshop on Algorithm Theory (SWAT)*, July 2002. To appear.
9. A. Krause and M. Vingron. A set-theoretic approach to database searching and clustering. *Bioinformatics*, 14:430–438, 1998.
10. O. Lehtinen, E. Sutinen, and J. Tarhio. Experiments on block indexing. In *Proc. 3rd South American Workshop on String Processing (WSP)*, pages 183–193. Carleton University Press, 1996.
11. G. Navarro. *Approximate Text Searching*. PhD thesis, Dept. of Computer Science, University of Chile, 1998.
12. G. Navarro. A guided tour to approximate string matching. *ACM Computing Surveys*, 33(1):31–88, 2001.
13. P. A. Pevzner and M. S. Waterman. Multiple filtration and approximate pattern matching. *Algorithmica*, 13(1/2):135–154, 1995.
14. E. Ukkonen. Approximate string matching with q-grams and maximal matches. *Theor. Comput. Sci.*, 92(1):191–212, 1992.
15. J. Weber and H. Myers. Human whole genome shotgun sequencing. *Genome Research*, 7:401–409, 1997.

Optimal Exact and Fast Approximate Two Dimensional Pattern Matching Allowing Rotations

Kimmo Fredriksson[1][*], Gonzalo Navarro[2][**], and Esko Ukkonen[1][* * *]

[1] Department of Computer Science, University of Helsinki
{kfredrik,ukkonen}@cs.helsinki.fi
[2] Department of Computer Science, University of Chile
gnavarro@dcc.uchile.cl

Abstract. We give fast filtering algorithms to search for a 2–dimensional pattern in a 2–dimensional text allowing any rotation of the pattern. We consider the cases of exact and approximate matching under several matching models, improving the previous results. For a text of size $n \times n$ characters and a pattern of size $m \times m$ characters, the exact matching takes average time $O(n^2 \log m / m^2)$, which is optimal. If we allow k mismatches of characters, then our best algorithm achieves $O(n^2 k \log m / m^2)$ average time, for reasonable k values. For large k, we obtain an $O(n^2 k^{3/2} \sqrt{\log m} / m)$ average time algorithm. We generalize the algorithms for the matching model where the sum of absolute differences between characters is at most k. Finally, we show how to make the algorithms optimal in the worst case, achieving the lower bound $\Omega(n^2 m^3)$.

1 Introduction

We consider the problem of finding the exact and approximate occurrences of a two–dimensional *pattern* of size $m \times m$ cells from a two–dimensional *text* of size $n \times n$ cells, when all possible rotations of the pattern are allowed. This problem is often called *rotation invariant template matching* in the signal processing literature. Template matching has numerous important applications in image and volume processing. The traditional approach [ ] to the problem is to compute the cross correlation between each text location and each rotation of the pattern template. This can be done reasonably efficiently using the Fast Fourier Transform (FFT), requiring time $O(Kn^2 \log n)$ where K is the number of rotations sampled. Typically K is $O(m)$ in the 2–dimensional (2D) case, and $O(m^3)$ in the 3D case, which makes the FFT approach very slow in practice. However, in many applications, "close enough" matches of the pattern are also accepted. To

* Work supported by ComBi and the Academy of Finland.
** Partially supported by the Millenium Center for Web Research.
* * * Work supported by the Academy of Finland.

A. Apostolico and M. Takeda (Eds.): CPM 2002, LNCS 2373, pp. 235–   , 2002.
© Springer-Verlag Berlin Heidelberg 2002

this end, the user may specify a parameter k, such that matches that have at most k differences with the pattern should be accepted.

Efficient two dimensional combinatorial pattern matching algorithms that do not allow rotations of the pattern can be found, e.g., in [, , ,]. Rotation invariant template matching was first considered from a combinatorial point of view in [ ]. In this paper, we follow this combinatorial line of work. If we consider the pattern and text as regular grids, then defining the notion of matching becomes nontrivial when we rotate the pattern: since every pattern cell intersects several text cells and vice versa, it is not clear what should match what. Among the different matching models considered in previous work [, ,], we stick to the simplest one in this paper: (1) the geometric center of the pattern has to align with the center of a text cell; (2) the text cells involved in the match are those whose geometric centers are covered by the pattern; (3) each text cell involved in a match should match the value of the pattern cell that covers its center.

Under this *exact matching* model, an online algorithm is presented in [ ] to search for a pattern allowing rotations in $O(n^2)$ average time.

The model (a 3D version) was extended in [ ] such that there may be a limited number k of mismatches between the pattern and its occurrence. Under this *mismatches* model an $O(k^4 n^3)$ average time algorithm was obtained, as well as an $O(k^2 n^3)$ average time algorithm for computing the lower bound of the distance; here we will develop a 2D version whose running time is $O(k^{3/2} n^2)$. This works for any $0 \le k < m^2$. For a small k, an $O(k^{1/2} n^2)$ average time algorithm was given in [ ].

Finally, a more refined model [, ,] suitable for gray level images adds up the absolute values of the differences in the gray levels of the pattern and text cells supposed to match, and puts an upper limit k on this sum. Under this *gray levels* model average time $O((k/\sigma)^{3/2} n^2)$ is achieved, assuming that the cell values are uniformly distributed among σ gray levels. Similar algorithms for indexing are presented in [ ].

In this paper we present fast filters for searching allowing rotations under these three models. Table   shows our main achievements (all are on the average). The time we obtain for exact searching is average-case optimal. For the k–mismatches model we present two different algorithms, based on searching for pattern pieces, either exactly or allowing less mismatches. For the gray levels model we present a filter based on coarsening the gray levels of the image, which makes the problem independent on the number of gray levels, with a complexity approaching that of the k–mismatches model.

2   Problem Complexity

There exists a general lower bound for d–dimensional exact pattern matching. In [ ] Yao showed that the one–dimensional string matching problem requires at least time $\Omega(n \log m/m)$ on average, where n and m are the lengths of the string and the pattern respectively. In [ ] this result was generalized for the

Table 1. The (simplified) average case complexities achieved for different models

Model	Previous result	Our results		
Exact matching	$O(n^2)$	$O(n^2 \log_\sigma m/m^2)$		
k Mismatches	$O(n^2 k^{3/2})$	$O(n^2 k \log_\sigma m/m^2),\ k < m^2/(3 \log_\sigma m)^2$		
		$O(n^2 m^3/\sigma^{m/\sqrt{k}}),\ k < m^2/(5 \log_\sigma m)$		
		$O(n^2 k^{3/2}\sqrt{\log m}/m),\ k < m^2\ (1 - \Theta(1/\sigma))$		
Gray levels	$O(n^2 (k/\sigma)^{3/2})$	$O(n^2 (k/\sigma) \log_\sigma m/m^2),\ k < m^2\sigma/(9e\ln^2 m)$		
		$O(n^2 (k/\sigma)^{3/2}\sqrt{\log m}/m),\ k < m^2\sigma/(5e\ln m)$		

d–dimensional case, for which the lower bound is $\Omega(n^d \log m^d/m^d)$ (without rotations).

The above lower bound also holds for the case with rotations allowed, as exact pattern matching reduces (as a special case) to the matching with rotations. To search for P exactly, we search it allowing rotations and once we find an occurrence we verify whether or not the rotation angle is zero. Since in 2D there are $O(m^3)$ rotations [ ], on average there are $O(n^2 m^3/\sigma^{m^2})$ occurrences. Each rotated occurrence can be verified in $O(1)$ average time (by the results of the present paper). Hence the total exact search time (et) is that of searching with rotations (rt) plus $O(n^2 m^3/\sigma^{m^2}) = o(n^2 \log_\sigma m/m^2)$ for verifications. Because of Yao's bound, $et = \Omega(n^2 \log_\sigma m/m^2) = rt + o(n^2 \log_\sigma m/m^2)$, and so $rt = \Omega(n^2 \log_\sigma m/m^2)$ as well. This argument can be easily generalized to d dimensions because there are $O(m^{O(d)}/d^m)$ matches to verify at $O(1)$ cost.

In Sec.   we give an algorithm whose expected running time matches this lower bound.

A lower bound for the k differences problem (approximate string matching with $\le k$ mismatches, insertions or deletions of characters) was given in [ ] for the one dimensional case. This bound is $\Omega(n(k + \log m)/m)$, where n is the length of the text string and m is the length of the pattern. This bound is tight; an algorithm achieving it was also given in [ ].

This lower bound can be generalized to the d–dimensional case also. By [ ], exact d–dimensional searching needs $\Omega(n^d \log m^d/m^d)$ time, and this is a special case of approximate matching. Following [ ], we have that at least $k+1$ symbols of a window of the size of P in T have to be examined to guarantee that the window cannot match P. So a second lower bound is $\Omega(kn^d/m^d)$. The lower bound $\Omega(n^d(k + \log m^d)/m^d)$ follows.

3   Definitions

Let $T = T[1..n, 1..n]$ and $P = P[1..m, 1..m]$ be arrays of unit squares, called *cells*, in the (x, y)–plane. Each cell has a value in ordered finite alphabet Σ. The size of the alphabet is denoted by $\sigma = |\Sigma|$. The corners of the cell for $T[i, j]$ are $(i - 1, j - 1), (i, j - 1), (i - 1, j)$ and (i, j). The center of the cell for $T[i, j]$ is $(i-\frac{1}{2}, j-\frac{1}{2})$. The array of cells for pattern P is defined similarly. The center of the whole pattern P is the center of the cell in the middle of P. Precisely, assuming for simplicity that m is odd, the center of P is the center of cell $P[\frac{m+1}{2}, \frac{m+1}{2}]$.

Assume now that P has been moved on top of T using a rigid motion (translation and rotation), such that the center of P coincides exactly with the center of some cell of T (the *center–to–center assumption*). The location of P with respect to T can be uniquely given as $((i,j),\theta)$ where (i,j) is the cell of T that matches the center of P, and θ is the angle between the x–axis of T and the x–axis of P. The (approximate) occurrence between T and P at some location is defined by comparing the values of the cells of T and P that overlap. We will use the centers of the cells of T for selecting the comparison points. That is, for the pattern at location $((i,j),\theta)$, we look which cells of the pattern cover the centers of the cells of the text, and compare the corresponding values of those cells.

More precisely, assume that P is at location $((i,j),\theta)$. For each cell $T[r,s]$ of T whose center belongs to the area covered by P, let $P[r',s']$ be the cell of P such that the center of $T[r,s]$ belongs to the area covered by $P[r',s']$. Then $M(T[r,s]) = P[r',s']$. So our algorithms compare the cell $T[r,s]$ of T against the cell $M(T[r,s])$ of P.

Hence the *matching function* M is a function from the cells of T to the cells of P. Now consider what happens to M when angle θ grows continuously, starting from $\theta = 0$. Function M changes only at the values of θ such that some cell center of T hits some cell boundary of P. It was shown in [ ] that this happens $O(m^3)$ times, when P rotates full 360 degrees. This result was shown to be also a lower bound in [ ]. Hence there are $\Theta(m^3)$ relevant orientations of P to be checked. The set of angles for $0 \le \theta \le \pi/2$ is

$$A = \{\beta, \pi/2 - \beta \mid \beta = \arcsin \frac{h + \frac{1}{2}}{\sqrt{i^2 + j^2}} - \arcsin \frac{j}{\sqrt{i^2 + j^2}};$$

$$i = 1, 2, \ldots, \lfloor m/2 \rfloor; j = 0, 1, \ldots, \lfloor m/2 \rfloor; h = 0, 1, \ldots, \lfloor \sqrt{i^2 + j^2} \rfloor\}.$$

By symmetry, the set of possible angles θ, $0 \le \theta < 2\pi$, is

$$\mathcal{A} = A \ \cup \ A + \pi/2 \ \cup \ A + \pi \ \cup \ A + 3\pi/2.$$

As shown in [ ], any match of a pattern P in a text T allowing arbitrary rotations must contain a so-called "feature", i.e. a one–dimensional string obtained by reading a line of the pattern in some angle and crossing the center. These features are used to build a filter for finding the position and orientation of P in T.

We now define a *set* of linear features (strings) for P (see Figure). The length of a particular feature is denoted by u, and the feature for angle θ and row q is denoted by $F^q(\theta)$. Assume for simplicity that u is odd. To read a feature $F^q(\theta)$ from P, let P be on top of T, on location $((i,j),\theta)$. Consider the cell $T[i - \frac{m+1}{2} + q, j - \frac{u-1}{2}], \ldots, T[i - \frac{m+1}{2} + q, j + \frac{u-1}{2}]$. Denote them as $t_1^q, t_2^q, \ldots, t_u^q$. Let c_i^q be the value of the cell of P that covers the center of t_i^q. The (horizontal) feature of P with angle θ and row q is now the sequence $F^q(\theta) = c_1^q c_2^q \cdots c_u^q$. Note that this value depends only on q, θ and P, not on T.

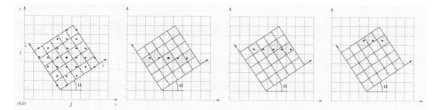

Fig. 1. Each text cell is matched against the pattern cell that covers the center of the text cell. For each angle θ, a set of features is read from P

The sets of angles for the features are obtained the same way as the set of angles for the whole pattern P. Note that the set of angles $\mathcal{B}^q$ for the feature set F^q is subset of $\mathcal{A}$, that is $\mathcal{B}^q \subset \mathcal{A}$ for any q. The size of $\mathcal{B}$ varies from $O(u^2)$ (the features crossing the center of P) to $O(um)$ (the features at distance $\Theta(m)$ from the center of P). Therefore, if a match of some feature $F^q(\theta)$ is found, there are $O(|\mathcal{A}|/|\mathcal{B}^q|)$ possible orientations to be verified for an occurrence of P. In other words, the matching function M can change as long as $F^q(\theta)$ does not change.

More precisely, assume that $\mathcal{B}^q = (\gamma_1, \ldots, \gamma_K)$, and that $\gamma_i < \gamma_{i+1}$. Therefore, feature $F^q(\gamma_i) = F^q(\theta)$ can be read using any θ such that $\gamma_i \leq \theta < \gamma_{i+1}$. On the other hand, there are $O(|\mathcal{A}|/|\mathcal{B}^q)|$ angles $\beta \in \mathcal{A}$ such that $\gamma_i \leq \beta < \gamma_{i+1}$. If there is an occurrence of $F^q(\theta)$, then P may occur with such angles β.

4   Exact Search Allowing Rotations

In [ ] only a set of features crossing the center of P and of length m is extracted from P, i.e. $q = \frac{m+1}{2}$ and $u = m$. The text is then scanned row–wise for the occurrence of some feature, and upon such an occurrence the whole pattern is checked at the appropriate angles.

The verification takes $O(m)$ time on average and $O(m^2)$ in the worst case in [ ]. The reason is that there are $O(m^2)$ cells in the pattern, and each one intersects $O(m)$ different text centers along a rotation of 360 degrees (or any other constant angle), so there are $O(m^3)$ different rotations for the pattern. The number of relevant rotations for a feature of $O(m)$ cells is, however, only $O(m^2)$, and therefore there are $O(m)$ different angles in which the pattern has to be tested for each angle in which a feature is found.

In [ ] the possibility of using features of length $u \leq m$ is considered, since it reduces the space and number of rotations. In what follows we assume that the features are of length $u \leq m$, and find later the optimal u.

We show now how to improve both search and verification time.

4.1   Faster Search

In [ ] a 2–dimensional search algorithm (not allowing rotations) is proposed that works by searching for all the pattern rows in the image. Only every mth row of the image needs to be considered because one of them must contain some pattern row in any occurrence.

We take a similar approach. Instead of taking the $O(u^2)$ features that cross the center of the pattern, we also take some not crossing the center. More specifically, we take features for q in the range $\frac{m-r}{2} + 1 \ldots \frac{m+r}{2}$, where r is an odd integer for simplicity. For each such q, we read the features at all the relevant rotations. This is illustrated in Fig. . This allows us to search only one out of r image rows, but there are $O(rum)$ features now. Figure   also shows that the features may become shorter than m when they are far away from the center and the pattern is rotated. On the other hand, there is no need to take features farther away than $m/2$ from the center, since in the case of unrotated patterns this is the limit. Therefore we have the limit $r \leq m$. If we take features from $r = m$ rows then the shortest ones (for the pattern rotated at 45 degrees) are of length $(\sqrt{2} - 1)m = \Theta(m)$. The features do not cross the pattern center now, but they are still fixed if the pattern center matches a text center.

The search time per character is independent on the number of features if an Aho–Corasick machine (AC) [ ] is used. Alternatively, we can use a suffix automaton (DAWG–MATCH algorithm) [ ] to get an optimal average search time. The worst case time for the suffix automaton is the same as for the AC automaton.

4.2   Faster Verification

We show how verifications can be performed faster, in $O(1)$ time instead of $O(m)$. Imagine that a feature taken at angle θ has been found in the text. Since the feature has length u and can be at distance at most m from the center, there at most $O(um)$ different angles, whose limits we call γ_1 to γ_K, and we have $\gamma_i \leq \theta < \gamma_{i+1}$.

We first try to extend the match of the feature to a match of the complete rotated row of the pattern. There are $O(m^2/(um))$ possible angles for the complete row, which lie between γ_i and γ_{i+1} (as the feature is enlarged, the matching angles are refined). However, we perform the comparison incrementally: first try to extend the feature by 1 cell. There are $O(((u+1)m)/(um)) = O((u+1)/u) = O(1)$ possible angles, and all them are tried. The probability that the $(u+1)$-th cell matches in some of the $O(1)$ permitted angles is $O(1/\sigma)$. Only if we succeed we try with the $(u+2)$-th cell, where there would be $O(((u+2)m)/((u+1)m))$ different angles, and so on.

In general, the probability of checking the $(u+i+1)$-th cell of the feature is that of passing the check for the $(u+1)$-th, then that of the $(u+2)$-th and so on. The average number of times it occurs is at most

$$\left(\frac{u+1}{u}\right)\frac{1}{\sigma} \times \left(\frac{u+2}{u+1}\right)\frac{1}{\sigma} \times \ldots \times \left(\frac{u+i}{u+i-1}\right)\frac{1}{\sigma} = \left(\frac{u+i}{u}\right)\frac{1}{\sigma^i}$$

and by summing for $i = 0$ to $\Theta(m) - u$ we obtain $O(1)$. This is done in both directions from the center, in any order.

The same scheme can be applied to extend the match to the rest of the pattern. Each time we add a new pattern position to the comparison we have only $O(1)$ different angles to test, and therefore an $O(1/\sigma)$ probability of success. The process is geometric and it finishes in $O(1)$ time on average.

Note that this result holds even if the cell values are not uniformly distributed in the range $1 \ldots \sigma$. It is enough that there is an independent nonzero probability p of mismatch between a random pattern cell and a random text cell, in which case $1/\sigma$ is replaced by $1 - p$.

4.3   Analysis

Using the suffix automaton the average search time is $O(n^2 \log_\sigma ru^2m/(r(u - \log_\sigma ru^2m)))$: there are $O(rum)$ features of length u, meaning that there are $O(ru^2m)$ suffixes to match, so the search enters to depth $O(\log_\sigma ru^2m)$ on average, we scan only every $O(r)$th row of the text, and the shift the automaton makes is on average $O(u - \log_\sigma ru^2m)$.

The verification time per feature that matches is $O(1)$ as explained, and there are $O(rum/\sigma^u)$ features matching each text position on average. This results in a total search cost

$$O\left(n^2 \frac{1}{r}\left(\frac{\log_\sigma ru^2m}{u - \log_\sigma ru^2m} + \frac{rum}{\sigma^u}\right)\right) = O\left(n^2\left(\frac{\log_\sigma ru^2m}{r(u - \log_\sigma ru^2m)} + \frac{um}{\sigma^u}\right)\right)$$

The optimum is at $r = u = \Theta(m)$, which leads to total average time

$$O(n^2(\log_\sigma m/m^2 + m^2/\sigma^m)) = O(n^2 \log_\sigma m/m^2).$$

which is optimal, so the exact matching problem can be solved in optimal average time $O(n^2 \log m/m^2)$. The space requirement of the suffix automaton is $O(m^4)$.

Again, this analysis is valid for non–uniformly distributed cell values, by replacing $1/\sigma$ by $1 - p$, where p is the probability of a mismatch.

5   Search Allowing Rotations and Mismatches

We first present a 2D version of the incremental algorithm of [ ] that runs in $O(k^{3/2}n^2)$ average time, to search for a pattern in a text allowing rotations and at most k mismatches.

Assume that, when computing the set of angles $\mathcal{A} = (\beta_1, \beta_2, \ldots)$, we also sort the angles so that $\beta_i < \beta_{i+1}$, and associate with each angle β_i the set $\mathcal{C}_i$ containing the corresponding cell centers that must hit a cell boundary at β_i. Hence we can evaluate the number of mismatches for successive rotations of P incrementally. That is, assume that the number of mismatches has been evaluated for β_i, then to evaluate the number of mismatches for rotation β_{i+1}, it suffices to re–evaluate the cells restricted to the set $\mathcal{C}_i$. This is repeated for each $\beta \in \mathcal{A}$. Therefore, the total time for evaluating the number of mismatches for P centered at some position in T, for all possible angles, is $O(\sum_i |\mathcal{C}_i|)$. This is $O(m^3)$ because each fixed cell center of T, covered by P, can belong to some $\mathcal{C}_i$ at most $O(m)$ times. To see this, note that when P is rotated the whole angle 2π, any cell of P traverses through $O(m)$ cells of T.

Then consider the k mismatches problem. The expected number of mismatches in N tests is $Np = N\frac{\sigma-1}{\sigma}$. Requiring that $Np > k$ gives that about $N > k/p$ tests should be enough in typical cases to find out that the distance must be $> k$.

This suggests an improved algorithm for the k–mismatches case. Instead of using the whole P, select the smallest subpattern P' of P, with the same center cell, of size $m' \times m'$ such that $m' \times m' > k/p$. Then search for P' to find if it matches with at most k mismatches. If so, then check with the gradually growing subpatterns P'' whether P'' matches, until $P'' = P$. If not, continue with P' at the next location of T. The expected running time of the algorithm is $O(m'^3 n^2)$ which is $O(k^{3/2}n^2)$.

Note that this algorithm assumes nothing of how we compare the cell values, any other distance measure than counting the mismatches can be also used.

We show now how to improve this time complexity.

5.1   Reducing to Exact Searching

The idea is to reduce the problem to an exact search problem. We cut the pattern into j pieces along each dimension, for $j = \lfloor\sqrt{k}\rfloor + 1$, thus obtaining j^2 pieces of size $(m/j) \times (m/j)$. Now, in each match with k differences or less necessarily one of those pieces is preserved without differences, since otherwise there should be at least one difference in each piece, for a total of $j^2 = (\lfloor\sqrt{k}\rfloor + 1)^2 > k$ differences overall. This fact was first utilized in [,]. So we search for all the j^2 pieces exactly and check each occurrence for a complete match.

Observe that this time the pieces cannot be searched for using the center to center assumption, because this holds only for the whole pattern. However, what is really necessary is not that the piece center is aligned to a text center, but just that there exists a fixed position to where the piece center is aligned. Once we fix a rotation for the whole pattern, the matching function of each pattern piece gets fixed too. Moreover, from the $O(m^3)$ relevant rotations for the whole pattern, only $O(mu)$ are relevant for each one-dimensional feature of length u. There is at most one matching function for each relevant rotation (otherwise we would have missed some relevant rotations). Hence we can work exactly as before when matching pieces, just keeping in mind that the alignment between the pattern center and the text center has to be shifted accordingly to the angle in which we are searching for the feature. The same considerations of the previous section show that we can do the verification of each matching piece in $O(1)$ time.

The search algorithm can look for all the features of all the j^2 patterns together. Since there are j^2 pieces of size $(m/j)^2$, there are $r = O(m/j)$ feature sets, which when considering all their rotations make up $O(j^2(m/j)mu) = O(jm^2u)$

features of length u that can match. So the suffix automaton takes time

$$O\left(n^2 \frac{\log_\sigma jm^2u^2}{\frac{m}{j}\left(\frac{m}{j} - \log_\sigma jm^2u^2\right)}\right).$$

The verification of the whole piece once each feature is found takes $O(1)$. Hence the total verification time is $O(n^2j^2(m/j)mu/\sigma^u)$. Note that for each piece of length $(m/j)^2$ there will be $O(m(m/j)^2)$ relevant rotations, because the piece may be far away from the pattern center.

Once an exact piece has been found (which happens with probability $O(m(m/j)^2/\sigma^{m^2/j^2})$) we must check for the presence of the whole pattern with at most k differences. Although after comparing $O(k)$ cells we will obtain a mismatch on average, we have to check for all the possible rotations. A brute force checking of all the rotations gives $m^3/(m(m/j)^2) = j^2$ checks, for a total $O(kj^2)$ verification time for each piece found.

We can instead extend the valid rotations incrementally, by checking cells farther and farther away from the center and refining the relevant rotations at the same time. Unlike the case of exact searching, we cannot discard a rotation until k differences are found, but the match will disappear on average after we consider $O(k)$ extra cells at each rotation. Hence, we stop the verification long before reaching all the $O(m^3)$ rotations.

Let K be a random variable counting the number of cells read until k differences are found along a fixed rotation. We know that $\overline{K} = O(k)$. Since we enlarge the match of the piece by reading cells at increasing distances from the center, by the point where we find k differences we will have covered a square of side R where $R^2 - (m/j)^2 = K$ (see Figure   ). The total number of rotations considered up to that point is $O(mR^2/(m(m/j)^2)) = O(1 + Kj^2/m^2)$. Since this is linear on K we can take the function on the expectation $\overline{K}$, so the average number of rotations considered until finding more than k differences is $O(1 + kj^2/m^2)$. We consider that we check all these rotations by brute force, making $\overline{K} = O(k)$ comparisons for each such rotation. Then the verification

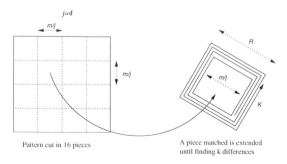

Fig. 2. On the left, the pattern is cut in $j^2 = 16$ pieces. On the right, a piece of width m/j found exactly is extended gradually until finding k differences

cost per piece found is $O(k + k^2j^2/m^2)$. This verification has to be carried out $O(j^2m(m/j)^2n^2/\sigma^{m^2/j^2}) = O(m^3n^2/\sigma^{m^2/j^2})$ times on average. Therefore the total search time is of the order of

$$
n^2 \left(\frac{\log_\sigma jm^2u^2}{\frac{m}{j}\left(\frac{m}{j} - \log_\sigma jm^2u^2\right)} + \frac{j^2(m/j)mu}{\sigma^u} + \frac{km^3}{\sigma^{m^2/j^2}} + \frac{k^2j^2m^3}{m^2\sigma^{m^2/j^2}} \right)
$$

where all the terms increase with j. If we select $j = \Theta(\sqrt{k})$, and $u = \Theta(m/j) = \Theta(m/\sqrt{k})$, the cost is

$$
n^2 \left(\frac{\log_\sigma m^4/\sqrt{k}}{\frac{m}{\sqrt{k}}\left(\frac{m}{\sqrt{k}} - \log_\sigma m^4/\sqrt{k}\right)} + \frac{m^3}{\sigma^{m/\sqrt{k}}} + \frac{km^3}{\sigma^{m^2/k}} + \frac{k^3m}{\sigma^{m^2/k}} \right)
$$

The first term of the expression dominates for $k < m^2/(3\log_\sigma^2 m)$, up to where the whole scheme is $O(n^2k\log_\sigma m/m^2)$ sublinear time. After that point the whole scheme is $O(n^2m^3/\sigma^{m/\sqrt{k}})$ time for $k < m^2/(4\log_\sigma m)$, and $O(n^2k^3m/\sigma^{m^2/k})$ time for larger k.

5.2   Reducing to Approximate Searching

Since the search time worsens with j we may try to use a smaller j, although this time the pieces must be searched for allowing some differences. More specifically, we must allow $\lfloor k/j^2 \rfloor$ differences in the pieces, since if there are more than $\lfloor k/j^2 \rfloor$ differences per piece then the total exceeds k.

The $O(k^{3/2}n^2)$ time incremental search algorithm can be used here. Since we search for j^2 pieces with k/j^2 differences, the total search cost for the pieces is $O(n^2j^2(k/j^2)^{3/2}) = O(n^2k^{3/2}/j)$.

However, the incremental algorithm assumes that the center of P coincides with some center of the cells of T, and this is not necessarily true when searching for pieces. We now present a filter that gives a lower bound for the number of mismatches.

Assume that P is at some location $((u, v), \theta)$ on top of T, such that $(u, v) \in T[i, j]$ is not a center–to–center translation, and that the number of mismatches is k for that position of P. Then assume that P is translated to $((i, j), \theta)$, that is, center–to–center becomes true while the rotation angle stays the same. As a consequence, some cell centers of T may have moved to the cell of P that is one of its eight neighbors. Now compute the number of mismatches such that $T[r, s]$ is compared against $M(T[r, s])$ and its eight neighbors as well. If any of those nine cells match with $T[r, s]$, then we count a match, otherwise we count a mismatch. Let the number of mismatches obtained this way be k'.

This means that $k' \leq k$, because all matches that contribute to $m^2 - k$ must be present in $m^2 - k'$ too. The value of k' can be evaluated with the incremental technique using β_s instead of θ where s is such that $\beta_s \leq \theta < \beta_{s+1}$, because the matching functions are the same for θ and β_s by our construction. Hence $k' \leq k$.

Hence we use the algorithm with the center–to–center assumption, but count a mismatch only when the text cells differs from all the 9 pattern cells that surround the one it matches with. The net result in efficiency is that the alphabet size becomes $\tau = 1/(1 - 1/\sigma)^9$, meaning that a cell matches with probability $1/\tau$.

For the verification cost of the pieces, we need to know the probability of a match with k differences. Since we can choose the mismatching positions and the rest must be equal to the pattern, the probability of a match is $\leq \binom{m^2}{k}/\tau^{m^2-k}$. By using Stirling's approximation to the factorial and calling $\alpha = k/m^2$, we have that the probability can be bounded by γ^{m^2}/m^2, where $\gamma = 1/(\alpha^{\alpha/(1-\alpha)}(1 - \alpha)\tau)^{1-\alpha} \leq (e/((1-\alpha)\tau))^{1-\alpha}$. This improves as m grows and α stays constant. On the other hand, $\alpha < 1 - e/\tau$ is required so that $\gamma < 1$.

If we are searching for a piece of size $(m/j)^2$, then the matching probability is $O(\gamma^{(m/j)^2})/(m/j)^2$, which worsens as j grows. On the other hand, we have to multiply this probability by $j^2 m(m/j)^2 = m^3$ to account for all the rotations of all the pieces. Once a piece matches we check the complete match, which as explained in Section   takes $O(k + k^2 j^2/m^2)$ time. The total cost is

$$n^2 \left(\frac{k^{3/2}}{j} + \gamma^{m^2/j^2} m j^2 \left(k + \frac{k^2 j^2}{m^2} \right) \right) \;=\; n^2 k(\sqrt{k}/j + \gamma^{m^2/j^2} j^2 (m + kj^2/m))$$

whose optimum is $j = m/\sqrt{4 \log_{1/\gamma} m + 1/2 \log_{1/\gamma} k}(1 + o(1))$, which can be achieved whenever it is smaller than $\sqrt{k}$, i.e. for $k > m^2/(5 \log_\sigma m)(1 + o(1))$ (for smaller k the scheme reduces to exact searching and the previous technique applies). For this optimum value the complexity is $O(n^2 k^{3/2} \sqrt{\log_{1/\gamma} m}/m)$.

This competes with the reduction to exact searching for high values of k. Reducing to approximate searching is indeed better for $k > m^2/(5 \log_\sigma m)$, i.e., wherever it can be applied. Recall that the scheme cannot be applied for $k > m^2(1 - e/\tau) = m^2(1 - \Theta(1/\sigma))$.

Figure   shows the complexities achieved.

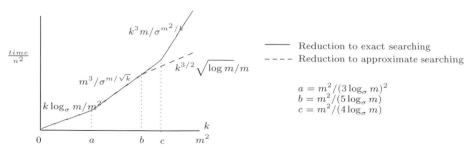

Fig. 3. The complexities obtained for the mismatches model depending on k

6   Searching under the Gray Levels Model

In principle any result for the mismatches model holds for the gray levels model as well, because if a pattern matches with total color difference k then it also matches with k mismatches. However, the typical k values are much larger in this model, so using the same algorithms as filters is not effective if a naive approach is taken.

In this case, we can improve the search by reducing the number of different colors, i.e. mapping s consecutive colors into a single one. The effect is that σ is reduced to σ/s and k is reduced to $1 + \lfloor k/s \rfloor = \Theta(k/s)$ too. For instance, if we consider reduction to exact searching, the scheme is $O(n^2 k \log_\sigma m/m^2)$ time for $k < m^2/(3 \log_\sigma^2 m)$. This becomes now $O(n^2 k/s \log_{\sigma/s} m/m^2)$ time for $k/s < m^2/(3 \log_{\sigma/s}^2 m)$. For example binarizing the image means $s = \sigma/2$ and gives a search time of $O(n^2 k/\sigma \log m/m^2)$ for $k < m^2/(3\sigma/\log_2^2 m)$.

This seems to show that the best is to maximize s, but the price is that now we have to check the matches found for potential matches, because some may not really satisfy the matching criterion on the original gray levels. After a match with reduced alphabet is found we have to check for a real match, which costs $O(1)$ and occurs $O(n^2 m^3 \delta^{m^2}/m^2) = O(n^2 m \delta^{m^2})$ times, where $\delta = 1/(\beta^{\beta/(1-\beta)}(1-\beta)\sigma/s)^{1-\beta} \le (e/((1-\beta)\sigma/s))^{1-\beta}$ and $\beta = \alpha/s = (k/s)/m^2$ (similar to γ in Section   ).

It is clear that this final verification is negligible as long as $\delta < 1$. The maximum s satisfying this is $(\sigma + \sqrt{\sigma^2 - 4e\sigma\alpha})/(2e) = \sigma/e(1 + O(1/\sqrt{\sigma}))$. The search cost then becomes $O(n^2 k/\sigma \log m/m^2)$ for $k < m^2\sigma/(9e \ln^2 m)$. This means that if we double the number of gray levels and consequently double k, we can keep the same performance by doubling s.

For higher k values, partitioning into exact searching worsens if we divide k and σ by s, so the scheme is applicable only for $k < m^2\sigma/(9e \ln^2 m)$. However, it is possible to resort to reduction to approximate matching, using the $O((k/\sigma)^{3/2} n^2)$ average time algorithm for this model. This cost improves as we increase s, and hence we can obtain $O(n^2 (k/\sigma)^{3/2} \sqrt{\log_{1/\delta} m}/m)$ time for $k < m^2\sigma/(5e \ln m)$.

7   Worst Case Optimal Algorithms

In [ ] it was shown that for the problem of the two dimensional pattern matching allowing rotations the worst case lower bound is $\Omega(n^2 m^3)$. Our efficient expected case algorithms above do not achieve this bound in the worst case. However, they are easily modified to do so. This can be done using the $O(m^3)$ time algorithm given in Sec.   for the verifications. Each time the filter suggests that there might be an occurrence in some position, we use the $O(m^3)$ time algorithm to verify it, if it is not verified before (which is possible because several features may suggest an occurrence at the same position). As each position is verified at most once, and the verification cost is $O(m^3)$, the total time is at most $O(n^2 m^3)$, which is optimal. This works for both the Hamming and gray levels cases. Moreover, this verification algorithm is very flexible, and can be adapted to many other distance functions.

8   Conclusions and Future Work

We have presented different alternatives to speed up the search for two dimensional patterns in two dimensional texts allowing rotations and differences.

The results can be extended to more dimensions. In three dimensions there are $O(m^{11})$ different rotations for P [   ], and $O(um^2)$ features of length u. However, the three–dimensional text must be scanned in two directions, e.g. along the x–axis and along the y–axis, to find out the candidate rotations for P. Only if two features are found (that suggest the same center position of P in T), we enter the verification, see Figure   . For the exact matching, the method works in $O(n^3 \log m / m^3)$ average time. The other results can be extended also.

It is also possible to make the verification probability lower by requiring that several pieces must match before going to the verification. This means smaller pieces or more differences allowed for the pieces. It is also possible to scan the text in two (in 2D) or in three (in 3D) ways instead of only one or two, using the same set of features than in the basic algorithm.

Note also that, until now, we have assumed that the center of P must be exactly on top of some center of the cells of T. It is also possible to remove this restriction, but the number of matching functions (and therefore the number of features) grows accordingly, see [   ]. This, however, does not affect the filtering time, but the verification for the approximate matching would be slower.

Finally, we have considered an error model where only "substitutions" are permitted, i.e. a cell value changes its value in order to match another cell, so we substitute up to k values in the text occurrence and obtain the pattern. More sophisticated error models exist which permit displacements (such as inserting/deleting rows/columns) in the occurrences, and search algorithms for those models (albeit with no rotations) have been developed for two and more dimensions [ ]. It would be interesting to combine the ability to manage those types of errors and rotations at the same time.

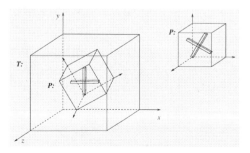

Fig. 4. Matching features in 3D

References

1. A. V. Aho and M. J. Corasick. Efficient string matching: an aid to bibliographic search. *Commun. ACM*, 18(6):333–340, 1975.
2. A. Amir, G. Benson, and M. Farach. An alphabet independent approach to two-dimensional pattern matching. *SIAM J. Comput.*, 23(2):313–323, 1994.
3. A. Amir, A. Butman, M. Crochemore, G.M. Landau, and M. Schaps. Two-dimensional pattern matching with rotations. Submitted for publication, 2002.
4. R. Baeza-Yates and G. Navarro. New models and algorithms for multidimensional approximate pattern matching. *J. Discret. Algorithms*, 1(1):21–49, 2000.
5. R. A. Baeza-Yates and M. Régnier. Fast two-dimensional pattern matching. *Inf. Process. Lett.*, 45(1):51–57, 1993.
6. L. G. Brown. A survey of image registration techniques. *ACM Computing Surveys*, 24(4):325–376, 1992.
7. W.I. Chang and T. Marr. Approximate string matching with local similarity. In *Proc. 5th Combinatorial Pattern Matching (CPM'94)*, LNCS 807, pages 259–273, 1994.
8. M. Crochemore, A. Czumaj, L. Gąsieniec, T. Lecroq, W. Plandowski, and W. Rytter. Fast practical multi-pattern matching. *Inf. Process. Lett.*, 71(3–4):107–113, 1999.
9. K. Fredriksson. Rotation invariant histogram filters for similarity and distance measures between digital images. In *Proc. 7th String Processing and Information Retrieval (SPIRE'2000)*, pages 105–115. IEEE CS Press, 2000.
10. K. Fredriksson and E. Ukkonen. A rotation invariant filter for two-dimensional string matching. In *Proc. 9th Combinatorial Pattern Matching (CPM'98)*, LNCS 1448, pages 118–125, 1998.
11. K. Fredriksson and E. Ukkonen. Combinatorial methods for approximate image matching under translations and rotations. *Patt. Recog. Letters*, 20(11–13):1249–1258, 1999.
12. K. Fredriksson and E. Ukkonen. Combinatorial methods for approximate pattern matching under rotations and translations in 3d arrays. In *Proc. 7th String Processing and Information Retrieval (SPIRE'2000)*, pages 96–104. IEEE CS Press, 2000.
13. G. Navarro K. Fredriksson and E. Ukkonen. An index for two dimensional string matching allowing rotations. In J. van Leeuwen, O. Watanabe, M. Hagiya, P.D. Mosses, and T. Ito, editors, *IFIP TCS2000*, LNCS 1872, pages 59–75, 2000.
14. J. Kärkkäinen and E. Ukkonen. Two- and higher-dimensional pattern matching in optimal expected time. *SIAM J. Comput.*, 29(2):571–589, 2000.
15. R. L. Rivest. Partial-match retrieval algorithms. *SIAM J. Comput.*, 5(1):19–50, 1976.
16. S. Wu and U. Manber. Fast text searching allowing errors. *Commun. ACM*, 35(10):83–91, 1992.
17. A. C. Yao. The complexity of pattern matching for a random string. *SIAM J. Comput.*, 8(3):368–387, 1979.

Statistical Identification
of Uniformly Mutated Segments within Repeats

S. Cenk Şahinalp[1][*], Evan Eichler[2], Paul Goldberg[3][**], Petra Berenbrink[4],
Tom Friedetzky[5], and Funda Ergun[6]

[1] Dept of EECS, Dept of Genetics and Center for Computational Genomics
CWRU, USA
cenk@cwru.edu
[2] Dept of Genetics and Center for Computational Genomics
CWRU, USA
eee@po.cwru.edu
[3] Dept of Computer Science, University of Warwick, UK
pwg@dcs.warwick.ac.uk
[4] Simon Fraser University, School of Computing, Canada
petra@cs.sfu.ca
[5] Pacific Institute of Mathematics, Simon Fraser University, Canada
tf@pims.math.ca
[6] NEC Research Institute and Dept of EECS, CWRU, USA
ergun@research.nj.nec.com

Abstract. Given a long string of characters from a constant size
(w.l.o.g. binary) alphabet we present an algorithm to determine whether
its characters have been generated by a single i.i.d. random source. More
specifically, consider all possible k-coin models for generating a binary
string S, where each bit of S is generated via an independent toss of one
of the k coins in the model. The choice of which coin to toss is decided by
a random walk on the set of coins where the probability of a coin change
is much lower than the probability of using the same coin repeatedly. We
present a statistical test procedure which, for any given S, determines
whether the *a posteriori* probability for $k = 1$ is higher than for any other
$k > 1$. Our algorithm runs in time $O(\ell^4 \log \ell)$, where ℓ is the length of S,
through a dynamic programming approach which exploits the convexity
of the *a posteriori* probability for k.

The problem we consider arises from two critical applications in analyzing long alignments between pairs of genomic sequences. A high alignment score between two DNA sequences usually indicates an evolutionary relationship, i.e. that the sequences have been generated as a result
of one or more copy events followed by random point mutations. Such
sequences may include functional regions (e.g. exons) as well as non-
functional ones (e.g. introns). Functional regions with critical importance
exhibit much lower mutation rates than non-functional DNA (or DNA

---

[*] Supported in part by an NSF Career Award and by Charles B. Wang Foundation.
[**] Partially supported by the IST Programme of the EU under contract number IST-
1999-14186 (ALCOM-FT).

A. Apostolico and M. Takeda (Eds.): CPM 2002, LNCS 2373, pp. 249–   , 2002.

with non-critical functionality) due to selective pressures for conserving
such regions. As a result, given an alignment between two highly similar
genome sequences, it may be possible to distinguish functional regions
from non-functional ones using variations in the mutation rate. Our test
provides means for determining variations in the mutation rate and thus
checking the existence of DNA regions of varying degrees of functionality.
A second application for our test is in determining whether two highly
similar, thus evolutionarily related, genome segments are the result of a
single copy event or of a complex series of copies. This is particularly an
issue in evolutionary studies of genome regions rich with repeat segments
(especially non-functional tandemly repeated DNA). Our approach can
be used to distinguish simple copies from complex repeats again by ex-
ploiting variations in mutation rates.

1 Introduction

As more of the human genome is sequenced and assembled, it is becoming apparent that it consists of numerous sequences "repeated" with various degrees
of similarity [, ,]. Such repeats are more likely to be generated as a result
of segmental copies during evolution rather than by chance.
Approximately 60% of the human genome appears to be repeated. Repeat segments are commonly classified into three main categories. Over 45% of the human genome comprises common repeats; one example is the $\sim$ 300bp *alu* element, occurring more than 1M times within a divergence rate of $5\% - 15\%$.
Another $\sim 5\%$ consists of the centromeric repeats, particularly the alpha satellite and microsatellite DNA. A final $\sim 7\%$ is made up of much longer repeat
segments (which include partial or complete genes) exhibiting small divergence
rates ($\leq 10\%$). These figures support the theory that copying followed by point
mutations provide the main process underlying genome evolution [ ].

Several biochemical mechanisms underlying segmental copies have been identified in the last 30 years (e.g. *unequal cross over* [ ], *replication slippage* and
retrotransposition); potentially many more are waiting to be discovered. The
task of identifying all copying mechanisms in the genome for a better understanding of the genome evolution process poses a number of algorithmic and
computational challenges. First and foremost, one needs to identify *a posteriori*
all pairs of repeat sequences which were generated as a result of a single copy

---

[1] Consider, for example, a 3×10^9 bp sequence (which is roughly of size similar to the human genome) generated by an i.i.d. random source on the four letter DNA alpha-
bet. Then the probability that two specified 100bp long segments have a Hamming distance of 5% or less is smaller than 2^{-150}. Thus, the probability that the whole genome (generated as above, without any copying events) includes a pair of 100bp long segments with Hamming distance of 5% or less is smaller than 2^{-75} (practically nil).

[2] Studies show that 60% of the nucleotides are contained in 1Kbp segments repeated with less than 30% divergence.

event during evolution. Note that a *repeat* can be a result of multiple complex *copy* events: for example a long tandemly repeated sequence S, S, S, S may be a result of tandemly copying the first S three times or copying the first S once to obtain S, S and copying this whole segment again to get S, S, S, S; there are many other possibilities. This is the subject of our concern.

Contributions. In this paper we present a statistical test for identifying whether a pair of genome sequences with a high similarity score are indeed a result of a single copy event. For this purpose we employ the "neutral hypothesis"; i.e. that point mutations occur independently at random with a fixed probability ($1.5 - 3 \times 10^{-9}$ per base pair per year for non-functional segments of humanoids and old world monkeys).

As mentioned above, a high similarity score between two sequences is an indication of an evolutionary relationship. One possible relationship between two such sequences is that one may have been copied from the other in a single copy event. Because after a copying event both copies would be subject to independent point mutations, a number of edit errors would be observed in their alignment. However, these mutations would have been applied to each character in an i.i.d. fashion; as a result, the normalized similarity score between the two sequences is expected to be uniform (allowing for the statistically expected amount of variation) throughout their alignment.

A common strategy for identifying copies between two genome regions is to iteratively locate pairs of sequences with the highest similarity score (e.g. via Smith-Waterman method). Shortcomings of this strategy in terms of "signal strength" are discussed in [ ], where an alternative "normalized" similarity measure based on [ ] is described along with an efficient algorithm for computing it. This approach is designed for identifying pairs of sequences with higher functional relationship rather than providing a tool for studying the evolution of repeat segments. As mentioned at the beginning of this section, a pair of sequences with a high alignment score may be a result of a number of complex copy events occurring at different points of evolutionary time. They may also involve segments with varying degrees of functionality which are subject to different rates of mutation; this is due to evolutionary pressures for conserving highly functional segments. As a result, a high overall alignment score (absolute or normalized) cannot be used (due to its consolidation of the individual alignment scores of smaller segments in the sequences) to measure the evolutionary time passed since the separation of two such sequences. For instance, in satellite DNA, which contains a large number of tandem repeats of the same subsequence, there are many possibilities as to the actual progression of the copying events, including their order, as well as the source and destination subsequences [ ]. Assuring that the sequences have been subject to independent point mutations only, rather than a complex series of copying events, is critical to the accuracy of phylogenetic analysis, especially based on distance comparisons (e.g. [ ]) involving these sequences.

To address the above issue we propose a new method for pairwise sequence comparison in the form of a *statistical test* to determine whether a given pair of sequences with high similarity score have been generated as the result of a single copy event. More specifically, we consider k-state Hidden Markov Models (HMMs) for generating the alignment sequence S (on which a 0 may represent a correct alignment and a 1 may represent a misalignment) between two highly similar sequences. In the models that we consider, the bit values of S are generated by independent tosses of biased coins (with output 0/1) which are fixed for each state of the HMM in consideration. (Thus each state represents a random process which imposes a fixed mutation rate on the segment it is applied upon.) The sequence of states which are responsible for the generation of S is decided by a random walk where the probability of a state change is much lower than that of remaining at a given state. We present an algorithm which determines *a posteriori* for any given S, whether among all possible k-state Hidden Markov Models, those for which $k = 1$ are more likely than any other for which $k > 1$ (we compare the aggregate likelihood of all 1-state HMMs with that of k-state HMMs for $k > 1$). Our algorithm runs in time $O(\ell^4 \log \ell)$, where ℓ is the length of S, through a dynamic programming approach which exploits the convexity of the probability function for k.

Similar problems have been considered earlier in [,   , , ,   ,   ]. In fact [ ] considered a two state HMM for identifying the cutoff point between one mutation rate and another for a *given* alignment sequence S. The *most likely* HMM is constructed through standard expectation maximization (EM) techniques. In contrast we focus on the *aggregate* effects of all possible 1 or 2 state HMMs rather than focusing on a single model for robustness purposes as it does not require specification of a cut-off point for differentiating one coin and two coin models. Furthermore our approach need not consider a single alignment between the pair of sequences considered: it is possible to generalize our method to aggregate over all possible alignments according to the likelihood of their occurrence.

2   Preliminaries

For the purposes of this paper, the *genome* is a long string of characters from the DNA alphabet $\{a, c, g, t\}$. A *genome segment* is a substring of the genome. We assume that we are given the correctly assembled genome (partially or as a whole) as part of the input.

Throughout the paper S and Q denote genome segments, $S[i]$ denotes the ith character of segment S, and $S[i : j]$ the substring between the ith and j^{th} characters (inclusive) of S. $|S|$ denotes the length of the segment S.

An alignment between two genome segments Q and S is the 2-tuple (Q', S'), where $Q', S' \in \{a, c, g, t, -\}^{\ell}$ for some $\ell = |Q'| = |S'|$, such that Q and S are obtained if all "$-$" are removed from Q' and S' respectively.

Given two characters x and y, $x \oplus y$ denotes the character-wise exclusive-OR (XOR) function; it evaluates to 1 if $x \neq y$ and to 0 otherwise. Given an alignment (Q', S'), where $|Q'| = |S'| = \ell$, we denote by $al(Q', S') = Q' \oplus S'$, the *alignment sequence* of (Q', S') whose i^{th} entry is $Q'[i] \oplus S'[i]$. We denote by $h(Q', S')$ the normalized Hamming distance between Q' and S', i.e. the number of 1's in $al(Q', S')$ divided by ℓ.

3 A Statistical Test for Detecting Simple Copies

The sequence comparison problem we consider can be formally described as follows. We are given two genome segments Q, S and their alignment (Q', S') for which $h(Q', S') \leq \delta$ for some predetermined threshold value $0 \leq \delta \leq 1$. Our goal is to determine whether the alignment sequence $al(Q', S')$ is more likely to have been generated by a single i.i.d. random source or a combination of k i.i.d. random sources, for some $k > 1$.

The underlying motivation for the above problem is the need to test whether the sequences Q and S have been generated by a single copying event, followed by independent point mutations only. Alternatively they could either be a result of more complex procession of multiple copying events, where the segments involved were subjected to mutations for different periods of time, or a result of a single copying event after which different subregions have been subjected to different mutation rates. When the latter possibility is indeed the case, one expects to observe varying mutation rates throughout the sequences, resulting in measurable variation in the normalized distance between aligned segments of Q' and S'. Thus a statistical test for determining whether the edit errors between Q' and R' are more likely to have been generated by a single i.i.d. random source than by multiple sources can be used as a tool for identifying pairs of sequences that have been a result of a single copying event. For such sequences an overall similarity score can be used to determine the evolutionary time passed since their separation.

3.1 Comparing Single and Multiple Coin Models

Given the alignment sequence $O = al(Q', S') = (O_1, \ldots, O_\ell)$ of length ℓ, we would like to compute the *a posteriori* probability that O has been generated by independent tosses of a single coin or by a procession of multiple, coins selected by performing a random walk in the set of coins. More formally, we define an n-coin model as an n-state Hidden Markov Model similar to many other applications of HMMs [ ]. Note that employing HMMs in the context of this paper is quite natural. Without any *a priori* information about which positions in the alignment sequence a coin switch is more likely, it is plausible to assume independent and identical distributions for the coin switch probabilities; this in turn defines an HMM.

Let $\mathcal{C} = \{C_1, \ldots, C_n\}$ denote a set of n coins, each with $0/1$ outcome. Let $p_i(b)$ denote the probability of outcome b, $b \in \{0, 1\}$ on a flip of coin C_i. Thus, $p_i(0) + p_i(1) = 1$. If $n = 1$, we will denote C_1 as C and $p_1(1)$ as p. We denote by q_j, the coin used in generating $O[j]$, $1 \le j \le \ell$. Note that for any $t = 1, \ldots, \ell$, $\Pr(O[t] = 1 \mid q_t = C_j) = p_j(1)$, since the outcome does not depend on the location itself but only on the "active" coin. Furthermore, let $a_{i,j}(t) = \Pr(q_{t+1} = C_j \mid q_t = C_i)$ denote the transition probability for from coin C_i to coin C_j between locations t and $t + 1$. Again, $a_{i,j}$ does not depend on the location t, hence in the following we will refer to $a_{i,j}$ only. Let $\pi_j = \Pr(q_1 = C_j)$ for $1 \le j \le n$ (the probability that the first location is generated by coin C_j).

With $A = \{a_{i,j} \; : \; 1 \le i, j \le n\}$, $P = \{p_j(1) \; : \; 1 \le j \le n\}$, and $\pi = \{\pi_j \; : \; 1 \le j \le n\}$, an n-coin model λ is now defined as $\lambda = (A, P, \pi)$. For $i \ge 1$, let Λ_i denote the set of all i-coin models. Denote by $\Lambda = \bigcup_{i \ge 1} \Lambda_i$ the set of all coin models.

Let $\Omega = \{0, 1\}^\ell \times \Lambda$ denote our probability space. Hence, an elementary event is an ordered pair (S, λ) where S is an ℓ-bit binary string, and λ a coin model. An experiment consists of the following steps. First select a coin model (by first choosing the number of coins, and then fixing the corresponding biases), next, using this model, generate an alignment sequence (a bit string of length ℓ).

For convenience, we define the following probabilities. For some coin model λ, let $\Pr(\lambda) = \Pr\left(\bigcup_{t \in \{0,1\}^\ell}(t, \lambda)\right)$. Similarly, for $S \in \{0, 1\}^\ell$, let $\Pr(S) = \Pr\left(\bigcup_{\mu \in \Lambda}(S, \mu)\right)$. Note that $\sum_{S \in \{0,1\}^\ell} \Pr(S) = \sum_{\lambda \in \Lambda} \Pr(\lambda) = 1$.

Let W_i denote the event that an i-coin model was chosen, i.e., $W_i = \Omega \cap (\{0, 1\}^\ell \times \Lambda_i)$. We are interested in the quantities $\Pr(W_i|O) = \Pr(W_i|\bigcup_{\lambda \in \Lambda}(O, \lambda))$ for all $i \ge 1$. Obviously,

$$\Pr(W_i|O) = \frac{\Pr(W_i \wedge O)}{\Pr(O)} = \frac{\Pr(O|W_i) \cdot \Pr(W_i)}{\Pr(O)},$$

where, as above, $\Pr(O|W_i) = \Pr(\bigcup_{\lambda \in \Lambda}(O, \lambda)|W_i)$ and $\Pr(W_i \wedge O) = \Pr(W_i \cap \bigcup_{\lambda \in \Lambda}(O, \lambda))$. Without any prior information, we assume that $\Pr(W_i) = \Pr(W_j)$ for all $i, j \ge 1$. Hence, we need to compute and compare all $\Pr(O|W_i)$ in order to compute the most probable number of coins to generate the sequence.

Single-coin model. A single-coin model λ is defined by p, the probability that a 1 is generated. For discrete valued p,

$$Pr(O \mid W_1) = \sum_{\forall p} Pr(O \mid p \wedge W_1) \times Pr(p \mid W_1).$$

For continuous valued p for which $Pr(p \mid W_1)$ is uniform,

$$Pr(O|W_1) = \int_{p=0}^{1} Pr(O|p \wedge W_1)dp = \int_{p=0}^{1} Pr(O[1] \cdot O[2] \ldots O[\ell] \mid p, W_1)dp$$

$$= \int_{p=0}^{1} p^k (1-p)^{\ell-k} \, dp$$

$$= \sum_{i=0}^{k+1} (\ell - k \ i) \frac{1}{\ell - i + 1}$$

where k is the number of 1's in O. Hence, in a single-coin model, provided that the number of 0's and 1's is fixed, their specific locations have no effect on $Pr(O \mid W_1)$, and thus on the likelihood of W_1 given O.

Multiple coin models. For any n-coin model λ and for any sequence O,

$$Pr(O|\lambda \wedge W_n) = \sum_{\forall q_1 \ldots q_\ell; \ k \ \text{s.t.} \ q_1 = C_k} \pi_k \cdot p_{q_1}(O[1])$$

$$\cdot a_{q_1,q_2} \cdot p_{q_2}(O[2]) \ldots a_{q_{\ell-1},q_\ell} \cdot p_{q_\ell}(O[\ell]),$$

which can be computed using the following recurrence relationship.

Let $\alpha_t(i) = Pr(\text{the coin model } \lambda \text{ generates } O[1:t] \wedge q_t = C_i \mid \lambda \wedge W_n)$. Now,

$$\alpha_1(i) = \pi_i \cdot p_i(O[1]) \quad \text{for all } i, \text{ and}$$

$$\alpha_{t+1}(j) = [\sum_{\forall i} \alpha_t(i) \cdot a_{i,j}] \cdot p_j(O[t+1]).$$

One can simply write

$$Pr(O \mid \lambda \wedge W_n) = \sum_{\forall i} \alpha_\ell(i).$$

To give an example, in a two-coin model $S = \{C_1, C_2\}$. Let $A = \{a_{1,2} = u, a_{1,1} = 1-u, a_{2,1} = v, a_{2,2} = 1-v\}$ and let $B = \{p_1(0) = r, \ p_1(1) = 1-r, \ p_2(0) = s, \ p_2(1) = 1-s\}$. Without any prior information, $\pi = \{\pi_1 = 1/2, \ \pi_2 = 1/2\}$. Thus

$$Pr(O \mid \lambda \wedge W_2) = \alpha_\ell(1) + \alpha_\ell(2)$$

$$= [\alpha_{\ell-1}(1) \cdot (1-u) + \alpha_{\ell-1}(2) \cdot v] \cdot (1-r)^{O[\ell]} \cdot r^{1-O[\ell]}$$

$$+ \ [\alpha_{\ell-1}(1) \cdot u + \alpha_{\ell-1}(2) \cdot (1-v)] \cdot (1-s)^{O[\ell]} \cdot s^{1-O[\ell]}.$$

Iteratively, we can replace the terms involving α_i with those involving α_{i-1}, finally replacing terms involving α_1 with the above definition of α_1, to obtain a multi-variate polynomial on u, v, r, s of total degree 2ℓ with $\frac{1}{4}(\ell^2 - \ell)^2 \leq \frac{\ell^4}{4}$ terms as follows.
Let

$$V_0 = \begin{bmatrix} (1-u) \cdot r & u \cdot s \\ v \cdot r & (1-v) \cdot s \end{bmatrix}$$

and

$$V_1 = \begin{bmatrix} (1-u) \cdot (1-r) & u \cdot (1-s) \\ v \cdot (1-r) & (1-v) \cdot (1-s) \end{bmatrix}.$$

Then one can simply write

$$Pr(O \mid u, v, r, s, W_2) = [1/2 \; 1/2] \cdot \{\prod_{i=1}^{\ell} V_{O[i]}\} \cdot \begin{bmatrix} 1 \\ 1 \end{bmatrix}$$

which can be evaluated by successive multiplications in $O(\sum_{i=1}^{\ell} i^4) = O(\ell^5)$ time. Thus for uniform $Pr(\lambda \mid W_2)$,

$$Pr(O \mid W_2) = \int_{v,u,r,s=0}^{1} Pr(O \mid v, u, r, s, W_2) \; dv \; dr \; du \; ds$$

$$= \int_{v,u,r,s=0}^{1} \sum_{\forall i} \alpha_\ell(i) \; dv \; dr \; du \; ds$$

and thus

$$Pr(O \mid W_2) = \int_{v,u,r,s=0}^{1} [1/2 \; 1/2] \cdot \{\prod_{i=1}^{\ell} V_{O[i]}\} \cdot \begin{bmatrix} 1 \\ 1 \end{bmatrix} \; dv \; dr \; du \; ds$$

Notice that one can impose bounds on coin transition probabilities by simply changing the range of integration. It is quite straightforward to determine and perform the symbolic integration of the above multivariate polynomial for $Pr(O|W_2)$ which involves $O(\ell^4)$ terms in $O(\ell^5)$ time.

We now show how to conclude whether the single coin explanation is the likeliest.

Lemma 1. *Knowledge of $Pr(O \mid W_1)$ and $Pr(O \mid W_2)$ is sufficient to conclude whether a single-coin model W_1 has the highest a posteriori probability among all W_i for creating the binary sequence O.*

An important observation about $Pr(O \mid W_i)$ as a function of i is that it has at most one local maximum; hence for any i-coin model, if $Pr(O \mid W_i) \geq Pr(O \mid W_{i-1})$ and $Pr(O \mid W_i) \geq Pr(O \mid W_{i-2})$, then $Pr(O \mid W_{i-1}) \geq Pr(O \mid W_{i-2})$. Thus if $Pr(O \mid W_1) \geq Pr(O \mid W_2)$, then for any i $Pr(O \mid W_1) \geq Pr(O \mid W_i)$, and thus W_1 is the most likely model for generating O.

The lemma implies that our test needs to compute and compare $Pr(O \mid W_1)$ and $Pr(O \mid W_2)$, which can be performed in $O(\ell^5)$ time. In the next section we show how to improve this running time to $O(\ell^4 \cdot \log \ell)$.

3.2 Improving the Running Time

The running time of our test is dominated by the time spent on computing $P(O \mid W_2)$. The multivariate polynomial evaluation in this step can be performed faster than $O(\ell^5)$ time via a divide and conquer approach: The multiplication of two k-variate polynomials where the degree of each variable in a term is bounded above by i can be done in $O(k \cdot i^k \cdot \log i)$ time using FFT. It is not difficult to see that the running time of the divide and conquer algorithm is dominated by that of the final step, which requires multiplying two 2×2 matrices where each entry is a 4-variate polynomial and the degree of each variable is at most ℓ. This leads to an overall running time of $O(\ell^4 \cdot \log \ell)$ for our test, which we state in the theorem below.

Theorem 1. *Given an alignment sequence O of length ℓ, it is possible to determine in $O(\ell^4 \log \ell)$ time whether a posteriori probability that O has been generated by a 1-coin model is higher than that for any other k-coin model for $k > 1$.*

3.3 Examples

For $O = 1\,1$ (or $O = 0\,0$)

$$Pr(O \mid W_1) = \int_0^1 (1-r)^2\, dr = \frac{8}{24}, \quad and$$

$$Pr(O \mid W_2) = \int_0^1 \frac{1}{2}[(1-r)^2(1-u) + (1-r)u(1-s)$$

$$+ (1-u)(1-s)^2 + (1-r)v(1-s)]\, dr\, du\, dv\, ds$$

$$= \frac{1}{2} \cdot (\frac{1}{3} \cdot \frac{1}{2} + \frac{1}{2}^3) \cdot 2 = \frac{7}{24}$$

thus it is more likely that O has been generated by a single coin model. This is quite intuitive as a very likely model for generating this sequence consists of a single coin with high bias.

For $O = 1\,0$ (or $O = 0\,1$)

$$Pr(O \mid W_1) = \int_0^1 r(1-r)\, dr = \int_0^1 r - r^2\, dr = \frac{4}{24}, \quad and$$

$$Pr(O \mid W_2) = \int_0^1 \frac{1}{2}[ru(1-r) + r(1-u)(1-s)$$

$$+ s(1-u)(1-r)^2 + sv(1-s)]\, dr\, du\, dv\, ds$$

$$= \frac{1}{2} \cdot (\frac{1}{6} \cdot \frac{1}{2} + \frac{1}{2}^3) \cdot 2 = \frac{5}{24}$$

thus it is more likely that O has been generated by a two-coin model. This is also intuitive as such a sequence can only be a result of a single coin which is not very biased; however one can think of both biased and unbiased two-coin models that could be responsible of its generation.

We programmed our algorithm to test the likelihood of $Pr(O \mid W_1)$ and $Pr(O \mid W_2)$ on a number of alignment sequences O.

The first table below provides some intuition on the likelihood of models on short sequences. It is interesting to note that the last sequence is much more likely to be generated by a two-coin model due to its periodic nature. The most likely model to generate this sequence would involve two coins which are highly and oppositely biased; the transition probabilities from one coin to the other should also be very high.

O	$Pr(O \mid W_1)$	$Pr(O \mid W_2)$	Likely model
101	0.0833	0.104	W_2
11100	0.0166	0.0208	W_2
111111	0.142	0.0822	W_1
1110111	0.0178	0.0156	W_1
1010101010	0.000360	0.00149	W_2

Here are some sequences which were generated with two coins of opposite biases switched exactly in the middle of each sequence. The test was able to successfully identify bias differences of 10% or more.

O	% 1's in 1st half	% 1's in 2nd half	$Pr(O \mid W_1)$	$Pr(O \mid W_2)$	Likely model
11101111111111110000001000000	93%	8%	$1.780 \cdot 10^{-9}$	$2.980 \cdot 10^{-8}$	W_2
11010110111111000010101000	77%	25%	$7.396 \cdot 10^{-9}$	$1.117 \cdot 10^{-8}$	W_2
010010110110010010101110	55%	45%	$6.163 \cdot 10^{-8}$	$8.450 \cdot 10^{-8}$	W_2

3.4   Extensions

It is possible to extend our statistical test by using a slightly larger alphabet $\{0, 1, -\}$ rather than the binary, where the character "$-$" represents a gap in only one of the sequences in the alignment. This increases the complexity of the problem as two new variables, r' and s', for representing the probabilities of generating a gap for each coin need to be incorporated into the algorithm. The corresponding increase in the number of variables in the multivariate polynomial from 4 to 6 leads to an $O(\ell^6 \log \ell)$ running time.

We also note that an alternative test, which compares $Pr(O \mid W_1, \lambda_1)$ and $Pr(O \mid W_2, \lambda_2)$, where λ_1 and λ_2 are the most likely one-coin and two-coin models respectively can be of use. It is easy to verify that obtaining λ_1 and λ_2

requires a differentiation of the respective univariate and multivariate polynomials and evaluating them at local maxima. This can be done in $O(\ell)$ time for the univariate polynomial, and in $O(\ell^4 \log \ell)$ time for the 4-variate polynomial.

3.5   Identifying All Copies of a Pattern in a Long Sequence

Given a long sequence S and a pattern Q, it is possible to extend our test to find all segments R of S for which the alignment (Q', R') obtained by an alignment algorithm of choice satisfies (1) $h(Q', R') \leq \delta$ for some threshold value $0 \leq \delta \leq 1$, and (2) the alignment sequence $al(Q', R')$ passes our statistical test. This generalizes available pattern matching algorithms for identifying segments of S that satisfies condition (1) only (some of the better known results in this direction include [, , , ,]). A simple implementation which slides Q through S takes $O(|S| \cdot |Q|^4 \log |Q|)$ time.

4   Open Problems and Discussion

An immediate open problem is whether it is possible to improve the running time of the pattern identification algorithm described above to $O(|S| \cdot |Q|^3 \log |Q|)$ for certain alignments. This raises the issue of generalizing our test, which considers a single alignment between a pair of sequences, to one which considers multiple possible alignments. Another important problem is how to apply this test to "discover" all repeats in a long genome segment, extending the work on sequence discovery algorithms available for non-tandem repeats [ ], and other motifs [, ,] under conventional measures of sequence similarity. One particularly interesting testbed is the identification of the exact boundaries of multi-layered tandemly repeated DNA segments. A practical approach to this problem is to slide a fixed size window across the sequence of interest, measuring the percentage similarity score of every window position w_i with every other w_j. It is expected that for those w_i and w_j for which $j - i + 1$ is a multiple of a *period* size, the percentage similarity score will be higher than other window positions; thus one can view each w_i, w_j pair whose similarity score is higher than a threshold as evidence that $k = j - i + 1$ is a candidate period size (usually on a 2-D plot). If the candidate period size k is supported by sufficient evidence, one may conclude that k is indeed the size of a period. Although this approach has been used in a number of applications, it raises a few issues.

(1) The widely accepted hypothesis for high order tandem repeat evolution (e.g. the high repeat alpha-satellite DNA) maintains that some early tandem copies at the monomeric level are followed by a k-mer copying event, after which almost all copying events occur at k-meric level [,]. In other words copying events occur hierarchically in time, and "larger period" sizes are always multiples of "smaller period" sizes.

However, one can imagine copies occurring in a number of different block sizes scattered over the sequence; this may lead the above strategy to fail to correctly identify the high order in the repeat pattern.

(2) Different window sizes may lead to different conclusions.

(i) if the window size is smaller than the size of a period, the method will not compare full periods against each other and the results derived can be misleading;

(ii) if the window size is much larger than the size of a period, then the variations in similarity between w_i, w_j pairs will be insignificant.

(3) The thresholds for (i) the similarity score and (ii) the number of evidences for identifying a potential period as an actual period play a significant role in the method. If the threshold values are too small, there will be too many periods to report; if they are too large, some of the periods may be ignored.

References

1. E. F. Adebiyi, T. Jiang, M. Kaufmann, An Efficient Algorithm for Finding Short Approximate Non-Tandem Repeats, *In Proceedings of ISMB 2001.*

2. A. N. Arslan, O. Egecioglu, P. A. Pevzner A new approach to sequence comparison: normalized sequence alignment, *Proceedings of RECOMB 2001.*

3. Bailey J. A., Yavor A. M., Massa H. F., Trask B. J., Eichler E. E., Segmental duplications: organization and impact within the current human genome project assembly, *Genome Research* 11(6), Jun 2001.

4. T. Bailey, C. Elkan. Fitting a mixture model by expectation maximization to discover motifs in biopolymers, *Proceedings of ISMB* 1994, AAAI Press.

5. J. Buhler and M. Tompa Finding Motifs Using Random Projections, *In Proc. of RECOMB 2001.*

6. J. Buhler Efficient Large Scale Sequence Comparison by Locality Sensitive Hashing, *Bioinformatics*17(5), 2001.

7. Richard Cole and Ramesh Hariharan, Approximate String Matching: A Simpler Faster Algorithm, *Proc. ACM-SIAM Symposium on Discrete Algorithms*, pp. 463-472, 25-27 January 1998.

8. Churchill, G. A. Stochastic models for heterogeneous DNA sequences, *Bulletin of Mathemathical Biology* 51, 79-94 (1989).

9. W. Chang and E. Lawler, Approximate String Matching in Sublinear Expected Time, *Proc. IEEE Symposium on Foundations of Computer Science*, 1990.

10. Fu, Y.-X and R. N. Curnow. Maximum likelihood estimation of multiple change points, *Biometrika* 77, 563-573 (1990).

11. Green, P. J. Reversible Jump Markov chain Monte Carlo Computation and Bayesian Model Determination *Biometrika* 82, 711-732 (1995)

12. A. L. Halpern Minimally Selected p and Other Tests for a Single Abrupt Changepoint in a Binary Sequence *Biometrics* 55, Dec 1999.

13. A. L. Halpern Multiple Changepoint Testing for an Alternating Segments Model of a Binary Sequence *Biometrics* 56, Sep 2000.

14. J. E. Horvath, L. Viggiano, B. J. Loftus, M. D. Adams, N. Archidiacono, M. Rocchi, E. E. Eichler Molecular structure and evolution of an alpha satellite/non-satellite junction at 16p11. *Human Molecular Genetics*, 2000, Vol 9, No 1.

15. Jackson, Strachan, Dover, *Human Genome Evolution, Bios Scientific Publishers*, 1996.

16. E. S. Lander et al., Initial sequencing and analysis of the human genome, *Nature*, 15:409, Feb 2001.
17. V. I. Levenshtein, Binary codes capable of correcting deletions, insertions and reversals, *Cybernetics and Control Theory*, 10(8):707-710, 1966.
18. T. Mashkova, N. Oparina, I. Alexandrov, O. Zinovieva, A. Marusina, Y. Yurov, M. Lacroix, L. Kisselev, Unequal crossover is involved in human alpha satellite DNA rearrangements on a border of the satellite domain, *FEBS Letters*, 441 (1998).

19. A. Marzal and E. Vidal, Computation of normalized edit distances and applications, *IEEE Trans. on PAMI*, 15(9):926-932, 1993.
20. L. Parida, I. Rigoutsos, A. Floratsas, D. Platt, Y. Gao, Pattern discovery on character sets and real valued data: linear bound on irredundant motifs and an efficient polynomial time algorithm, *Proceedings of ACM-SIAM SODA, 2000*.
21. S. C. Sahinalp and U. Vishkin, Approximate and Dynamic Matching of Patterns Using a Labeling Paradigm, *Proc. IEEE Symposium on Foundations of Computer Science*, 1996.
22. George P. Smith Evolution of Repeated DNA Sequences by Unequal Crossover, *Science*, vol 191, pp 528–535.   ,
23. J. D. Thompson, D. G. Higgins, T. J. Gibson, Clustal-W: improving the sensitivity of progressive multiple sequence alignment through sequence weighting, position specific gap penalties and weight matrix choice, *Nucleic Acid Research* 1994, Vol. 22, No. 22.
24. E. Ukkonen, On Approximate String Matching, *Proc. Conference on Foundations of Computation Theory*, 1983.
25. Venter, J. and Steel, S. Finding multiple abrupt change points. *Computational Statistics and Data Analysis* 22, 481-501. (1996).
26. C. Venter et. al., The sequence of the human genome, *Science*, 16:291, Feb 2001.

Simple and Practical Sequence Nearest Neighbors with Block Operations

S. Muthu Muthukrishnan[1] and S. Cenk Şahinalp[2][*]

[1] Dept of CS, Rutgers University and AT& T Labs – Research
Florham Park, NJ, USA
`muthu@cs.rutgers.edu`
[2] Dept of EECS, Dept of Genetics, Cntr for Computational Genomics
CWRU, Cleveland, OH, USA
`cenk@cwru.edu`

Abstract. Sequence nearest neighbors problem can be defined as follows. Given a database D of n sequences, preprocess D so that given any query sequence Q, one can quickly find a sequence S in D for which $d(S,Q) \leq d(S,T)$ for any other sequence T in D. Here $d(S,Q)$ denotes the "distance" between sequences S and Q, which can be defined as the minimum number of "edit operations" to transform one sequence into the other. The edit operations considered in this paper include single character edits (insertions, deletions, replacements) as well as block (substring) edits (copying, uncopying and relocating blocks).

One of the main application domains for the sequence nearest neighbors problem is computational genomics where available tools for sequence comparison and search usually focus on edit operations involving single characters only. While such tools are useful for capturing certain evolutionary mechanisms (mainly point mutations), they may have limited applicability for understanding mechanisms for segmental rearrangements (duplications, translocations and deletions) underlying genome evolution. Recent improvements towards the resolution of the human genome composition suggest that such segmental rearrangements are much more common than what was estimated before. Thus there is substantial need for incorporating similarity measures that capture block edit operations in genomic sequence comparison and search. Unfortunately even the computation of a block edit distance between two sequences under any set of non-trivial edit operations is NP-hard.

The first efficient data structure for approximate sequence nearest neighbor search for any set of non-trivial edit operations were described in [ ];

[*] Supported in part by an NSF Career Award and by Charles B. Wang foundation.
[1] Note that more "accurate" distance measures for evolutionary time between two copies of a genome segment which have been subject to such segmental rearrangements may incorporate information about "likeliness" of of each block edit operation occurring at any position. The block edit distance considered in this paper can provide a first approximation to such "weighted block edit distances" similar to what Levenshtein edit distance provided for the "weighted edit distances" used by BLAST and other sequence search and comparison tools.

A. Apostolico and M. Takeda (Eds.): CPM 2002, LNCS 2373, pp. 262–   , 2002.
© Springer-Verlag Berlin Heidelberg 2002

the measure considered in this paper is the block edit distance. This method achieves a preprocessing time and space polynomial in size of D and query time near-linear in size of Q by allowing an approximate factor of $O(\log \ell (\log^* \ell)^2)$. The approach involves embedding sequences into Hamming space so that approximating Hamming distances estimates sequence block edit distances within the approximation ratio above.

In this study we focus on simplification and experimental evaluation of the [ ] method. We first describe how we implement and test the accuracy of the transformations provided in [ ] in terms of estimating the block edit distance under controlled data sets. Then, based on the hamming distance estimator described in [ ] we present a data structure for computing approximate nearest neighbors in hamming space; this is simpler than the well-known ones in [,]. We finally report on how well the combined data structure performs for sequence nearest neighbor search under block edit distance.

1   Introduction

The *sequence nearest neighbor (SNN) problem* is as follows. We are given a database D of sequences for preprocessing; given an on-line query sequence Q, our goal is to return a sequence S in D whose distance $d(S, Q)$ to Q is no more than that of any other sequence T to Q. Distance between two sequences is defined to be the minimum number of edit operations needed to transform one to another (all edit operations of interest are reversible so that $d(S, T) = d(T, S)$ for any two sequences T and S). The nature of SNN problem depends on the edit operations permitted which correspond to the notion of similarity between sequences one may wish to capture for an application. We will come back to this issue later in this section.

Let the number of sequences in D be n and the maximum length of the sequences in D be ℓ. Under nearly any set of edit operations, known algorithms for the SNN problem face the "dimensionality bottleneck" in the worst case; that is, they cannot use subexponential $(o(2^\ell))$ preprocessing cost and yet obtain query time bounds better than that obtained by comparing the query sequence to each in the database (say, sublinear – $o(n\ell)$ – time). Overcoming this dimensionality bottleneck under nontrivial edit operations has been an important open issue in Combinatorial Pattern Matching.

A similar bottleneck exists in the vector nearest neighbor problem (VNN) as well. Here the objects in the database are vectors and the distance between two objects is defined by a vector measure (such as Hamming, Euclidean, or L_∞). Recently, substantial progress was made on these problems: in [,], authors present polynomial time preprocessing algorithms that with near-linear time processing of the query vector, return an "approximate" nearest neighbor under vector distances considered.

Although SNN problems are related to VNN problems (a sequence of length ℓ may be thought of as an ℓ dimensional vector) they offer distinct challenges.

Edit operations allow nontrivial alignments between two sequences (position i in one sequence may align with position $j \neq i$ in the other) while vector distance functions allow only the trivial alignment of the ith position of one vector with the same in the other. In presence of certain "block edit operations" even computing the distance between two sequences can be NP-hard (allowing arbitrary block copies is a sufficient condition for hardness [ ]); in contrast, the vector distance measure can be computed in linear time.

Recently a number of approximation algorithms have been developed for distances allowing a subset of these block edits. For example Bafna and Pevzner [ ] studied the case where each string is a permutation of characters and the only edit operations allowed are transpositions; they provide a constant approximation for this problem. An indexing structure for this measure was later given in [ ]. The restriction on permutations were relaxed by a recent work of Shapira and Storer [ ], where general string distances under transpositions are considered and a logarithmic factor approximation is provided.

Independently, the first progress towards overcoming the dimensionality bottleneck in SNN problem was given in [ ] which describes the first known efficient algorithm for "approximate" nearest neighbor search with preprocessing time and space polynomial in size of D and query time near-linear in size of Q. This result holds for *block edit distance* $d(S,T)$ between two sequences S and T defined to be the the minimum number of single character insertions, deletions, replacements and block (substring) copies, uncopies (deleting one of the two copies of a block), relocations and reversals needed to transform one to the other.

Our Contributions. In this study we focus on simplification and experimental evaluation of the data structure presented in [ ]. The original [ ] data structure is constructed in two steps. The first step transforms each sequence S in D to a binary vector $T(S)$ such that the hamming distance between two such binary vectors provide an $O(\log \ell (\log^* \ell)^2)$ approximation to the block edit distance between their corresponding sequences; this is based on the *distance preserving transformation* provided in [ ]. The second step constructs a data structure for computing an approximate $(1+\epsilon)$ nearest neighbor of $T(Q)$ among the transformed sequences in D under hamming distance; this is based on [,].

The combined data structure "approximately" solves the SNN problem. Let D be a database involving n sequences $S_1, \ldots, S_n$ where $\ell = \max_i |S_i|$ and $|D| = \sum_i |S_i|$. The data structure of [ ] can be constructed in time and space polynomial in $|D|$ and ℓ. A query with sequence Q takes time $O(|Q|\ polylog(n\ell))$ and returns a sequence S in D such that $d(S,Q) = O(\log \ell (\log^* \ell)^2) d(T,Q)$ for any other sequence T in D, with high probability (at least $1 - 1/n$). Here, $d(S,Q)$ denotes the minimum number of character edits (insertions, deletions and replacements) and block edits (relocations, copies, uncopies and reversals) needed to transform S to Q (or vice versa).

In this paper we first describe how we implement and test the accuracy of the *distance preserving transformation* described in [ ] in terms of how well it approximates the block edit distance, under controlled data sets. Then, based on the hamming distance estimator described in in [ ] we present a simpler data structure for computing approximate nearest neighbors in hamming space than those described in [,]. We finally report on how well the combined data structure performs for sequence nearest neighbor search under block edit distance.

Map. In Section , we provide a somewhat detailed motivation for studying block edit distances drawn from processing biological sequences. We describe how to implement two key steps of [ ] in Sections and respectively in a simple way. In Section , we describe our experiments and present some observations.

2   Motivating Block Edit Distances

As mentioned earlier the distance between two sequences is defined as the minimum number of allowed edit operations needed to transform one to another, which are imposed by the specific considerations in the problem domain. In this paper we particularly focus on applications in computational genomics where the similarity between DNA sequences are of key importance. Specifically we focus on sequence similarity measures that can help us to identify mechanisms for structural rearrangement in the human genome, particularly genomic duplications, which provide the key to the understanding of the causes of several *genomic diseases.*

The human genome sequence consists of many segments which are *repeated* along the genome within a *small divergence rate* (i.e. small *normalized edit distance* [ ]). Preliminary studies predict that more that 60% of the human genome is duplicated [ ]. Most of these repeat sequences are *common repeats* [ ] which are usually short and tend to occur several thousands of times throughout the genome (for example, the well known *alu* segment consists of approximately 300 nucleotides and is repeated along the genome for more than 10^6 times within a divergence rate of $5\% - 15\%$). For capturing similarities between these *short* common repeats, distance functions which only allow *single nucleotide/character edit operations* are sufficiently powerful.

As the human genome project gets close to completion, it is becoming apparent that the genome consists of much longer segments comprising portions of genes and even entire gene segments which are duplicated with higher similarity/lower divergence [ ]. Some of those long duplicative segments are known to cause recurrent *structural rearrangements* (i.e. edit operations on whole segments/blocks) in chromosomes, such as segmental *deletions, translocations,* and *reversals,* as well as local *duplications.* A main mechanism that causes such rearrangements (particularly segmental deletions and repeats) is what is known as *unequal crossover* [ ]. Other mechanisms that cause such structural rearrangements include *replication slippage* and *retrotransposition* [ ]. These segmental rearrangements may occur hierarchically and in multiple layers. (For example, a long block of tandemly repeated DNA may have a complete duplicate elsewhere; this may indicate that the tandem duplication event is evolutionarily older than the long segmental duplication). Some of these rearrangements occur frequently, and may result in the deletion or duplication of a developmentally important gene segment, causing a genetic defect [ ]. In order to understand mechanisms underlying genome wide structural rearrangements so as to (i) help diagnosis and treatment of genetic defects, as well as (ii) improve understanding of the evolution of human genome, the sequence similarity/distance measure used should allow edit operations on genome segments/blocks, as well as single nucleotides/characters.

Thus the edit distance we consider in this study involves (1) *character edits* which include inserting or deleting a single character or replacing a single character by another, (2) *block edits* which involve moving a block (any consecutive set of characters) from one location to another or copying blocks from one place to another within a sequence, or reversing an entire block.

This motivation has been articulated for the first time here; even though block edit distances have been studied for quite some time in algorithmic literature, the discussion above motivates the specific block edit operations of interest in Computational Genomics, in particular, the copying or reversing block operations. In this paper, we will not consider the block reversal operations but all our discussions apply equally well to them, see [ ] for details.

In the rest of the paper we denote sequences by $P, Q, R, S..$, integers and binary numbers by $i, j, k..$ and constants by $\alpha, \beta, \gamma...$ All sequences have characters drawn from the DNA alphabet $\sigma = \{a, c, g, t\}$.

3   Implementation of Distance Preserving Transformations

In this section we review how to compute "signatures" of a set of binary sequences to approximate block edit distance [ ]. Given binary sequences S and Q, the [ ] method transforms them to bit strings $T(S)$ and $T(Q)$ which are used as signatures of S and Q. The hamming distance between $T(S)$ and $T(Q)$ approximates the block edit distance between the original sequences S and Q.

To compute the signatures, we first compute what [ ] calls core blocks (substrings) of S and Q. Here we present a simplified version of the core block computation which we employ in our implementations.

(1) **Computing the core blocks.** The core blocks of S are computed hierarchically as follows: At the base level every substring of the form 10^k1 and

---

2   Such defects, usually referred as *genomic disorders*, occur at a rate of approximately 1 in every 1000 births. Several birth defects are a result of these disorders, including *spina bifida* and *cleft lip/cleft palate*, as well as a number of adult diseases such as *cardiovascular disease* and *osteoporosis*.

3   or binary representations of sequences from the four letter DNA alphabet

01^l0 for all $k > 1$, $l > 2$ is a core block. Core blocks at the *next level* are concatenation of a number of successive core blocks at the current level - this is done iteratively for each level through the locally consistent parsing (LCP) (see [,] for details). We describe a simplified, practical implementation of LCP below. The key properties of the core blocks are:

   (i) there are $O(\log |S|)$ levels of core blocks.
  (ii) if V and W are two identical substrings of S, and V is a core block, then W is necessarily a core block as well,
 (iii) core blocks at a given level cover the whole sequence S with the exception of a few characters in the left and right margins, however,
  (iv) core blocks do not partition S as there may be overlaps between successive core blocks.

Each unique core block has a unique label. When processing a set of sequences, identical core-blocks get identical labels no matter in which sequence they occur at.

(2) **Computing signatures.** Consider the set of unique core blocks in all sequences, $\{C_1, C_2, \ldots, C_k\}$. Then the signature $T(S)$ is simply the string of bit flags where the ith bit is 1 if C_i is a core block of S and is 0 if C_i is not a core block of S. For improving space efficiency, we simply represent $T(S)$ as a list labels of all core blocks in S.

The space efficient representation of signature $T(S)$ can be computed in time $O(|S| \log |S|)$, and requires $O(|S| \log |S|)$ space. It is proven in [ ] that the hamming distance between the signatures of two sequences provides an approximation to the block edit distance. More specifically $\Omega(d(Q, S))/\log^*(|Q|+|S|) = h[(T(Q), T(S)] = O(d'(Q, S) \log(|Q| + |S|) \log^*(|Q| + |S|))$ where $h[T(Q), T(S)]$ is the hamming distance between $T(Q)$ and $T(S)$ which is equal to $\sum_i T(Q[i]) \oplus T(S[i])$.

Implementation Issues. Our implementation for computing signatures were done in C++ using STL (standard template library). The implementation requires the input genomic sequences to be added in the database be in the standard FASTA format. Once a sequence from D is read, it is converted to a binary sequence, two bits per character. We relax some of the theoretical conditions on core block computations of [ ] to improve efficiency. In our implementation, base level core blocks are designated to be all one-character sequences; to identify level-1 cores, repeating bits bracketed by opposite bits (e.g. 0110, 10001), are designated with the exception of 101. These cores are stored in a vector with their starting and ending positions in the original binary sequences.

To compute level-i cores from level-$i-1$ cores iteratively, each level-$i-1$ core C is compared to its right neighbor as per LCP [ ] and $p(C)$, the least significant bit position at which the C differs from its right neighbor, is computed. Then $p(C)$ is concatenated with the value of the $p(C)^{th}$ bit of C which gives $tag(C)$. If in the sequence of tags, C's tag is a local maxima, C is concatenated with two the two neighboring cores to its left and the two neighboring cores to its right (after their overlap is eliminated). This gives a new core in level-i which is added to the list of cores in level-i. The core computation is stopped once less than 5 cores remain at a level. After all cores of each of the sequences in the database are computed, they are sorted in lexicographic order, while multiple occurrences of the identical cores are eliminated.

Each core is labeled with its lexicographic ordering in the list of unique cores. Finally for each sequence S its signature $T(S)$ is computed as the list of core labels that are present in S, sorted in lexicographically.

4   Nearest Neighbor Search under Hamming Metric

In this section we describe a randomized data structure for approximate nearest neighbor search in the hamming space, simplifying the results of [ ] and [ ]. Given a set D of n bit vectors we first describe a random mapping $m : \{0,1\}^{\ell} \rightarrow \{0,1\}^{\log_{\beta} \ell}$ for some small constant $\beta > 1$ so that given $X, Y \in \{0,1\}^{\ell}$, one can estimate $h(X,Y)$ within a factor of β through $m(X)$ and $m(Y)$, following [ ].

For any X in D, a random mapping $m(X)$ is computed in ℓ iterations. In the ith iteration we uniformly and independently pick a sequence of β^i locations $r_1, \ldots, r_{\beta^i}$ in X. We hash the sequence of bit values specified by this sample, $\hat{X}_i = X[r_1], \ldots, X[r_{\beta^i}]$, to $m_i(X) = \bigoplus_{j=1,\ldots,\beta^i} X[r_j]$. The mapping of X is then simply $m(X) = m_1(X), \ldots, m_{\log_{\beta} \ell}(X)$.

Lemma 1. *It is possible to compute an estimate $\hat{h}(X,Y)$ of $h(X,Y)$ as a function of the longest common prefix of $m(X)$ and $m(Y)$ such that (1) the probability that $\hat{h}(X,Y) \leq \beta h(X,Y)$ is at most a user specified parameter ϵ, provided $\frac{1}{\epsilon}\frac{\beta-1}{} \leq \frac{\sqrt{5}-1}{2}$ (2) the probability that $h(X,Y) \leq \alpha \hat{h}(X,Y)$ for $\alpha = \frac{3\beta \ln 1/\epsilon}{2 \ln 2} \geq 1$. is at most $1/2$.*

Proof. The proof is mostly provided in [ ]. We only highlight some of the main points here. Observe that $m_i(X) \oplus m_i(Y) = \bigoplus_{j=1,\ldots,\beta^i} X[r_j] \oplus Y[r_j]$. Therefore $\Pr[m_i(X) \oplus m_i(Y) = 1]$ is equal to the probability of getting an odd parity out of β^i random i.i.d. bits, each of which has value 1 with probability $p = h(X,Y)/\ell$. This can easily be bounded as follows. Let $b_1, \ldots, b_k$ be independent Boolean random variables with expectation p. Then one can easily verify that $\Pr[\sum b_i$ is even $] = (1 + (1 - 2p)^k)/2$. Denote by $C(k,p) = (1 + (1 - 2p)^k)/2$. Then, for all $k \geq 2$, and $0 \leq p \leq 1$: (i) $e^{-kp} \leq C(k,p) \leq (1 + e^{-2kp})/2$; (ii) if $e^{-\frac{2}{3}kp} \geq \frac{\sqrt{5}-1}{2}$, i.e. $kp \leq 0.7218\ldots$, then $C(k,p) \leq e^{-\frac{2}{3}kp}$.

Now consider a sequence of k bits where the ith bit represents the parity of β^i i.i.d. coin flips each with bias p. Then $Z(k,p,\beta)$, the probability that all of the k bits are zero, is $\prod_{i \geq 0}^{u} C(k/\beta^i, p)$, where $u = \log_{\beta} k$. It is not difficult to check

---

[4] Note that these results do not work only for the Hamming space but also for Euclidean spaces; our simplification is only for the Hamming space with a relaxed set of constraints.

that $Z(k, p, \beta) \geq e^{-kp/(1-1/\beta)}$; and $Z(k, p, \beta) \leq e^{-\frac{2}{3}kp(1-1/\beta)}$ provided $e^{-\frac{2}{3}kp} \geq \frac{\sqrt{5}-1}{2}$.

Let $k = |pref(m(Y), m(Y))|$, the size of the longest common prefix of X and Y. Then for any given value of $\epsilon > 0$, we can estimate $p\ (= h(X, Y)/\ell)$ by choosing $\hat{p}$ so that $Z(k/\beta, \hat{p}, \beta) \leq \epsilon$, since the probability of not getting odd parity before k is at most ϵ if $p \geq \hat{p}$. We can satisfy $Z(k/\beta, \hat{p}, \beta) \leq \epsilon$ by setting $e^{\frac{-2k\hat{p}}{3(\beta-1)}} = \epsilon$, i.e. by setting $\hat{p} = \frac{3(\beta-1)\ln 1/\epsilon}{2k}$, and ensuring $e^{\frac{-2k\hat{p}}{3}} \geq \frac{\sqrt{5}-1}{2}$. This holds provided $\beta \leq 1 + \frac{1}{\ln 1/\epsilon} \ln(\frac{\sqrt{5}-1}{2})$.

Thus given $k = |pref(m(X), m(Y))|$, one can observe that $\hat{h}(X, Y) = \hat{p} \cdot \ell = \ell \cdot \frac{3(\beta-1)\ln 1/\epsilon}{2k} \leq \beta h(X, Y)$ with probability $\leq \epsilon$. It is not difficult to show that $\Pr[\hat{h}(X, Y) \geq \alpha h(X, Y)] \leq 1/2$, for $\alpha \geq \frac{3\beta \ln 1/\epsilon}{2\ln 2} \geq 1$ (see [ ] for details).

Thus given sequences X, Y and a query sequence Q where $h(Q, X) \leq \alpha h(Q, Y)$ (so that Q is closer to X than Y) for which $\alpha \geq \frac{3\beta \ln 1/\epsilon}{2\ln 2} \geq 1$ and $\frac{1}{\epsilon}^{\beta-1} \leq \frac{\sqrt{5}-1}{2}$, one can observe that $\Pr[|pref(m(Q), m(Y))| > |pref(m(Q), m(X))|] = \Pr[\hat{h}(Q, Y) < \hat{h}(Q, X)]$ is upper bounded by $(1 + \epsilon)/2$. Thus the probability that some $\log n$ independent random mappings $m_1, \ldots m_{\log n}$ all estimate Q to be closer to Y is $O(1/n)$. It is trivial to verify that given n sequences $Y_1, \ldots, Y_n$ for which $h(Q, X) \leq \alpha h(Q, Y)$ the probability that some $\log^2 n$ independent random mappings all fail to estimate that X is closer to Q is $O(1/n)$ as well.

Our data structure for approximate nearest neighbors thus consists of $O(\log^2 n)$ binary tries T_i each composed of an independent random mapping $m_i(X)$ of each of the bit vectors X in D. One can compute the mapping $m_i(X)$ in $O(\ell)$ time, thus the time needed for computing the data structure is simply $O(|D| \cdot \log^2 n)$.

To search for a query sequence Q in D, we simply find in each trie T_i one sequence S_i whose corresponding mapping has the longest common prefix with that of Q; this can be done in an overall time of $O(\ell \log^2 n)$. Testing which of these $\log^2 n$ sequences is a "closer" match to Q can again be performed in $O(\ell \log^2 n)$ time. Notice that this will again be done by computing the hamming distance between the transformed sequences $T(S_i)$ with $T(Q)$; which will only give an approximate answer to the SNN query.

5   Experimental Results

We used a variety of data sets with multiple settings to verify the practical value of our implementations. There are two key issues that we test in our experiments. (i) The accuracy of the approximate block edit distance computation: the worst case upper bound is $O(\log \ell (\log^* \ell)^2)$; we would like to test the approximation ratio in an experimental setting. (ii) Our data structure returns a number of nearest neighbor candidates for each query sequence Q, depending on the number of tries employed. A measure of success for a data structure of this nature is the ratio of the number of sequences returned by the data structure and the total number of sequences in the database. We first determine the number of tries needed to "guarantee" that the data structure returns the correct nearest neighbor among the candidates (in order to avoid "false negatives"). Then under the settings imposed by these figures, we determine the average and maximum number of candidate sequences returned by the data structure for a given query sequence.

To test our methods we randomly generate sets of sequences in which we control the block edit distances. Key parameters in our data sets are (i) the size of the sequences, (ii) the number of sequences, (iii) the type of edit operations involved (we also test how well our data structure approximates the Levenshtein [ ] edit distance), (iv) the distribution of the sequences in sequence space, i.e. the distances of the sequences to a given query sequence, (v) the locations of edit operations, (vi) the size of the block operations.

To generate each data set we start with a single sequence (of some specified length) which is later used as the query sequence. This query sequence is generated by picking for each position a character from the DNA alphabet "uniformly at random". All of the other sequences for a particular data set are generated as "mutations" of the query sequence, with mutations following a prescribed model. We start by applying a small number of edit operations on the query sequence to generate its nearest neighbor. We increase the number of edit operations (linearly or geometrically) to generate more sequences in the data set. More specifics of the data set generation is provided below.

5.1   Data Sets

Random sequences under Levenshtein Edit Distance. This data set is intended to help measure how well the transformations of [ ] approximate the Levenshtein edit distance in the absence of block edit operations when creating one sequence from the other. This scenario is inspired by measuring the evolutionary relationship between shorter duplicated segments in the genome on which only point mutations apply during evolution. To model this "molecular clock" phenomena, we generate sets of sequences with uniformly random character edit operations at uniformly random locations. Each data set is generated by first creating a *query sequence* that will serve as the template for all mutated sequences. The lengths of these sequences chosen in our experiments were 1000 and 10000 characters. Other sequences in the data set were created by performing a specified number of random edit operations to the query sequence. Each edit operation were chosen uniformly at random among the three character edit operations, namely character insertion, deletion or replacement. The location of each edit operation and the character to be used in case the edit operation is an insertion or a replacement, was also chosen uniformly at random.

---

[5] We use UNIX rand() for pseudo-random number generation throughout this study.

The prescribed distance from the query sequence grows either linearly or geometrically. For the linear rate, the number of mutations was chosen to start at 5 (the closest to the original sequence). For query sequences of length 1000, the growth rate was by 5 for each successive mutated sequence (i.e. 5, 10, 15, ...), and for query sequences of length 10000, the mutations increased by 50 for each successive sequence. For geometrical growth rate, the number of edit operations applied for each successive sequence was either doubled or increased by a factor of 1.2.

Random sequences under Block Edit Distance. While character edit operations represent point mutations, block edit operations represent genome wide segmental rearrangements. We emulate this phenomena by adding two operations to the character edit model: block relocation and block copy. In our data sets, we pick each edit operation uniformly at random among character edits and block edits and further among the specific edit operation in the selected category. The location to which the edit operation is applied is also selected uniformly at random. The size of the block to be edited is determined through binomial distribution; i.e. for each block, we start with a single character and iteratively extend it by one character with probability of 2^{-i}; we generated data sets with $i = 1, \ldots, 5$ to verify how the block size affects the accuracy of the transformations.

5.2   Tests on Data Sets

Accuracy of the transformations. We compare the estimated block edit distance with the number of edit operations applied, which provides a good approximation to the block edit distance. The lengths of the original sequences, the rates of growth of the number of edit operations, and the number of derived sequences were all chosen so that the chance of an overlap between operations were relatively low. A good measure of accuracy is how well signatures preserve the order of the sequences in the data set in terms of their original and estimated distances to the query sequence which we demonstrate for each data set in the tables below.

We start by interpreting the results for sequences with a linear growth factor with lengths 1Kb and 10Kb. In both cases, the ordering results matched that of the correct order well, although the 10Kb sequences were somewhat more accurate. For the 1Kb sequences, since the growth factor was only 5, the sequence differences could be quite small, as opposed to the longer 10Kb sequences with a growth factor of 50. By examining the actual values of the linearly increasing generated sequences, we can infer some more intuition about each distance computation. (1) The hamming distance between signatures clearly exhibit a tendency to "level off" as the sequence size increases, rather than increasing linearly with the number of actual mutations. This should be at least in part a result of the clustering effect of our method: When the edit operations are sparse each such operation will tend to modify some logarithmic number of cores When the edit operations are dense then several closely located operations will cluster in the modification of high level cores resulting in somewhat lesser number of modifications *per edit*. (2) Also apparent is the dependence of the hamming distances between transformations on the average size of the edited blocks.

We also make a few observations on the data set with geometrical growth rates. Here we can see how the larger difference between each sequence contributes to perfect orderings for the 1Kb length sequences, and near perfect for the 10Kb length sequences. With the 10Kb sequences, the orderings for the first few sequences generated from the query sequence have incorrect order primarily because the growth rate was small (1.2), thus the sequences did not differ by more than 15 mutations (still relatively close) until 12 successive sequences had been generated from the original (i.e. the first few sequences are all very close to each other and original).

Effectiveness of the tries. We tested the tries of transformed sequences on larger data sets. Our first goal was to determine the "typical" number of tries needed to ensure that the correct nearest neighbor was one of the sequences returned for a given query. We used three data sets in which the size of the sequences were approximately 250 characters and the number of sequences in the data sets were 111, 251, and 2551 respectively. A linear growth in the induced block edit distance from 5 to 45 with a factor of 3 were used with increasing number of sequences within each group.

We also tested multiple sampling factor growth rates (i.e. β values) for creating fingerprints, factors of 1.2, 1.5, and 2 were tested with each of the data sets. The results are quite as expected as demonstrated in the last three tables: As the data set size increases one needs to decrease the sampling factor growth rate. For data set size 111, the most effective sampling factor growth rate is 1.5 both in terms of the number of tries to be examined and the number of unique matches. For data set size 311, the most effective sampling factor growth rate becomes 1.2 in terms of both measures for most of the runs. For data set size 2551, the sampling factor growth rate of 1.2 clearly outperforms the other two. We leave the determination of the best sampling factor growth rate as a function of data size to further studies.

6   Concluding Remarks

We have shown simple, effective ways to implement the sequence nearest neighbors result from [ ]. Our experimental study was motivated by processing genomic sequences, and we hope to further strengthen our implementations.

Acknowledgements

We gratefully acknowledge the contributions of Greg Cox and Jai Macker in the programming and testing of the algorithms described in the paper.

Table 1. Signature computation for psuedo random strings, size 1k, with character edit distance from query increasing linearly

"Actual"		Estimated (Run 1)		Estimated (Run 2)		Estimated (Run 3)	
Order	Edit Distance	Order	Hamming Dist	Order	Hamming Dist	Order	Hamming Dist
1	5	1	99	1	71	1	82
2	10	3	146	2	137	2	146
3	15	2	148	3	191	3	207
4	20	4	243	4	252	4	264
5	25	5	295	5	307	5	296
6	30	6	318	6	310	6	330
7	35	7	325	8	366	7	355
8	40	8	374	7	369	8	379
9	45	9	431	9	403	9	407
10	50	12	442	10	414	10	444
11	55	10	448	11	432	12	473
12	60	11	459	14	467	11	500
13	65	14	505	12	468	13	524
14	70	13	520	13	520	14	556
15	75	16	533	15	523	15	558
16	80	15	557	16	532	19	578
17	85	17	559	17	551	18	595
18	90	18	574	18	559	17	596
19	95	19	596	20	586	20	602
20	100	20	601	19	616	16	609

Table 2. Signature computation for psuedo random strings, size 1k, with character edit distance from query increasing linearly

"Actual"		Estimated (Run 1)		Estimated (Run 2)		Estimated (Run 3)	
Order	Edit Distance	Order	Hamming Dist	Order	Hamming Dist	Order	Hamming Dist
1	5	1	99	1	71	1	82
2	10	3	146	2	137	2	146
3	15	2	148	3	191	3	207
4	20	4	243	4	252	4	264
5	25	5	295	5	307	5	296
6	30	6	318	6	310	6	330
7	35	7	325	8	366	7	355
8	40	8	374	7	369	8	379
9	45	9	431	9	403	9	407
10	50	12	442	10	414	10	444
11	55	10	448	11	432	12	473
12	60	11	459	14	467	11	500
13	65	14	505	12	468	13	524
14	70	13	520	13	520	14	556
15	75	16	533	15	523	15	558
16	80	15	557	16	532	19	578
17	85	17	559	17	551	18	595
18	90	18	574	18	559	17	596
19	95	19	596	20	586	20	602
20	100	20	601	19	616	16	609

Table 3. Signature computation for psuedo random strings, size 1k, with block edit distance from query increasing linearly. BED is the induced Block Edit Distance; HD is the observed Hamming Distance from signature comparison. The run i determines for block operations the chance of extending a block by $(1 - 1/2^i)$ chance of another character being added to the block

"Actual"		Estimated (Run i = 1)		Estimated (Run i = 2)		Estimated (Run i = 3)		Estimated (Run i = 4)		Estimated (Run i = 5)	
Order	BED	Order	HD	Order	HD	Order	HD	Order	HD	Order	HD
1	5	1	71	1	95	1	104	1	110	1	128
2	10	2	151	3	220	2	173	2	208	2	175
3	15	3	214	2	224	3	250	3	239	3	314
4	20	4	290	4	298	4	250	4	283	4	340
5	25	6	350	5	316	5	322	5	352	5	354
6	30	5	364	6	392	6	359	6	364	6	443
7	35	7	403	8	409	7	440	7	371	7	475
8	40	8	408	7	424	8	456	9	468	8	530
9	45	9	434	9	477	9	460	8	481	9	540
10	50	10	471	10	504	10	498	10	543	10	572
11	55	11	518	11	534	12	557	12	549	11	614
12	60	12	533	13	568	11	564	11	558	12	646
13	65	15	550	15	579	14	585	14	571	13	670
14	70	13	566	14	580	13	586	15	611	14	699
15	75	14	597	12	587	16	598	13	613	16	705
16	80	17	619	18	654	15	608	16	638	17	715
17	85	20	632	16	655	17	683	17	667	18	750
18	90	16	648	17	660	18	691	18	691	15	755
19	95	18	657	20	683	20	696	20	693	19	805
20	100	19	659	19	699	19	731	19	715	20	814

Table 4. Signature computation for psuedo random strings, size 10k, with block edit distance from query increasing linearly. BED is the induced Block Edit Distance; HD is the observed Hamming Distance from signature comparison. The run i determines for block operations the chance of extending a block by $(1 - 1/2^i)$ chance of another character being added to the block

"Actual"		Estimated (Run i = 1)		Estimated (Run i = 2)		Estimated (Run i = 3)		Estimated (Run i = 4)		Estimated (Run i = 5)	
Order	BED	Order	HD	Order	HD	Order	HD	Order	HD	Order	HD
1	50	1	1032	1	998	1	980	1	1066	1	1159
2	100	2	1612	2	1640	2	1716	2	1747	2	1797
3	150	3	2044	3	2382	3	2423	3	2372	3	2435
4	200	4	2680	4	2618	4	2703	4	2835	4	2867
5	250	5	2823	5	3008	5	3011	5	3165	5	3404
6	300	6	3161	6	3256	6	3431	6	3624	6	3809
7	350	7	3476	7	3659	7	3761	7	3827	7	3935
8	400	8	3698	8	4056	8	4063	8	4039	8	4124
9	450	9	4034	9	4091	9	4285	9	4462	9	4561
10	500	10	4233	10	4343	10	4512	10	4585	10	4727
11	550	11	4309	11	4614	11	4663	11	4758	11	4912
12	600	12	4390	12	4672	13	4905	13	5038	12	5251
13	650	13	4661	13	4810	12	5050	12	5053	13	5431
14	700	15	4920	15	5060	14	5288	15	5371	14	5661
15	750	14	4936	14	5082	15	5299	14	5402	15	5724
16	800	16	5101	16	5277	16	5544	16	5626	16	5865
17	850	17	5286	17	5394	17	5601	17	5850	18	6112
18	900	18	5358	18	5430	18	5745	18	5913	17	6245
19	950	19	5406	20	5560	19	5810	19	5980	19	6268
20	1000	20	5502	19	5654	20	5858	20	6093	20	6387

Table 5. Signature computation for psuedo random strings, size 10k, with block edit distance from query increasing linearly. BED is the induced Block Edit Distance; HD is the observed Hamming Distance from signature comparison. The run i determines for block operations the chance of extending a block by $(1 - 1/2^i)$ chance of another character being added to the block

"Actual"		Estimated (Run i = 1)		Estimated (Run i = 2)		Estimated (Run i = 3)		Estimated (Run i = 4)		Estimated (Run i = 5)	
Order	BED	Order	HD	Order	HD	Order	HD	Order	HD	Order	HD
1	50	1	1032	1	998	1	980	1	1066	1	1159
2	100	2	1612	2	1640	2	1716	2	1747	2	1797
3	150	3	2044	3	2382	3	2423	3	2372	3	2435
4	200	4	2680	4	2618	4	2703	4	2835	4	2867
5	250	5	2823	5	3008	5	3011	5	3165	5	3404
6	300	6	3161	6	3256	6	3431	6	3624	6	3809
7	350	7	3476	7	3659	7	3761	7	3827	7	3935
8	400	8	3698	8	4056	8	4063	8	4039	8	4124
9	450	9	4034	9	4091	9	4285	9	4462	9	4561
10	500	10	4233	10	4343	10	4512	10	4585	10	4727
11	550	11	4309	11	4614	11	4663	11	4758	11	4912
12	600	12	4390	12	4672	13	4905	12	5038	12	5251
13	650	13	4661	13	4810	12	5050	12	5053	13	5431
14	700	15	4920	15	5060	14	5288	15	5371	14	5661
15	750	14	4936	14	5082	15	5299	14	5402	15	5724
16	800	16	5101	16	5277	16	5544	16	5626	16	5865
17	850	17	5286	17	5394	17	5601	17	5850	18	6112
18	900	18	5358	18	5430	18	5745	18	5913	17	6245
19	950	19	5406	20	5560	19	5810	19	5980	19	6268
20	1000	20	5502	19	5654	20	5858	20	6093	20	6387

Table 6. Signature computation for psuedo random strings, size 10k, with block edit distance from query increasing exponentially by a factor of 1.2. BED is the induced Block Edit Distance; HD is the observed Hamming Distance from signature comparison. The run i determines for block operations the chance of extending a block by $(1 - 1/2^i)$ chance of another character being added to the block. Note that the starting BED from the query is 10 due to the fact that the sampling rate growth factor of 1.2 does not allow us to differentiate distances smaller than 0.1% of the sequence size

"Actual"		Estimated (Run i = 1)		Estimated (Run i = 2)		Estimated (Run i = 3)		Estimated (Run i = 4)		Estimated (Run i = 5)	
Order	BED	Order	HD	Order	HD	Order	HD	Order	HD	Order	HD
10	10	10	266	11	255	10	284	10	233	10	253
11	12	12	329	10	277	11	357	11	312	11	343
12	15	13	335	12	379	12	371	12	347	12	450
13	18	11	352	13	411	13	454	13	395	13	458
14	21	14	438	14	529	14	553	14	644	14	459
15	26	16	635	15	571	15	627	15	648	15	648
16	31	15	700	16	598	16	740	16	673	16	784
17	37	17	710	17	842	17	829	17	763	17	855
18	44	18	840	18	946	18	981	18	960	18	958
19	53	19	1064	19	1151	19	1091	19	1129	19	1189
20	64	20	1125	20	1210	20	1201	20	1224	20	1307
21	77	21	1304	21	1265	21	1481	21	1457	21	1653
22	92	22	1543	22	1510	22	1606	22	1705	22	1715
23	110	23	1734	23	1834	23	1916	23	1886	23	1874
24	132	24	1821	24	2001	24	2104	24	2166	24	2414
25	159	25	2303	25	2216	25	2355	25	2423	25	2536
26	191	26	2403	26	2469	26	2754	26	2662	26	2664
27	229	27	2875	27	2874	27	3041	27	3062	27	2988
28	275	28	3067	28	3167	28	3359	28	3267	28	3477
29	330	29	3531	29	3474	29	3714	29	3599	29	3868
30	396	30	3832	30	3881	30	4202	30	4181	30	4251
31	475	31	4123	31	4244	31	4423	31	4497	31	4601
32	570	32	4454	32	4577	32	4766	32	4830	32	5143
33	684	33	4718	33	4969	33	5162	33	5317	33	5494
34	820	34	5121	34	5218	34	5579	34	5594	34	6091
35	984	35	5448	35	5551	35	5837	35	6033	35	6249

Table 7. Testing SNN I

Run	Sampling Factor					
	1.2		1.5		2	
	Tries	Unique Matches	Tries	Unique Matches	Tries	Unique Matches
1	11	12	4	4	5	16
2	4	6	2	2	16	21
3	2	3	3	4	2	4
4	7	7	2	5	22	35
5	1	2	6	9	21	31
6	12	12	5	9	11	10
7	5	5	5	10	21	37
8	3	4	1	1	7	10
9	1	18	4	7	26	31
10	1	2	2	3	31	36
Average	4.7	7.1	3.4	5.4	16.2	23.1
Maximum	12	18	6	10	31	37
Avg % Seq Elim	93.60%		95.10%		79.20%	
Worst % Seq Elim	83.70%		90.90%		66.70%	

This table summarizes the results for SNN search tests on a data set of 111 sequences. The results for each of the ten runs at each of the sampling factors are shown. The numbers shown for the run are the total number of tries it took to find the correct nearest neighbor in that run and the unique matches found in that run that must be further examined

Table 8. SNN Tests II

Run	Sampling Factor					
	1.2		1.5		2	
	Tries	Unique Matches	Tries	Unique Matches	Tries	Unique Matches
1	3	4	3	8	7	13
2	18	27	49	60	1	6
3	1	1	11	23	30	42
4	9	12	5	5	15	26
5	21	25	22	33	45	56
6	2	2	65	73	73	88
7	12	17	6	7	1	1
8	1	20	3	7	91	118
9	17	27	6	7	56	68
10	3	4	3	5	23	40
Average	8.7	13.9	17.3	22.8	34.2	45.8
Maximum	21	27	65	73	91	118
Avg % Seq Elim	95.50%		92.40%		84.70%	
Worst % Seq Elim	91%		75.70%		60.70%	

This table summarizes the results from SNN search tests on a data set of 301 sequences. The results for each of the ten runs at each of the sampling factors are shown. The numbers shown for the run are the total number of tries it took to find the nearest neighbor in that run and the unique matches found in that run that must be further examined

Table 9. SNN Tests III

| | Sampling Factor | | | | | |
| | 1.2 | | 1.5 | | 2 | |
Run	Tries	Unique Matches	Tries	Unique Matches	Tries	Unique Matches
1	5	6	42	62	16	54
2	3	4	30	50	11	30
3	9	12	46	69	26	81
4	7	9	15	23	60	160
5	2	2	4	6	17	55
6	2	2	3	3	17	79
7	6	9	4	5	5	21
8	11	13	47	63	6	25
9	1	1	6	7	210	620
10	4	6	28	44	16	62
Average	5	6.4	22.5	33.2	38.4	119
Maximum	11	13	47	69	210	620
Avg % Seq Elim	94.2%		88.9%		95.3%	
Worst % Seq Elim	88.2%		77.0%		75.7%	

This table summarizes the results from the SNN search tests on a data set of 2551 sequences. The results for each of the ten runs at each of the sampling factors is shown. The numbers shown for the run are the total number of tries it took to find the nearest neighbor in that run and the unique matches found in that run that must be further examined

References

1. A. N. Arslan, O. Egecioglu, P. A. Pevzner *A new approach to sequence comparison: normalized sequence alignment, Proceedings of RECOMB 2001*.
2. Bailey J.A., Yavor A.M., Massa H.F., Trask B.J., Eichler E.E., *Segmental duplications: organization and impact within the curren t human genome project assembly, Genome Research* 11(6), Jun 2001.
3. G. Cormode, M. Paterson, S. C. Sahinalp and U. Vishkin. Communication Complexity of Document Exchange. *Proc. ACM-SIAM Symp. on Discrete Algorithms*, 2000.   ,   ,   ,   ,
4. G. Cormode, S. Muthukrishnan, S. C. Sahinalp. Permutation editing and matching via Embeddings. *Proc. ICALP*, 2001.
5. Feng D.F., Doolittle R.F., *Progressive sequence alignment as a prerequisite to correct phylogenetic trees, J Mol Evol.* 1987;25(4):351-60.
6. P. Indyk and R. Motwani. Approximate Nearest Neighbors: Towards Remving the Curse of Dimensionality. *Proc. ACM Symp. on Theory of Computing*, 1998, 604–613.   ,   ,   ,
7. Jackson, Strachan, Dover, *Human Genome Evolution, Bios Scientific Publishers*, 1996.
8. Y. Ji, E. E. Eichler, S. Schwartz, R. D. Nicholls, *Structure of Chromosomal Duplications and their Role in Mediating Human Genomic Disorders, Genome Research* 10, 2000.
9. E. Kushilevitz, R. Ostrovsky and Y. Rabani. Efficient search for approximate nearest neighbor in high dimensional spaces. *Proc. ACM Symposium on Theory of Computing*, 1998, 614–623.   ,   ,   ,
10. D. Lopresti and A. Tomkins. Block edit models for approximate string matching. *Theoretical Computer Science*, 1996.
11. S. Muthukrishnan and S. C. Sahinalp, *Approximate nearest neighbors and sequence comparison with block operations Proc. ACM Symposium on Theory of Computing*, 2000.   ,   ,   ,   ,   ,   ,   ,

12. V. I. Levenshtein, Binary codes capable of correcting deletions, insertions and reversals, *Cybernetics and Control Theory*, 10(8):707-710, 1966.
13. V. Bafna, P. A. Pevzner, Sorting by transpositions. *SIAM J. Discrete Math*, 11, 224-240, 1998.
14. D. Shapira and J. Storer, Edit distance with move operations,t Proceedings of *CPM*, (2002).
15. S. C. Sahinalp and U. Vishkin, Approximate and Dynamic Matching of Patterns Using a Labeling Paradigm, Proceedings of *IEEE Symposium on Foundations of Computer Science*, (1996).
16. George P. Smith *Evolution of Repeated DNA Sequences by Unequal Crossover*, *Science*, vol 191, pp 528–535.
17. J. D. Thompson, D. G. Higgins, T. J. Gibson, *Clustal-W: improving the sensitivity of progressive multiple sequence alignment through sequence weighting, position specific gap penalties and weight matrix choice*, *Nucleic Acid Research* 1994, Vol. 22, No. 22.
18. L. Wang and T. Jiang, *On the complexity of multiple sequence alignment*, *Journal of Computational Biology*, 1:337–348, 1994.
19. C. Venter et. al., *The sequence of the human genome*, *Science*, 16:291, Feb 2001.

Constructing NFAs by Optimal Use of Positions in Regular Expressions

Lucian Ilie* and Sheng Yu**

Department of Computer Science, University of Western Ontario
N6A 5B7, London, Ontario, CANADA
ilie,syu@csd.uwo.ca

Abstract. We give two new algorithms for constructing small nondeterministic finite automata (NFA) from regular expressions. The first constructs NFAs with ε-moves (εNFA) which are smaller than all the other εNFAs obtained by similar constructions. Their size is at most $\frac{3}{2}|\alpha| + \frac{5}{2}$, where α is the regular expression. The second constructs NFAs. It uses ε-elimination in the εNFAs we just introduced and builds a quotient of the well-known position automaton. Our NFA is always smaller and faster to compute than the position automaton. It uses optimally the information from the positions of a regular expression.

Keywords: regular expressions, nondeterministic finite automata, regular expression matching, algorithms, positions, partial derivatives, quotients, right-invariant equivalence, ε-elimination

1 Introduction

The importance of regular expressions for applications is well known. They describe lexical tokens for syntactic specifications and textual patterns in text manipulation systems. Regular expressions have become the basis of standard utilities such as scanner generators (lex), editors (emacs, vi), or programming languages (perl, awk), see [,   ]. While regular expressions provide an appropriate notation for regular languages, their implementation is done using finite automata. The size of the automata is crucial for the efficiency of the algorithms using them; e.g., for regular expression matching. Since the deterministic finite automata obtained from regular expressions can be exponentially larger in size, in many cases nondeterministic finite automata (NFA) are used instead. Minimization of NFAs is PSPACE-complete, see [   ], so other methods need to be used to obtain small NFAs. Probably the most famous such constructions are the ones of Thompson [   ] which builds an NFA with ε moves (εNFA) and the one of Glushkov and McNaughton-Yamada [,   ] which outputs an NFA (called position automaton). While Thompson's automaton has linear size (in terms of the size of the regular expression), the position automaton has

---

* Research partially supported by NSERC grant R3143A01.
** Research partially supported by NSERC grant OGP0041630.

A. Apostolico and M. Takeda (Eds.): CPM 2002, LNCS 2373, pp. 279–   , 2002.
© Springer-Verlag Berlin Heidelberg 2002

size between linear and quadratic and can be computed in quadratic time, as proved by Brügemann-Klein [   ]. We notice that throughout the paper the size of automata will include both transitions and states.

Antimirov [   ] generalized Brozozowski's derivatives and built the partial derivative automata. Champarnaud and Ziadi [,   ] improved very much Antimirov's $\mathcal{O}(n^5)$ algorithm for the construction of such NFA; their algorithm runs in quadratic time. They proved also that the partial derivative automaton is a quotient of the position automaton and so it is always smaller.

The best worst case is in the construction of Hromkovič et al. [   ]; their NFA has size at most $\mathcal{O}(n(\log n)^2)$ and, by the algorithm of Hagenah and Muscholl [   ], it can be computed in time $\mathcal{O}(n(\log n)^2)$. However, the number of states is artificially increased to reduce the number of transitions.

In spite of these improvements, the position automaton is still very much used probably because of its simplicity, but maybe also because none of the proposed constructions proved to be clearly better. In this paper, we propose new algorithms to construct very small nondeterministic finite automata, with or without ε-moves, from regular expressions. Our NFAs are shown to be better in all respects than the position automata.

Our first algorithm constructs εNFAs which are smaller than all the others obtained by similar constructions; e.g., the one of Thompson [   ] or the one of Sippu and Soisalon-Soininen [   ] (which builds a smaller εNFA than Thompson's). Given a regular expression α, the size of our εNFA for α is at most $\frac{3}{2}|\alpha| + \frac{5}{2}$. This is very close to the optimal; we prove a lower bound of $\frac{4}{3}|\alpha| + \frac{5}{2}$.

We give then a method for constructing NFAs. It uses ε-elimination in the εNFA newly introduced. Although the construction of this NFA has, apparently, nothing to do with positions, it turns out, unexpectedly, that the NFA is a quotient of the position automaton with respect to the equivalence given by the follow relation; therefore giving the name of follow automaton. The follow automaton is always smaller and faster to compute than the position automaton. Therefore, it is a very good candidate for replacing the position automaton in applications. Moreover, it uses optimally the information from the positions of the regular expression and thus it cannot be improved this way. It is in general uncomparable with the partial derivative automaton. As a by-product of our construction of the follow εNFA, we obtain also a new algorithm to compute the transition mapping of the position automaton. The new algorithm has the same worst case as the one of Brügemann-Klein [   ] but it is usually faster.

Due to limited space, all proofs are omitted.

2   Regular Expressions and Automata

We recall here the basic definitions we need throughout the paper. For further details we refer to [   ] or [   ].

Let A be an alphabet and A^* the set of all words over A; ε denotes the empty word and the length of a word w is denoted $|w|$. A *language* over A is a subset of A^*. A *regular expression* over A is $\emptyset$, ε, or $a \in A$, or is obtained from these applying the following rules finitely many times: for two regular expressions α and β, the *union*, $\alpha + \beta$, the *catenation*, $\alpha \cdot \beta$, and the *star*, α^* are regular expressions. The regular language denoted by a regular expression α is $L(\alpha)$. Also, we define $\varepsilon(\alpha)$ to be ε if $\varepsilon \in L(\alpha)$ and $\emptyset$ otherwise. The *size* of α is denoted $|\alpha|$ and represents the number of symbols in α when written in postfix (parentheses are not counted).

A *finite automaton* is a quintuple

$$M = (Q, A, q_0, \delta, F),$$

where Q is the set of states, A is the input alphabet, $q_0 \in Q$ is the initial state, $F \subseteq Q$ is the set of final states, and $\delta \subseteq Q \times (A \cup \{\varepsilon\}) \times Q$ is the transition mapping; we shall denote, for $p \in Q, a \in A \cup \{\varepsilon\}$, $\delta(p, a) = \{q \in Q \mid (p, a, q) \in \delta\}$. The automaton M is called
   - *deterministic* (DFA) if $\delta : Q \times A \to Q$ is a (partial) function,
   - *nondeterministic* (NFA) if $\delta \subseteq Q \times A \times Q$, and
   - *nondeterministic with ε-moves* (εNFA) if there are no restrictions on δ.

The language recognized by M is denoted $L(M)$. The *size* of a finite automaton M is $|M| = |Q| + |\delta|$; we count both states and transitions.

Let $\equiv \subseteq Q \times Q$ be an equivalence relation. For $q \in Q$, $[q]_\equiv$ denotes the equivalence class of q w.r.t. $\equiv$ and, for $S \subseteq Q$, $S/_\equiv$ denotes the quotient set $S/_\equiv = \{[q]_\equiv \mid q \in S\}$. We say that $\equiv$ is *right invariant* w.r.t. M iff (i) $\equiv \subseteq (Q - F)^2 \cup F^2$ (that is, final and non-final states are not $\equiv$-equivalent) and (ii) for any $p, q \in Q$ and any $a \in A$, if $p \equiv q$, then $\delta(p, a)/_\equiv = \delta(q, a)/_\equiv$. If $\equiv$ is right invariant, the *quotient automaton* $M/_\equiv$ is constructed by

$$M/_\equiv = (Q/_\equiv, A, [q_0]_\equiv, \delta_\equiv, F/_\equiv)$$

where $\delta_\equiv = \{([p]_\equiv, a, [q]_\equiv) \mid (p, a, q) \in \delta\}$; notice that $Q/_\equiv = (Q - F)/_\equiv \cup F/_\equiv$, so we do not merge final with non-final states. Notice that $L(M/_\equiv) = L(M)$.

3   Small εNFAs from Regular Expressions

We give in this section an algorithm to construct an εNFA from a regular expression. The regular expression is assumed *reduced* such that 'many' redundant $\emptyset$'s, ε's, and $*$'s are eliminated. The precise way to do such reductions is omitted.

Algorithm 1 Given a regular expression α, the algorithm constructs an εNFA for α inductively, following the structure of α, and is shown in Figure .

The steps should be clear from the figure but we bring further improvements at each step which, rather informally, are described as follows:
   (a) if $p \xrightarrow{\varepsilon} q$ and the outdegree of p (or indegree of q) is one, then p and q can be merged,
   (b) any cycle of ε-transitions can be collapsed (all states merged and transitions removed),
   (c) multiple transitions (the same source, target, and label) are removed.

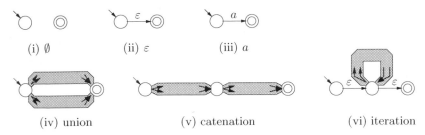

(i) ∅   (ii) ε   (iii) a

(iv) union   (v) catenation   (vi) iteration

Fig. 1. The construction of $\mathbf{A}_f^\varepsilon$

We call the obtained nondeterministic finite automaton with ε-moves *follow* εNFA (the reason for this name will be clear later) and denote it

$$\mathbf{A}_f^\varepsilon(\alpha) = (Q_f^\varepsilon, A, 0_f, \delta_f^\varepsilon, q_f).$$

Example 2 Consider the regular expression $\tau = (a + b)(a^* + ba^* + b^*)^*$. The εNFA for τ constructed by Algorithm   for τ, that is, $\mathbf{A}_f^\varepsilon(\tau)$, is shown in Figure . We shall use the same regular expression to show various other automata recognizing the same language. The automaton $\mathbf{A}_f^\varepsilon(\tau)$ is shown again in Figure to compare with the NFA obtained from it by removing ε-transitions.

The next theorem gives the upper bound on the size of our εNFA.

Theorem 3 *For any reduced regular expression α we have:*
(i) $L(\mathbf{A}_f^\varepsilon(\alpha)) = L(\alpha)$,
(ii) $\mathbf{A}_f^\varepsilon(\alpha)$ *can be computed in time* $\mathcal{O}(|\alpha|)$, *and*
(iii) $|\mathbf{A}_f^\varepsilon(\alpha)| \le \frac{3}{2}|\alpha| + \frac{5}{2}$.

In fact, the upper bound in Theorem (iii) is very close to the lower bound in Lemma   and hence very close to optimal.

Lemma 4 *Let* $\alpha_n = (a_1^* + a_2^*)(a_3^* + a_4^*) \cdots (a_{2n-1}^* + a_{2n}^*)$. *Every εNFA accepting* $L(\alpha_n)$ *has size at least* $8n - 1 = \frac{4}{3}|\alpha| + \frac{1}{3}$.

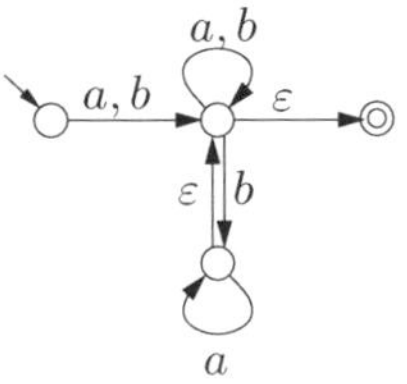

$$a, b$$

$$a, b \qquad \varepsilon$$

$$\varepsilon \quad b$$

$$a$$

Fig. 2. $\mathbf{A}_f^\varepsilon(\tau)$ for $\tau = (a + b)(a^* + ba^* + b^*)^*$

4   Positions and Partial Derivatives

We recall in this section two well-known constructions of NFAs from regular expressions. The first is the *position automaton*, discovered independently by Glushkov [   ] and McNaughton and Yamada [   ].

Let α be a regular expression. The set of *positions* in α is

$$\mathsf{pos}(\alpha) = \{1, 2, \ldots, |\alpha|_A\}.$$

Denote also $\mathsf{pos}_0(\alpha) = \mathsf{pos}(\alpha) \cup \{0\}$. All letters in α are made different by marking each letter with its position in α; denote the obtained expression $\overline{\alpha} \in \overline{A}^*$, where $\overline{A} = \{a_i \mid a \in A, 1 \le i \le |\alpha|_A\}$. For instance, if $\alpha = a(baa + b^*)$, then $\overline{\alpha} = a_1(b_2 a_3 a_4 + b_5^*)$. Notice that $\mathsf{pos}(\alpha) = \mathsf{pos}(\overline{\alpha})$. The same notation will also be used for removing indices, that is, for unmarked expressions α, the operator $\bar{\ }$ adds indices, while for marked expressions $\overline{\alpha}$ the same operator $\bar{\ }$ removes the indices: $\overline{\overline{\alpha}} = \alpha$. We extend the notation for arbitrary structures, like automata, in the obvious way. It will be clear from the context whether $\bar{\ }$ adds or removes indices.

Three mappings first, last, and follow are then defined as follows. For any regular expression α and any $i \in \mathsf{pos}(\alpha)$, we have:

$$\begin{aligned}
\mathsf{first}(\alpha) &= \{i \mid a_i w \in L(\overline{\alpha})\} \\
\mathsf{last}(\alpha) &= \{i \mid w a_i \in L(\overline{\alpha})\} \\
\mathsf{follow}(\alpha, i) &= \{j \mid u a_i a_j v \in L(\overline{\alpha})\}
\end{aligned} \tag{1}$$

For future reasons, we extend $\mathsf{follow}(\alpha, 0) = \mathsf{first}(\alpha)$. Also, let $\mathsf{last}_0(\alpha)$ stand for $\mathsf{last}(\alpha)$ if $\varepsilon(\alpha) = \emptyset$ and $\mathsf{last}(\alpha) \cup \{0\}$ otherwise.

The *position automaton* for α is

$$\mathbf{A}_{\mathrm{pos}}(\alpha) = (\mathsf{pos}_0(\alpha), A, \delta_{\mathrm{pos}}, 0, \mathsf{last}_0(\alpha))$$

with $\delta_{\mathrm{pos}} = \{(i, a, j) \mid j \in \mathsf{follow}(\alpha, i), a = \overline{a_j}\}$.

As shown by Glushkov [   ] and McNaughton and Yamada [   ], we have the following result.

Theorem 5 *For any regular expression α, $L(\mathbf{A}_{\mathrm{pos}}(\alpha)) = L(\alpha)$.*

The first part of the following result concerns the size of the position automaton and is clear from its construction. Its optimality is easily seen for expressions like $(a_1 + \varepsilon)(a_2 + \varepsilon) \cdots (a_n + \varepsilon)$. The second part has been proved by Brügemann-Klein [   ]; she gave the fastest algorithm known to compute the position automaton.

Theorem 6 *For any regular expression α, we have*
  (i) $|\mathbf{A}_{\mathrm{pos}}(\alpha)| = \mathcal{O}(|\alpha|^2)$ (can be quadratic in the worst case),
  (ii) $\mathbf{A}_{\mathrm{pos}}(\alpha)$ can be computed in time $\mathcal{O}(|\alpha|^2)$.

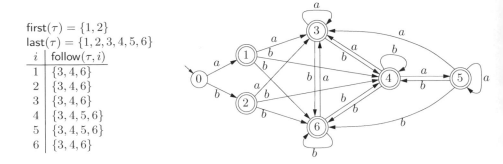

first$(\tau) = \{1, 2\}$
last$(\tau) = \{1, 2, 3, 4, 5, 6\}$

i	follow(τ, i)
1	$\{3, 4, 6\}$
2	$\{3, 4, 6\}$
3	$\{3, 4, 6\}$
4	$\{3, 4, 5, 6\}$
5	$\{3, 4, 5, 6\}$
6	$\{3, 4, 6\}$

Fig. 3. $\mathbf{A}_{\mathrm{pos}}(\tau)$ for $\tau = (a + b)(a^* + ba^* + b^*)^*$

Example 7 For the regular expression $\tau = (a + b)(a^* + ba^* + b^*)^*$ from Example , the marked version is $\bar{\tau} = (a_1 + b_2)(a_3^* + b_4 a_5^* + b_6^*)^*$. The values of the mappings first, last, and follow for τ and the corresponding position automaton $\mathbf{A}_{\mathrm{pos}}(\tau)$ are given in Figure .

The second construction we recall in this section is the *partial derivative automaton*, introduced by Antimirov [   ]. Recall the notion of partial derivative introduced by Antimirov [   ]. For a regular expression α and a letter $a \in A$, the set $\partial_a(\alpha)$ of partial derivatives of α w.r.t. a is defined inductively as follows:

$$\partial_a(\varepsilon) = \partial_a(\emptyset) = \emptyset$$
$$\partial_a(b) = \begin{cases} \{\varepsilon\}, & \text{if } a = b \\ \emptyset, & \text{otherwise} \end{cases}$$
$$\partial_a(\alpha + \beta) = \partial_a(\alpha) \cup \partial_a(\beta) \tag{2}$$
$$\partial_a(\alpha\beta) = \begin{cases} \partial_a(\alpha)\beta, & \text{if } \varepsilon(\alpha) = \emptyset \\ \partial_a(\alpha)\beta \cup \partial_a(\beta), & \text{if } \varepsilon(\alpha) = \varepsilon \end{cases}$$
$$\partial_a(\alpha^*) = \partial_a(\alpha)\alpha^*$$

The definition of partial derivatives is extended to words by $\partial_\varepsilon(\alpha) = \{\alpha\}$, $\partial_{wa}(\alpha) = \partial_a(\partial_w(\alpha))$, for any $w \in A^*$, $a \in A$.

The set of all partial derivatives of α is denoted $\mathrm{PD}(\alpha) = \{\partial_w(\alpha) \mid w \in A^*\}$. Antimirov [   ] showed that the cardinality of this set is less than or equal to $|\alpha|_A + 1$ and constructed the *partial derivative automaton*

$$\mathbf{A}_{\mathrm{pd}}(\alpha) = (\mathrm{PD}(\alpha), A, \delta_{\mathrm{pd}}, \alpha, \{q \in \mathrm{PD}(\alpha) \mid \varepsilon(q) = \varepsilon\}),$$

where $\delta_{\mathrm{pd}}(q, a) = \partial_a(q)$, for any $q \in \mathrm{PD}(\alpha), a \in A$. Antimirov proved also the following result.

Theorem 8 *For any regular expression* α, $L(\mathbf{A}_{\mathrm{pd}}(\alpha)) = L(\alpha)$.

$\partial_a(\tau) = \{\tau_1\}$ $\tau_1 = (a^* + ba^* + b^*)^*$
$\partial_b(\tau) = \{\tau_1\}$
$\partial_a(\tau_1) = \{\tau_2\}$ $\tau_2 = a^*\tau_1$
$\partial_b(\tau_1) = \{\tau_2, \tau_3\}$ $\tau_3 = b^*\tau_1$
$\partial_a(\tau_2) = \{\tau_2\}$
$\partial_b(\tau_2) = \{\tau_2, \tau_3\}$
$\partial_a(\tau_3) = \{\tau_2\}$
$\partial_b(\tau_3) = \{\tau_2, \tau_3\}$

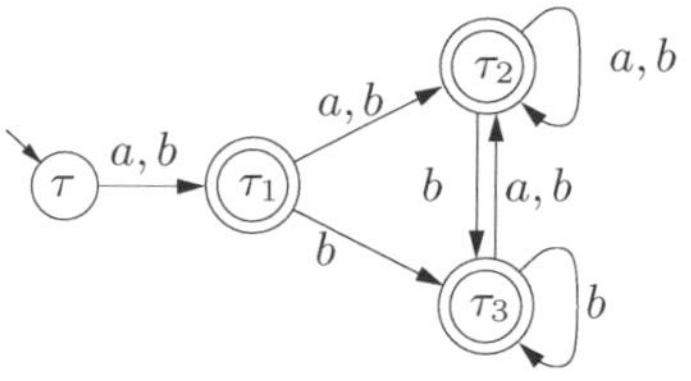

Fig. 4. $\mathbf{A}_{\mathrm{pd}}(\tau)$ for $\tau = (a + b)(a^* + ba^* + b^*)^*$

Example 9 Consider the regular expression τ from Example . The partial derivatives of τ are computed in Figure   where also its partial derivative automaton $\mathbf{A}_{\mathrm{pd}}(\tau)$ is shown.

The first part of the following result concerns the size of the partial derivative automaton and has been proved by Antimirov [   ]. Its optimality is proved also by the expressions $(a_1 + \varepsilon)(a_2 + \varepsilon) \cdots (a_n + \varepsilon)$. The second part has been proved by Champarnaud and Ziadi [,   ]. They also improved very much Antimirov's $\mathcal{O}(n^5)$ algorithm for the construction of the position automaton; their algorithm runs in quadratic time (third part in the theorem below).

Theorem 10 *For any regular expression α, we have:*
(i) $|\mathbf{A}_{\mathrm{pd}}(\alpha)| = \mathcal{O}(|\alpha|^2)$ (can be quadratic in the worst case),
(ii) $\mathbf{A}_{\mathrm{pd}}(\alpha)$ is a quotient of $\mathbf{A}_{\mathrm{pos}}(\alpha)$,
(iii) $\mathbf{A}_{\mathrm{pd}}(\alpha)$ can be computed in time $\mathcal{O}(|\alpha|^2)$.

5   Follow Automata

In this section we construct NFAs from regular expressions. This is done by eliminating ε-transitions from the $\mathbf{A}_f^\varepsilon(\alpha)$. Essentially, for any path labelled ε, $p \overset{\varepsilon}{\rightsquigarrow} q$, and any transition $q \overset{a}{\rightarrow} r$, we add a transition $p \overset{a}{\rightarrow} r$. The obtained automaton is called *follow* NFA, denoted

$$\mathbf{A}_f(\alpha) = (Q_f, A, 0_f, \delta_f, F_f).$$

Define then the equivalence $\equiv_f \subseteq \mathsf{pos}_0(\alpha)^2$ by

$i \equiv_f j$ iff   (i) both i, j or none belong to $\mathsf{last}(\alpha)$ and
(ii) $\mathsf{follow}(\alpha, i) = \mathsf{follow}(\alpha, j)$

Notice that we restrict the equivalence so that we do not make equivalent final and non-final states in $\mathbf{A}_{\mathrm{pos}}(\alpha)$. Also, $\equiv_f$ is a right-invariant equivalence, so we can make the quotient. The next result says that $\mathbf{A}_f(\alpha)$ is also a quotient of $\mathbf{A}_{\mathrm{pos}}(\alpha)$. This is unexpected because the construction of $\mathbf{A}_f(\alpha)$ does not have, apparently, anything to do with positions. However, the consequences of this result, discussed in the next sections, are very important.

Theorem 11 $\mathbf{A}_f(\alpha) \simeq \mathbf{A}_{\mathrm{pos}}(\alpha)/_{\equiv_f}$.

We notice that the restriction we imposed on $\equiv_f$ so that final and non-final states in $\mathsf{pos}_0(\alpha)$ cannot be $\equiv_f$-equivalent is essential, as shown by the expression $\alpha = (a^*b)^*$. Here $\mathsf{follow}(\alpha, i) = \{1, 2\}$, for any $0 \leq i \leq 2$. However, merging all three states of $\mathbf{A}_{\mathrm{pos}}(\alpha)$ is an error as the resulting automaton would accept the language $(a + b)^*$.

We remark that both $\mathbf{A}_f^\varepsilon$ and $\mathbf{A}_f$ can be shown to be very good data structures for fast computation of the values of the mappings first, last, and follow. This gives a simpler proof for the results of Brüggeman-Klein [   ] concerning the computation of the position automaton in quadratic time (see Theorem   (ii)). The idea is to construct first the automaton $\mathbf{A}_f^\varepsilon(\alpha)$ and then compute the three mappings (and so, implicitly, the transition mapping of $\mathbf{A}_{\mathrm{pos}}(\alpha)$) using the same properties of $\mathbf{A}_f^\varepsilon$ which implied the unexpected result in Theorem   . The worst case of this algorithm is the same as for the algorithm of Brügemann-Klein [   ] but in many cases ours is faster.

Example 12 We give an example of an application of Theorem   . For the same regular expression $\tau = (a+b)(a^*+ba^*+b^*)^*$ from Example   , we build in Figure   the $\mathbf{A}_f^\varepsilon(\tau)$ and then give the equivalence classes of $\equiv_f$ and the automaton $\mathbf{A}_f(\tau)$.

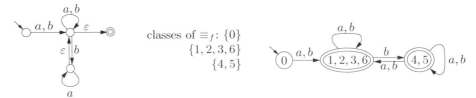

classes of $\equiv_f$: $\{0\}$
$\{1, 2, 3, 6\}$
$\{4, 5\}$

Fig. 5. $\mathbf{A}_f^\varepsilon(\tau)$ and $\mathbf{A}_f(\tau) = \mathbf{A}_{\mathrm{pos}}(\tau)/_{\equiv_f}$ for $\tau = (a + b)(a^* + ba^* + b^*)^*$

6   Comparisons with other Constructions

A first consequence of our result in Theorem   is that $\mathbf{A}_f$ is always smaller or equal to $\mathbf{A}_{\mathrm{pos}}$. Also, it can be computed faster. Thus it is a good candidate for replacing $\mathbf{A}_{\mathrm{pos}}$ in applications.

We give below several examples showing that $\mathbf{A}_f$ can be much smaller than either of $\mathbf{A}_{\mathrm{pos}}$ and $\mathbf{A}_{\mathrm{pd}}$ and that $\mathbf{A}_f$ and $\mathbf{A}_{\mathrm{pd}}$ are incomparable in general.

Example 13 Consider $\alpha_1 = (a_1 + \varepsilon)^*$ and define inductively, for all $i \geq 1$, $\alpha_{i+1} = (\alpha_i + \beta_i)^*$, where β_i is obtained from α_i by replacing each a_j by $a_{j+|\alpha_i|_A}$. For instance,

$$\alpha_3 = (((a_1 + \varepsilon)^* + (a_2 + \varepsilon)^*)^* + ((a_3 + \varepsilon)^* + (a_4 + \varepsilon)^*)^*)^*.$$

We have then
- $|\mathbf{A}_{\mathrm{pos}}(\alpha_n)| = |\mathbf{A}_{\mathrm{pd}}(\alpha_n)| = \Theta(|\alpha_n|^2)$ and
- $|\mathbf{A}_{\mathrm{f}}(\alpha_n)| = \Theta(|\alpha_n|)$.

Example 14 Consider the regular expression

$$\alpha_n = a_1(b_1 + \cdots + b_n)^* + a_2(b_1 + \cdots + b_n)^* + \ldots + a_n(b_1 + \cdots + b_n)^*.$$

We have
- $|\mathbf{A}_{\mathrm{pos}}(\alpha_n)| = \Theta(|\alpha_n|^{3/2})$,
- $|\mathbf{A}_{\mathrm{f}}(\alpha_n)| = \Theta(|\alpha_n|)$, and
- $|\mathbf{A}_{\mathrm{pd}}(\alpha_n)| = \Theta(|\alpha_n|^{1/2})$.

Example 15 Consider the regular expression (identifiers in programming languages)

$$\alpha_n = (a_1 + a_2 + \cdots + a_n)(a_1 + a_2 + \cdots + a_n + b_1 + b_2 + \cdots + b_m)^*.$$

We have
- $|\mathbf{A}_{\mathrm{f}}(\alpha_n)| = |\mathbf{A}_{\mathrm{pd}}(\alpha_n)| = \Theta(|\alpha_n|)$ and
- $|\mathbf{A}_{\mathrm{pos}}(\alpha_n)| = \Theta(|\alpha_n|^2)$.

We notice that we did not compare our construction with the one of Chang and Paige [   ] since we do not work with compressed automata.

7   Optimal Use of Positions

Finally, we show that the follow automaton $\mathbf{A}_{\mathrm{f}}(\alpha)$ uses the whole information which comes from positions of α. Indeed, the follow automaton for marked expressions cannot be improved. $\mathbf{A}_{\mathrm{f}}(\overline{\alpha})$ is a deterministic automaton and let the minimal automaton equivalent to it be $\mathsf{min}(\mathbf{A}_{\mathrm{f}}(\overline{\alpha}))$. Then $\overline{\mathsf{min}(\mathbf{A}_{\mathrm{f}}(\overline{\alpha}))}$ is an NFA accepting $L(\alpha)$ which can be computed in time $\mathcal{O}(|\alpha|^2 \log |\alpha|)$ using the minimization algorithm of Hopcroft [   ]. This is, in fact, another way of using positions to compute NFAs for regular expressions. However, it is interesting to see that $\overline{\mathsf{min}(\mathbf{A}_{\mathrm{f}}(\overline{\alpha}))}$ brings no improvement over $\mathbf{A}_{\mathrm{f}}(\alpha)$.

Theorem 16 $\overline{\mathsf{min}(\mathbf{A}_{\mathrm{f}}(\overline{\alpha}))} \simeq \mathbf{A}_{\mathrm{f}}(\alpha)$.

Notice also that computing $\mathbf{A}_{\mathrm{f}}(\alpha)$ by ε-elimination in $\mathbf{A}_{\mathrm{f}}^{\varepsilon}(\alpha)$ is faster than using Hopcroft's algorithm [   ] plus unmarking.

References

[ASU86]   Aho, A., Sethi, R., Ullman, J., *Compilers: Principles, Techniques, and Tools*, Addison-Wesley, MA, 1988.

[An96]   Antimirov, V., Partial derivatives of regular expressions and finite automaton constructions, *Theoret. Comput. Sci.* **155** (1996) 291–319.   ,

288   Lucian Ilie and Sheng Yu

[BeSe86]   Berry, G, Sethi, R., From regular expressions to deterministic automata, *Theoret. Comput. Sci.* **48** (1986) 117–126.
[BrK93]   Brüggemann-Klein, A., Regular expressions into finite automata, *Theoret. Comput. Sci.* **120** (1993) 197–213.   ,   ,
[Br64]   Brzozowski, J., Derivatives of regular expressions, *J. ACM* **11** (1964) 481–494.
[ChZi01a]   Champarnaud, J.-M., Ziadi, D., New finite automaton constructions based on canonical derivatives, *Proc. of CIAA 2000*, LNCS 2088, Springer, 2001, 94 –104.   ,
[ChZi01b]   Champarnaud, J.-M., Ziadi, D., Computing the equation automaton of a regular expression in $\mathcal{O}(s^2)$ space and time, *Proc. of 12th Combinatorial Pattern Matching (CPM 2001)*, LNCS 2089, Springer, 2001, 157–168.   ,

[ChPa97]   Chang, C.-H., Paige, R., From regular expressions to DFA's using compressed NFA's, *Theoret. Comput. Sci* **178** (1997) 1–36.
[CrHa97]   Crochemore, M., Hancart, C., Automata for pattern matching, in: G. Rozenberg, A. Salomaa, eds., *Handbook of Formal Languages, Vol. II*, Springer-Verlag, Berlin, 1997, 399–462.
[Fr98]   Friedl, J., *Mastering Regular Expressions*, O'Reilly, 1998.
[Gl61]   Glushkov, V. M., The abstract theory of automata, *Russian Math. Surveys* **16** (1961) 1–53.   ,
[HaMu00]   Hagenah, C., Muscholl, A., Computing ϵ-free NFA from regular expressions in $O(n \log^2(n))$ time, *Theor. Inform. Appl.* **34** (4) (2000) 257–277.

[Ho71]   Hopcroft, J., An $n \log n$ algorithm for minimizing states in a finite automaton, *Proc. Internat. Sympos. Theory of machines and computations*, Technion, Haifa, 1971, Academic Press, New York, 1971, 189–196.
[HoUl79]   Hopcroft, J. E., Ullman, J. D., *Introduction to Automata Theory, Languages, and Computation*, Addison-Wesley, Reading, Mass., 1979.
[HSW01]   Hromkovic, J., Seibert, S., Wilke, T., Translating regular expressions into small ϵ-free nondeterministic finite automata, *J. Comput. System Sci.* **62** (4) (2001) 565–588.
[McNYa60]   McNaughton, R., Yamada, H., Regular expressions and state graphs for automata, *IEEE Trans. on Electronic Computers* **9** (1) (1960) 39–47.   ,

[SiSo88]   Sippu, S., Soisalon-Soininen, E., *Parsing Theory: I Languages and Parsing*, EATCS Monographs on Theoretical Computer Science, Vol. 15, Springer-Verlag, New York, 1988.
[Th68]   Thompson, K., Regular expression search algorithm, *Comm. ACM* **11** (6) (1968) 419–422.   ,
[Yu97]   Yu, S., Regular Languages, in: G. Rozenberg, A. Salomaa, eds., *Handbook of Formal Languages, Vol. I*, Springer-Verlag, Berlin, 1997, 41–110.   ,

Author Index